ISWA 88
PROCEEDINGS
of the
5th International Solid Wastes Conference
International Solid Wastes

ISWA 88 PROCEEDINGS of the 5th International Solid Wastes Conference

International Solid Wastes
and Public Cleansing Association

September 11–16th, 1988
Copenhagen, Denmark

VOLUME I
ORAL PRESENTATIONS

Edited by
Lizzi Andersen and Jeanne Møller
Vester Farimagsgade 29
DK 1606 Copenhagen V
Denmark

1988

AP

ACADEMIC PRESS

Harcourt Brace Jovanovich, Publishers
London San Diego New York Berkeley
Boston Sydney Tokyo Toronto

ACADEMIC PRESS LIMITED
24/28 Oval Road,
London NW1 7DX

United States Edition published by
ACADEMIC PRESS INC.
San Diego, CA 92101

ISBN 0-12-058451-4

Printed in Great Britain by St Edmundsbury Press Limited,
Bury St Edmunds, Suffolk

PREFACE

The ISWA Quadrennial Conference and Exhibition was established to provide an international forum for current topics in the field of solid wastes management, technology and research.

With its wide range of topics, the ISWA 88 Conference offers unique facilities for the presentation and exchange of novel technology and thoughts both in a wide context and on more specific aspects.

These Proceedings contain papers selected from 254 submitted abstracts by the programme committee. The ISWA 88 General Secretariat acknowledges the 140 authors for their cooperation and timely submittal of camera-ready manuscripts.

Lizzi Andersen
Secretary General

Jeanne Møller
Administrative Secretary

ISWA 88 General Secretariat
Vester Farimagsgade 29
DK 1606 Copenhagen V, Denmark

ISWA 88

Thank you for registering for the ISWA 88 Congress. We would like to welcome you and we hope that the content of the conference programme and the leisure activities will match your expectations.

Apart from the speeches by representatives from the World Bank, the World Health Organization, and the United Nations Environmental Programme, you will also hear contributions from specialists from 28 countries. They will deal with the most far-reaching areas of concern within ISWA's domain. I sincerely hope that each contribution will be followed by discussions with frank exchanges of views.

Although the contributors only have a limited time to speak, each one has put in a great deal of time and effort in the preparations. We would like to thank all the speakers for their participation in the success of this major ISWA event.

J. DEFÈCHE
President of ISWA

Contents

Session 1

ISWA 88 Opening Session

Session 2

Public Attitudes Towards Solid Waste Management

PACKINGTON ESTATE ENTERPRISES LTD

A Paper for

The 1988 Conference of

INTERNATIONAL SOLID WASTES ASSOCIATION

by

C.A.R. Biddle Ll.B.,F.R.S.A,M.Inst.W.M.

Group Managing Director

"ENCOURAGING PUBLIC CONFIDENCE IN WASTE DISPOSAL"

ABSTRACT: A simplified explanation of how history, supported by enlightened self-interest, has lead to an integrated management package in an attempt to provide an acceptable perception of waste disposal activities carried out in sensitive Green Belt locations. Social responsibility and teamwork are stressed together with more about practice than preaching.

INTRODUCTION: For Packington Estate Enterprises Ltd (PEEL) the 23rd February 1988 has brought another connotation to the term "D DAY." Not D for DUMP for that is a four letter word in the English language never used at Packington other than to describe a competitor operation (usually abroad) for the dictionary definition is: "To discard haphazardly, usually of rubbish." There is nothing haphazard in PEEL's operations.

In this case, "D DAY" stands for DEMONSTRATION DAY. PEEL's Managing Director stood before HRH The Prince of Wales to receive, at his hands, one of the six "Better Environment Awards for Industry 1987." This was partly sponsored by the European Commission for their European Year of the Environment. Extracts from the opening paragraphs of the speech by the Company's Managing Director forms a suitable introduction to this Paper.

ISWA PROCEEDINGS
VOLUME 1 ISBN 0-12-058451-4

OBJECTIVE: The Brief requires an outline of the Packington project. Technically, it is "Controlled Landfill" where, each year, Packington now receives more than 1 million tonnes of "controlled wastes." Packington works on the principle that "all wastes are toxic: but some more than others" since it is known as a "Co-Disposal" site by accepting large tonnages of so called "Hazardous Wastes."

PACKINGTON'S ACHIEVEMENTS: Packington possesses a sound record of operational disciplined innovation among which are: 1: Above Ground Landfill which necessitated the development of 2: Cell Tipping so providing 3: Natural Litter Curtains and also 4: The High Density laying of waste so 5:Reducing Settlement while these four together led to 6: Progressive Restoration which has resulted in 7: Combined Use of landfill with agricultural use as well as leisure and recreational activities. Of far greater significance is an eighth factor pioneered by Packington.

THE ROYAL COMMISSION ON ENVIRONMENTAL POLLUTION: The R.C.E.P. has commended Packington as "one landfill operator who has obtained (environmental impairment) insurance on favourable terms because he has been able to demonstrate a high level of professionalism backed by monitoring by an independent consultant."

It is the maintenance of this Insurance Policy which is the corner-stone towards the practice (and thus the recognition) of the high environmental standards achieved at Packington.

THE HISTORY OF PACKINGTON: The Packington Estate is situated almost precisely mid-way between Birmingham and Coventry in the centre of England. The Estate comprises 5,000 acres (2,200 hectares). It can trace continuous family ownership, by the Earl of Aylesford and his forbears, since the early

part of the 16th Century.
There are now numerous activities on the Estate as a whole which today finds itself in a sensitive Green Belt location between two major conurbations. Among these many diverse uses of land are 1: The joint domestic as well as commercial use of a Stately Home: 2: Deer Ranching in the Home Park: 3: 11 agricultural tenant farmers: 4: Land leased to the Forestry Commission: 5: Sporting enterprises including: a) Fishing - both fly (trout) and Coarse, b) Shooting c) A Golf Club with i) 9 hole and ii) 18 hole Championship Courses iii) 120 bedroom Hotel (under construction). There is also 6: Extensive mineral extraction for a) Sand and Gravel b) Brick earth (clay) together with 7: Controlled wastes management especially: a) Engineered landfill, b) Resource recovery comprising:- i) Salvage of recyclable materials, ii) Renewable energy ie. landfill gas, c)Combined agricultural and leisure use conjointly with wastes disposal.
POLICY FOR ENGINEERED WASTES MANAGEMENT: The landfill site at Packington, like 3 others in the immediate vicinity controlled by PEEL, is very close to the River Blythe. This river is a (potable) drinking water supply for 300,000 people some 3.5 miles (8 kilometres) downstream of the site. Thus the self-same policy which applies for the 15 foregoing activities was set by Lord Aylesford and his son, Lord Guernsey, who is Chairman of PEEL. In short, this policy is "to develop a profitable operation but, **UNDER NO CIRCUMSTANCES**, should there be built-up any liabilities for future generations."
THE PRACTICE OF POLICY: It is the caveat in this policy which is so important, especially when considering disposal of wastes to land. The liabilities, if any, might only reveal themselves long after "the suitcase" (or void) is full with no revenue earnings let alone profits to carry out

rectification. The first priority therefore was to seek some form of protection through insurance.

Here, the prior history of trading at Packington proved essential in the light of the misguided perceptions of the insurance industry in the U.K. These appeared to PEEL to be based on the irrelevant experiences found as much in America as Europe itself.

HISTORICAL DIFFERENCES: The Commission of the European Communities give no scientific justification for their intentions to proscribe MOST Controlled Wastes going directly to landfill by the year 2000. Even worse, the CEC has ignored the fundamentally different historical and environmental developments in the United Kingdom as compared with either Europe or the U.S.A. It should not be necessary to remind everyone that we in the U.K. have had successful detailed central (and especially local) statutory environmental as well as planning controls since the middle of the last Century; a time when both those Continents were solving their internal political difficulties, sometimes ferociously!

It may come as a surprise to Delegates to learn that we at Packington (like all our competitors in the U.K.) are supervised, inspected and controlled by no less than 61(sixty one) separate Officers of District, County, Regional and National Authorities; the vast majority appointed by statutes of our Parliament long before the Treaty of Rome was envisaged, let alone ratified!

TEAMWORK: This prior history of trading at Packington was effectively under the same Management with the self-same team of employees together with professional advisors from many scientific, engineering and other disciplines. Fortunately also most of the 61 Inspecting Officers mentioned above also remained constant. Their co-operation and contribution has been invaluable throughout which the Company is only too grateful to acknowledge.

From the beginning the whole objective was to create a credible as much as reliable track record to establish the minimum adverse environmental impact. Thus no one aspect took priority but rather attention was paid to the detailed engineering throughout for practice appeared more important than mere theory.

ENVIRONMENTAL IMPACT INSURANCE: The relevant quotation from Chapter 9 of the 11th Report of the Royal Commission on Environmental Pollution has already been given in the Introduction above.

Every aspect of PEEL's operations on any of the six Licensed landfill sites (there are two others elsewhere) is designed to sustain the Insurance Policies. These allow the Company to claim up to £3 million for rectification works in any one event as much within the boundaries as in adjoining areas.

There are many who have argued that this form of protection admits a degree of risk. This PEEL acknowledges "freely" in that the annual premiums are remarkably low at less than £20,000 in all ie. about half of 1% of turnover.

ASSOCIATED COSTS: The real true costs in obtaining this level of cover, at such a low premium, is in all the supporting work carried out by PEEL. For instance, there are 122 existing boreholes together with surface water and other monitoring points on the principal site alone comprising some 385 acres (162 hectares). Some 4,000 separate analyses have been and continue to be undertaken each year apart from any corrective measures which might be indicated. Taken together, PEEL estimates the total cost of sustaining the policies is of the order of 50 pence per tonne.

ANNUAL ENVIRONMENTAL AUDIT CERTIFICATE: At this stage, what might at first appear a slight digression seems relevant. As much under the Companies' Acts as the Finance Acts in the U.K., there are formal statutory requirements to the effect that: ".... there should be an independent annual audit of the books of account to reflect a true and accurate statement of the affairs of the business at a point in time." Presumably this is to ensure that the management are neither defrauding either the Shareholders and/or the Revenue Authorities.

What a regrettable state of affairs that there is no similar requirement on all business activities and especially waste management for a similar Independent Audit Certificate as to environmental impact (if any).

CONTROL OF POLLUTION ACT 1974: In the U.K. one Local Authority has set a magnificent legal precedent however. Every other Local Authority ought now to apply the principle established by Shropshire County Council through Judicial Review so as to require at least environmental impairment insurance (if not also bonding) as they did when granting a Licence to Wrights Waste Re-cycling at Bridgnorth. The concluding paragraph of Section 6(2) of CoPA certainly gives these powers.

More pertinently, in the context of influencing public perception, reference ought also to be made to Section 5 where a valid Planning Permission must be in force which itself must be dependent upon decisions of Elected Representatives who, no doubt, are likely to reflect public opinion. The experience at Packington certainly substantiates the statement that a continual environmental monitoring programme can have a marked effect if only at the stage of a Planning Appeal.

ENLIGHTENED SELF INTEREST: It is only natural to question whether such costs as 50 pence per tonne to implement the Packington's philosophy can be justified; especially when it is neither a mandatory requirement nor incurred by competitors in spite of the fact that it truly implements the recommendations of the Gregson Report that the "Polluter must Pay."

The proof of PEEL's policy can be so easily demonstrated in several ways. In business terms by the success of the recent Planning Appeal. This allows not only an extension of the base area but, more particularly, the substantial construction above natural contour so as to provide a hill some 50 metres or more in height. At the rates of intake at that time, this increased "the life" of the void by some 65 years.

With such a projected "life" it then became economically feasible to sustain a project as costly as £2 million(albeit with some Government Grant Aid) so as to control the hazards of landfill gas by harnessing the properties of the carbon dioxide as much as the methane content by generating 28 million units of electricity per annum through a gas turbine and alternator set.

Awards in themselves can be attractive. The Company is proud to have received the Commemorative Award of the Institute of Wastes Management as well as the E.Y.E. Award itself presented by HRH The Prince of Wales. These Awards seem to provide a credibility for other business enterprises to approach PEEL for conjoint operations both at home and overseas.

So we conclude where we began. Public recognition of the high standards prove not only an advantage in reducing the nation's overheads (while showing minimal impact upon the environment) but, in addition, the creation of both pride and wealth on a continuing basis: enlightened self interest!

RONALD POHORYLES & WOLF-ERICH ECKSTEIN
(Interdisciplinary Centre for Comparative Research in Technology and Social Policy, ICCR - Vienna, Austria)

THE PUBLIC DEBATE ON HAZARDOUS WASTE DISPOSAL: CONFLICTS AND HOW TO ACHIEVE CONSENSUS FOR A RATIONAL STRATEGY IN POLITICAL ACTION

This study aims to explain, by comparing a conflict- (i.e. problem facing) and a control population, the failure so far, due to civic protest, of establishing a special waste disposal depot. Important clues are supplied on the one hand by the course of past conflicts around the installation of such or comparable facilities, as documented from periodicals'and documentation analyses and from expert discussion. Supplementing them, and, more specifically, the findings of population polls may be perused.

1. Experience gained in conflicts past

It is one of the central experiences in past conflicts that their theme was never the concept of well regulated special waste disposal: the bone of contention was always the _type_ of facility, or its potential _operator_.

This clue is decisive if one remembers that the handling of special waste is considered a grave burden by the communities concerned - and rightly so. In this context a rational political strategy must be conceptually unassailable, and in more than one sense:

* On the one hand it must be clearly demonstrated that in an higher interest the installation of special waste disposal facilities is a necessity and that careful consideration has been given to presenting a cohesive overall concept of disposal; and
* evidence must be offered that the choice of location and the technology selected are the result of that concept. Choice of location is merely one element of a complex disposal system. The dispute about the isolated subject of location will quickly raise emotions, impeding rational conflict resolution - if not making it impossible.

On the international plane, too, it can be seen that the special waste disposal depots are only the last link in a complex disposal strategy. With the presentation and discussion of the complex system as a whole it becomes clear that special waste disposal poses a more intricate problem than the mere establishment of a depot. By demonstrating the existence of a systematically planned concept it should be

ISWA PROCEEDINGS
VOLUME 1 ISBN 0-12-058451-4

possible to channel the discussion in a direction away from bandying arguments about the particular depot. Initially, and quite justifiably, the population sees special waste disposal as a political task. The environmental policy content of the problem is evident and requires proof of competent problem solution. If it can be made clear that the depot is but one element in a complex disposal chain, public discussion might take quite a different course.

But it is decisive that proof be given in this discussion that the decision-takers and those charged with effecting the disposal are competent. This competence is decidedly twodimensional, encompassing one objective dimension as well as one of good will and good intentions. Past conflicts in the disposal debate offer plain clues that the difficulties were due to conceptual faults and shortcomings. The projects presented to the population appeared insufficiently thought out, and that was underlined by the spectacle of numerous experts coming forward at various times with a multitude of contradictory information on these very projects. Fragmentary information is hardly suited to dispel public mistrust.

Responsibility in the disposal debate should not be delegated to the communal (or municipal) level. Local politics were the decisive factor for the failure of these projects, while overriding higher interests were at stake. In fact, local authorities and politicians are not so very deeply interested in the establishment of special waste disposal depots within the community that they would be able to withstand substantial popular pressure. And why should one assume that they have the necessary factual and technical competence, both in respect of the overall concept and of the planned facilities? Thus civic protest proved most effective by way of pressure on local politicians.

In this context there is also the question of the operator. The proposal to argue the overall concept in the disposal debate even when the building of a singe facility (e.g., a depot) is under consideration, makes it plain that the operator is expected to act as a subordinate and tightly controlled recognisable carrier of responsibility. Irrespective of whether the facility is run by a private or public operator, the aspect of transparency must be assured by active, competent and trustworthy control. To achieve a breakthrough it will be necessary to create reliable controls. This presupposes that the controlling bodies are known to the public and that their trustworthiness is acknowledged.

Control must be open to public scrutiny and the controlling bodies must be aware of the need to keep in touch with active citizens. It merits careful consideration to what extent the active citizens should be made to share in the control of the facilities. Involvement of citizens' initiatives in this process may help to establish public confidence.

2. The population and the active citizen

A comparison of the directly affected ("conflict"-) and the control population suggests that in the special waste debate complex attidudes are, at least latently, present, making protest likely.

While attitudes to disposal in the home region are less pronounced in the control community, argumentation patterns are nevertheless very similar in both communities. Conflict arises not from the activities of ad hoc created pressure groups but from latent anxieties.

Opposition to the installation of special waste disposal facilities is fairly widely extant. But this is by no means to say that appropriate projects are politically unrealisable. Compared with other political tasks the decision-making process concerning such a facility does not turn on the relative strength of advocacy contra opposition. It is obviously contrary to the structure of this problem to expect direct advocacy of such facilities on the part of the population affected by this issue. For with the installation of such a facility the population must undoubtedly accept burdens without the expectation of any directly recognisable benefits - such as might, for example, result from the building of a major industrial installation. Thus the political problem is not to win a majority for the establishment of a facility but acquiescence to accepting the burdens and disadvantages involved. The approach to this end is, surely, comprehensive information.

We have demonstrated that those citizens are most concerned about the unsolved problem of waste disposal who have already gone into the matter and have acquired relevant information. Since this was the case only among one fifth of the control community and a third of the conflict community, improving the level of popular information is clearly an essential political task.

The argument that the installation of a special waste facility is not a question of the relative strength of advocates and opponents but only a question of acceptance is also borne out by the fact that a majority of respondents does not, by any means, demand a plebiscite on the erection of such a facility. Preference is given rather to an expert decision (and the right to decide is thereby relinquished by the respondents). But what is wanted is not necessarily experts in the accepted sense: a very heterogeneous range of experts is variously named by respondents in this debate.

It is notable that the more active and more conflict inclined groups are more ready to recognise the competence of politicians in this question. To them conflict seems mainly the instrument for urging the authorities to put forward arguments - here especially those of technical safety and adequate controllability - that testify to their claimed competence; and to take up the arguments that were formulated and put forward by the citizens' initiatives. This argument is also supported by the fact that the engaged active groups are the ones who are more convinced of the feasibility of technologically adequate, safe disposal. However, proof is demanded that such safe solution is in fact to be provided in the planned facility and that clear responsibilities are laid down for controlling the facility and its operation. It is precisely in this sense that the active citizen is also the natural partner - be he, at this initial stage, conflict partner.

On the basis of attitudes to the various dimensions of the special waste disposal problem it could be shown that arguments of environmental policy for the installation of depots must be complex to assure the necessary trust and confidence.

We see that the more participatory groups do not have to be convinced of the technological feasibility of environmentally innocuous special waste disposal per se, but of the honest intention to apply it. Contrary to relevant theories, the more active groups are less suspicious of industry and technology. What these active groups insist on is the transparency of the overall concept as a prerequisite for efficient control and the granting of control facilities for their safe running.

But if these demands are not adequately met, the more active groups can count on, and arouse, latent anti-industrialism, existing scepticism concerning technical feasibility, and suspicion of politicians, rife among less active groups. This, ultimately, is also the basis of the citizens' initiatives in past conflicts: the interplay of active citizens, whose control demands were not sufficiently or not convincingly regarded and the latent fears of the population, for which public policy has not made adequate allowance in the past.

3. Conclusions for political action

(1) For a special waste disposal facility to be accepted it is necessary to present a clear overall concept. It must incorporate information on the overall organisation of disposal, as well as about the type of facility planned, the intended operator and the controls envisaged.

(2) Concerning control in particular, it must be made clearly apparent that the ministry concerned is aware of its responsibility. That is in accordance with the legal position, but it is also a widely shared demand of the population. It would surely be sensible for the ministry to name a responsible superintendent not directly connected with the potential operator, and acting in a clearly discernible control function.

(3) In the context of optimal control, the appointment of independent experts should be considered. This will be in the interests of the objective need for tackling the safety problem of the disposal facility, as well as to create the necessary trust and confidence. In selecting the experts it will be advisable to give due consideration to the expectations of the population. In associating the public in the selection of independent experts and by publicly debating the expert reports it will be possible to make a substantial contribution to defusing and rationalising the debate around the installation of a dangerous waste disposal facility.

(4) It must be made clear that the subject matter of expert opinion is the ecological compatibility and operational safety of the facility, whereas the location and/or choice of location is not negotiable. Other themes might be problems of the infrastructure (e.g. Transport, Access), warning systems and assured optimal control. In all these matters the wishes of the population should be

met to the utmost possible degree.

(5) A concept that encompasses the active citizens' direct participation as well as their indirect association by way of the experts involved, can no doubt increase the acceptability of the special waste disposal facility; but it would be an illusion to suppose that no remnant of popular dissention would remain: some part of the population would vote against such a facility under any circumstances.

The measures here proposed aim at minimising that opposition and at preventing an association of this latter remnant with those who choose conflict strategies in order to achieve adequate controls and elicit credible answers on ecological problems.

For it seems not unlikely that - given the appropriate prerequisites - citizens will be prepared to accept the burdens connected with the disposal of noxious waste. It is the finding of this project that this may be confidently expected - provided the active citizen is accepted and taken seriously as a conflict partner.

A NEGOTIATION SUPPORT SYSTEM FOR WASTE MANAGEMENT

Jean Simos, Lucien Y. Maystre, Lausanne

1. The problem

The new trends in waste management are against the traditional principle of "everything into the garbage bag for incineration". Fighting the continuous increase of waste production, the resulting increase of environmental pollution and the squandering of resources, these new trends advocate separate collection practices, differentiated treatment schemes and a more responsible attitude of the population and enterprises, according to the "polluter-pays" principle.
The authorities of Geneva (Switzerland) have set up an expert committee to investigate all the feasible alternatives for an extension of the present waste management system.

2. The situation

From 1971 to 1986, the per capita production of household and industrial solid waste of Geneva has increased from 404 to 652 $kg.y^{-1}$ and the total production from 138'000 to 238'000 $to.y^{-1}$. The existing incineration plant will soon not be able to swallow this huge quantity of waste: the government has recommended an extension of the plant with larger units, but the local parliament has allocated a fund for the studies, provided that a parallel study is made, concerning new alternatives for the year 2000. This parallel study was in part entrusted to the Institute for environmental engineering (IGE) of the Federal Polytechnic of Lausanne (EPFL). The expert committee was created with 8 members representing the WWF, a national association for the protection of the environment, a regional consumer's organisation, the suburban municipalities, the waste collection service of Geneva, the incineration plant operation, the state energy commission, the public works department.

ISWA PROCEEDINGS
VOLUME 1 ISBN 0-12-058451-4

3. The composition of waste

Crushed garbage is impossible to sort correctly: a sorting campaign has been carried out on samples of garbage bags (or bins) prior to loading into collector trucks. 52 tons of waste were hand-sorted during one year, divided among 8 housing types and 11 economic activities, over 5 weeks during every season, on different days of the week and following the rules of a random stratified sampling. Only household waste and waste of similar nature collected through the municipal services, i.e. about 80 % of the total solid waste, were thus analysed. [1].

It was possible to identify 47 classes of waste: in Fig. 1, the 47 classes have been grouped into 11 "macro-classes" and the weighted yearly average of the composition is indicated.

More specific investigations have been carried out to determine the quantities of PVC in plastics [2], the composition of heterogeneous packaging material, of bulk waste, of batteries.

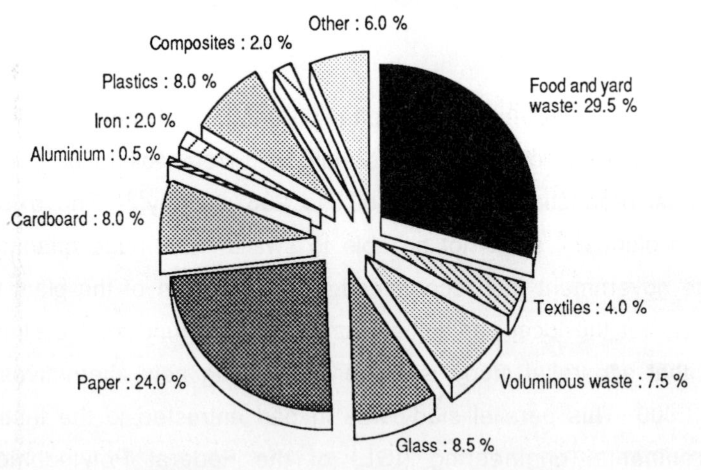

Figure 1: The composition of waste by "macro-classes"

4. The calorific value

Content of hydrogen, moisture and LCV were determined for each of the 47 classes of waste, using own analytical results and data from the technical literature. Thus the average LCV could be calculated for each housing type and economic activity. In 1986 the average moisture content was 32 % and the average LCV was 2300 kcal.kg^{-1} [3]. The predicted LCV for the year 2000 is around 2900 kcal.kg^{-1}.

5. Flows of chemical elements

Eleven chemical elements were selected, either because they are themselves pollutants or because they are good representatives of the behaviour of more complex organic combinations polluting the environment. These are: C, S, N, F, Cl, Fe, Cu, Zn, Hg, Pb. The aim was to compare the global flow of these elements, as output (solid, liquid and gaseous) of the incineration plant, to the input from the 47 classes. The concentration of each of the 11 elements in each class was determined either through analyses or with the informations from production companies or from the literature. This investigation produced graphs as the one shown in Fig. 2.

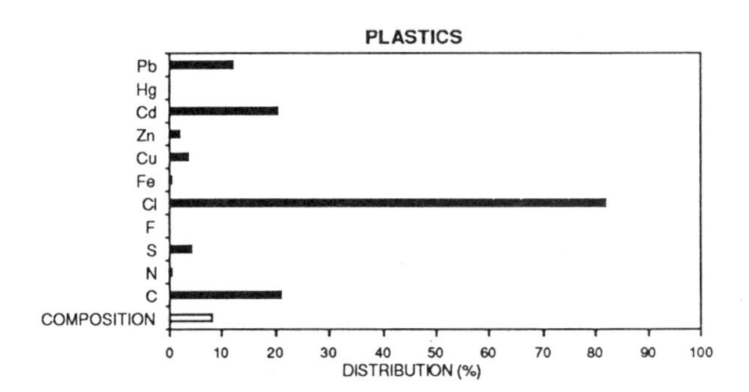

Figure 2: The relative importance of the 11 elements in the macro-class "plastics"

6. Technical investigations

The present collection scheme and recycling system, the up-to-date techniques of waste treatment and of waste collection and sorting, the market situation for recycled products have also been analysed [4]. Every collection and treatment system has been described through its "transfer functions" i.e. the partition of inflow into solid, liquid and gaseous outflows.

7. Inputs of the negotiation procedure

Every feasible situation, solution or point of view supported by any member of the expert committee was accepted for comparison. Four aspects have been considered:

- the socio-economic and political chances of success of separate collection of waste (i.e. non-mixing at the source), depending on the participation of consumers, enterprises and public services: the members of the expert committee, plus 14 members of the local parliament and 15 enterprises were asked 13 questions. This resulted into an expected "span of success" between 10 and 70 %, depending on the action and the class of waste [5].
-alternatives: 14 alternative collection and treatment schemes (so-called "actions") were retained for comparison. These actions are shown in Fig. 3.
-criteria: 11 criteria of comparison have been selected by the expert committee: costs, reliability, flexibility, necessary final disposal volume, recycled materials, energy recovery, marketing of the recycled materials, pollution resulting from the waste treatment, pollution resulting from the use of the recycled materials, nuisances (dirt, noise, traffic),consumer's education and participation [6].
-weighting: each member of the expert committee has given his personal weighting of these 11 criteria.

8. Output of the negotiation procedure

All comparisons were made pairwise between 14 actions, using 11 criteria weighted by 8 experts under 5 ratings of the chances of success, using the outranking and partial aggregation method Electre III [7] [8] [9].

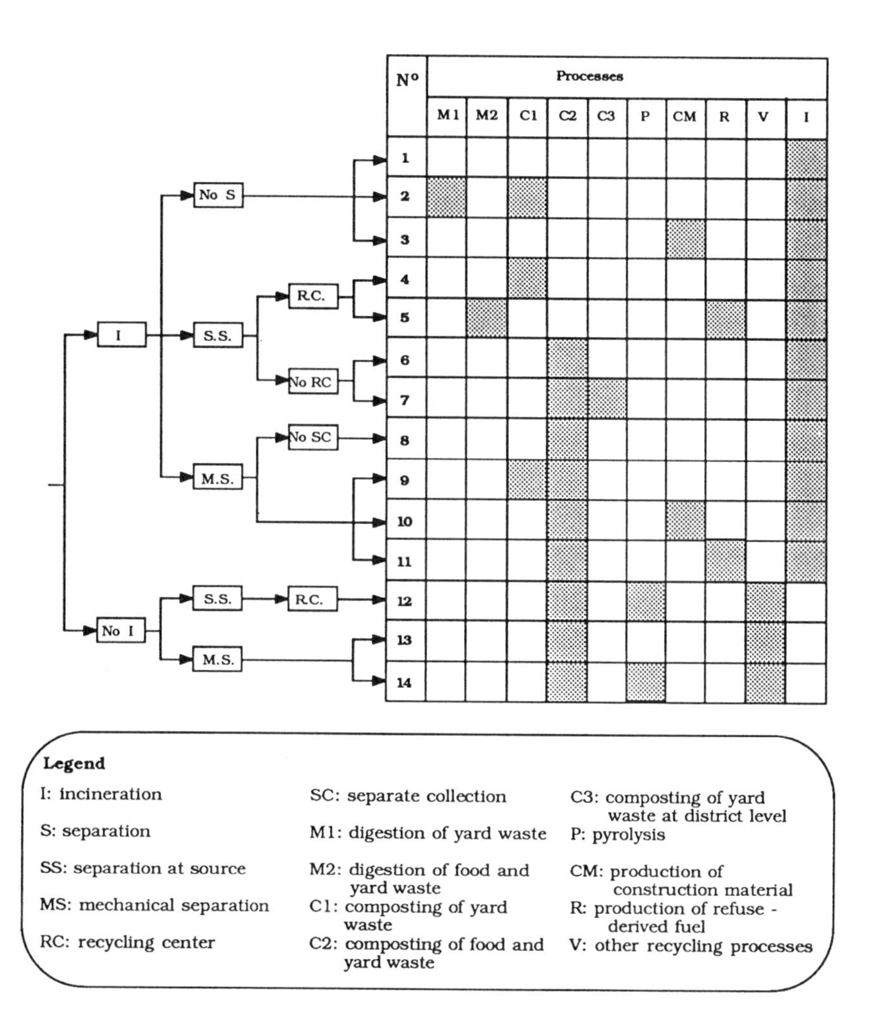

Figure 3: Generation of actions

The comparisons considered three thresholds for "indifference", "clear preference" and "veto". The final result of this analysis is one table for each alternative, indicating how many other alternatives are "better", "worse", "incomparable" according to each member of the expert committee. 5 alternatives are rated "worse" by all experts, 4 have mixed and unreliable ratings, 5 are "better". Of the latter, 2 are definitely rated as the most attractive and deserve further, more thorough and detailed analysis.

They both include incineration combined with composting or digestion of food and yard waste, together with waste collection centres and a minor RDF production plant. Location and number of those other treatment facilities are still to be investigated; the most attractive solution could be a combination of these two better alternatives (no 5 and no 7). The alternative of incineration alone (no 1) is considered among the worst: this is not surprising at a time of new ideas for both a more economical and more ecological, therefore differentiated waste management policy.

References:

[1] Diserens, Th. (1987). Composition des déchets ménagers et assimilés - situation actuelle, IGE, Lausanne.

[2] Diserens, Th., Maystre, L.Y. (1988). Les emballages plastiques dans les déchets: cas du canton de Genève, Swiss plastics 1-2, Zürich.

[3] Viret, F. (1987). Estimation du PCI des déchets ménagers et assimilés du canton de Genève, IGE, Lausanne.

[4] Maystre, L.Y., Leroy, D. (1987). Analyse du coût de la gestion des déchets solides - la récupération dans le canton de Genève, IGE, Lausanne.

[5] Simos, J. (1987). Mesures à la source pour stabiliser ou diminuer la production des déchets, IGE, Lausanne.

[6] Simos, J. (1988). Critères d'appréciation et analyse multicritère des scénarios, IGE, Lausanne.

[7] Maystre, L.Y., Simos, J. (1987). In L'aide à la décision dans l'organisation (Ed. Afcet), 253-258, Association française pour la cybernétique économique et technique, Paris.

[8] Simos, J. et al. (1986). Les méthodes ELECTRE, IGE, Lausanne.

[9] Roy, B. (1985). Méthodologie multicritère d'aide à la décision, Economica, Paris.

PUBLIC ATTITUDES TO RECYCLING

Ole Kaysen, cand.scient.
gendan a/s, Copenhagen

INTRODUCTION

This paper discusses public attitudes to recycling including sorting of waste at the source. The paper does not comment on the attitudes of trades and industries.

The conclusions of the paper are based on an opinion poll and a number of surveys in connection with selected recycling experiments during the period 1975-1987. The main stress is laid on the two latest experiments.

The main conclusion is that the public seem to take up a more sympathetic attitude to recycling after having participated in waste-sorting experiments than before being involved. Consequently, it is important not to be deterred in advance by prejudices against recycling before practical measurements of, e.g. turnout and systems efficiency, have been appraised.

Another conclusion is that the public are both willing to and capable of sorting their waste, but a high sorting efficiency is dependent on continuous information and repeated calling attention to the rules of the sorting.

ATTITUDE TO RECYCLING

In Denmark, real experiments with sorting of waste at the source first took place in the town of Bogense in Funen. In 1975, well over 1,000 private households tried to sort their waste into fractions (1). At that time, such matters as waste problems, utilization of materials in the waste, and development of resources could not in any way be commented on, neither in newspapers nor in other news media.

Being asked whether all households were participating in the experiment, the waste collector answered that three households were not, but it would not be long before they were too. He would see to that! This incident shows the importance of personal engagement for the success of an experiment. Today, how-

ISWA PROCEEDINGS
VOLUME 1 ISBN 0-12-058451-4

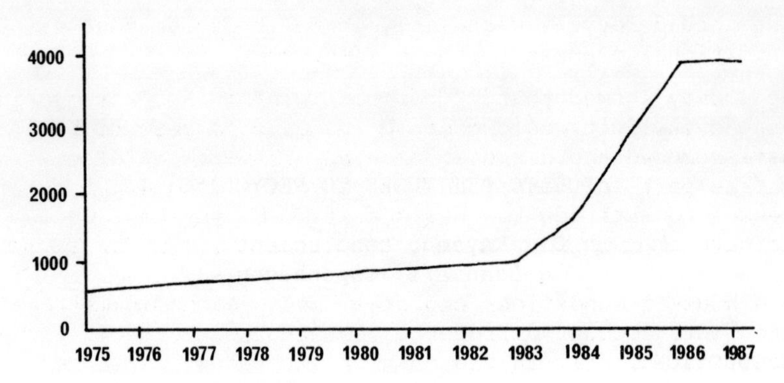

Number of newspaper cuttings concerning waste and/or recycling (in thousands).

ever, idealism on the part of the waste collector is not sufficient for the success of recycling arrangements; the extent, the consequences, and the demands are often too great for that.

Since the above-mentioned experiment was carried out in Bogense the interest in recycling has increased dramatically. This is illustrated by the figure showing the number of newspaper cuttings in which the key words waste and/or recycling were used.

The pronounced increase taking place in 1984 was primarily due to a growing realization of the necessity of an alternative utilization of our waste, but increasing problems as regards the disposal of waste was of importance too.

The debate about waste disposal and alternative methods of waste treatment including recycling should not be seen as an isolated phenomenon; it is related with the "green wave" and the growing realization among the public of the need to strengthen and save our environment.

So, what do we know about the public attitude to recycling? How do we find out whether the recycling duties we impose on the citizens are carried out effectively? How do we keep the motivation of the individual citizen?

The first Danish opinion poll with the purpose of measuring the attitude to recycling took place in 1983 when gendan a/s made the opinion poll institute Observa carry out a survey sending questionnaires to a representative section of the population numbering 1,300 persons over 18 years of age (2). Some of the questions to be answered read as follows:

Do you think that
- we should recycle the waste from Danish households, even if this would cause a 50 kr. rise of the annual rate on removal and disposal of waste?

- we should recycle waste only where it can be done without additional expense to the household?

The survey showed that 79 per cent went in for a recycling of waste even if it would cost 50 kr. extra a year to have the waste removed and disposed of. Only 17 per cent did not want to take part in the paying for recycling, and 4 per cent expressed no attitude to this question. 87 per cent would participate if recycling would not cause expense to the household.

When housing conditions and other socio-economic matters are taken into account differences in attitude to recycling are seen. Most clearly in the case of retirement pensioners, who do not think they have sufficient physical and financial resources, but also independent farmers and fishermen show less interest.

ATTITUDE TO SORTING AT THE SOURCE

It is one thing to be asked to express one's attitude to recycling and another to be involved in a recycling project having to sort one's waste.

In 1983, six municipalities in the county of Frederiksborg in North Zealand participated in a recycling project for the collection of glass and paper. Beforehand, the attitude of the households was registered by means of questionnaires sent to 1,200 households (3). Later in the course of the project the outcome of the collection was analysed in detail by a comparison with potentials.

Asked what the households did with their newspapers and magazines they answered that nearly half of them were collected for recycling (49 per cent), but the subsequent measurements showed that less than one third (27 per cent) was actually delivered for recycling. Recycling experiments have established the fact that people's opinion of how much they deliver for recycling does not correspond to what is actually delivered. Similarly, waste and sorting analyses have shown discrepancies between people's estimate of the weight of their waste and check weighings.

It is important that this discrepancy be stressed because it makes it clear that it is not sufficient to test people's attitude in order to estimate their willingness to contribute to recycling; it is necessary also to include measurements of results from practical experiments.

CURRENT EXPERIMENTS WITH SORTING AT THE SOURCE

Recycling arrangements that are instituted today confront the households with demands beyond that of a mere sorting out of conventional materials such as paper and glass. In preparation for further treatment the reusable products that are sorted out must be clean and they must be sorted very carefully, i.e. glass, paper, waste food, etc. separately. Consequently, it is crucial that the sorting takes place at the source.

In two areas in Denmark, experiments are in progress with a rough sorting at the source, which means division of the household waste into mixed fractions. Both experiments involve a large number of private households where the waste is divided up into a dry and a wet fraction.

In one of these areas (AFAV) the wet fraction is being composted while the dry portion is being burnt or dumped. In the other area (Søndersø) the dry portion is taken to a central sorting plant where the sorting out of reusable paper, cardboard, plastic, etc. is in focus. Thus, in both experiments households must place their waste in bags in two holders indoors (e.g. under the sink) and afterwards take the bags to two holders or bins outdoors, which are emptied by the waste collector.

Both experiments have been in progress for about a year (1986-1987); they affect 8,500 households in eight municipalities (AFAV) and 3,000 households (Søndersø) respectively. In conclusion of the sorting experiments, i.e. before the experiments were made permanent, the attitude of the households to the experiments and to the information given was investigated by means of surveys. In AFAV, 50 households in each municipality received a questionnaire and after a reminder procedure an average of 88 per cent of the letters had been answered. The survey shows a high degree of support and only minor differences between the municipalities are seen.

	Always	Sometimes	Never	Total
Do you sort your waste into wet and dry fractions?	92%	6%	2%	100%
	Easy	Difficult	Impossible	Total
Do you like the system?	88%	10%	2%	100%
	Yes	No	Do not know	Total
Are you satisfied with the information given?	89%	4%	7%	100%

The replies to the question as to what people thought of the system were compared with sorting analyses, which show that

90 percent of the households sort their waste correctly. This is in agreement with people's impression of the system as being easy. A more thorough examination of the survey shows, however, that a number of matters concerning the technical part of the experiment will have to be adjusted (not sufficient room in the bags, too few bags delivered for use indoors). These deficiencies, however, do not seem to be decisive for the acceptance of the system.

Furthermore, in a number of cases the survey shows that discontentedness with the system, e.g. with too little room in the bags, is related to the size of the household, households with three persons and upwards feeling such things distinctly. A similar relation has been ascertained in the Søndersø survey (5).

In Søndersø the experiment was concluded - before being made permanent - by a sample inquiry in the form of questionnaires sent to 400 households (column B). Also, while the experiment was still in progress the households were asked their attitude, 200 households being randomly selected (column A). As the inquiry was held anonymously it was not possible to press for replies. In both inquiries well over 60 per cent replied.

		A	B
I.	Did the sorting of the waste into dry and wet fractions mean extra labour to you?		
	no or very little	67%	71%
	little	24%	23%
	some	5%	3%
	much	2%	1%
	no reply	2%	2%
II.	Are you sometimes in doubt as to which bag to use for a specific type of waste?		
	no	90%	79%
	yes	9%	17%
	no reply	1%	4%
III.	Is the extra work of taking the bin to the pavement		
	without importance	79%	76%
	somewhat inconvenient	18%	14%
	very inconvenient	2%	7%
	no reply	1%	3%

In Søndersø the degree of acceptance is high, which is also the case in the AFAV area. According to both Søndersø surveys the sorting into wet and dry fractions is considered to mean no or little extra work by more than 90 per cent (I). There is no relation between the inconvenience experienced and the size of the household.

Although a direct comparison between the two Søndersø surveys is not possible, these surveys do present interesting features. For example, the problems as to which bag should be used for the dry and the wet waste respectively remain unchanged, they may even be slightly increasing (II). Half of the persons who are in doubt do think, however, that the information they received at the outset was sufficient. Information was given only once.

In Søndersø and in a single AFAV municipality the outdoor bin must be taken to the pavement for collection. In Søndersø more than 75 per cent consider this extra effort to be unimportant and a similar result was obtained in AFAV. There seems to be a relation between the size of household and the feeling of an extra effort - the more persons in the household, the less inconvenience.

ATTITUDE TO THE SORTING OUT OF WASTE FOOD AT THE SOURCE

Furthermore, the households in Søndersø were asked to comment on an extension of the sorting that would include batteries, problem waste, and waste food. 44 per cent are willing to participate in a collection of batteries, and 51 per cent are interested in a collection of problem waste, while only 15 per cent want to contribute to a separate collection of waste food.

The Tarup experiment in 1985 (6) demonstrated that people's interest in sorting out waste food depend on whether they had tried it before. Beyond doubt, the interest was significantly greater among those who had tried it before than among those who had not.

A comparison of the very low potential participation in Søndersø with the AFAV area's acceptance of the experiment where waste food was to be sorted out with a view to composting shows that apparently there is a very great difference between the imaginated and the actual experience of the sorting out of waste food. Many people's notion of the sorting out of waste food is probably based on an idea of something unappetizing or unhygienic as compared with the sorting out of dry materials; this is confirmed by the fact that the willingness to sort out problem waste and batteries is considerably greater, although, in a sense, these types of waste may be said to be just as unsavoury or to imply risks of another kind.

Besides, it should be noted that in Søndersø, people's interest in a further sorting out of, e.g. waste food is quite independent of whether or not it is found easy to sort into wet and dry.

FUTURE ATTITUDE AND MOTIVATION

Of course, it is not possible to predict anything definite about future treatment and sorting of waste by the individual households. In a sorting experiment in the Copenhagen suburb of Albertslund, waste is divided into paper, plastic, metal, glass, and waste food fractions. The reaction of the households to this experiment is not yet known, but it will probably be an indication of how far it will be possible to go in terms of extending the sorting at the source.

The implementation of an effective sorting at the source depends exclusively on the willingness of the households to sort the waste and on their opinion of when the limit has been reached for what seems to be meaningful and reasonably easy. Whether this limit is reached with three, four or five different fractions remains to be seen in the immediate future.

It is important to maintain people's motivation and their interest in recycling. As mentioned before, it is a condition that the demands made on the households are felt to be meaningful.

Indeed, the outcome of the survey and the professional expertise involved in the experiments do demonstrate that to secure an acceptable level of recycling it is necessary continuously to inform about results obtained so far, to rectify mistakes, and to encourage people to a continued willingness to sort.

Previously, gendan a/s developed the idea of a recycling consultant who was supposed to visit all kinds of enterprises in order to suggest and arrange things that might contribute to increased recycling. Without any problems, this practice could be applied to private households. The recycling consultant could pay a visit, not as a control measure, but in order to instruct the households, induce them to sort the waste correctly, and thereby encourage them to contribute to an increased recycling.

REFERENCES

(1) Resultater fra Genbrug 75, Bogense, H.J. Hansen A/S et al.
(2) Genbrug 1983, meningsmålingsundersøgelse, Observa.
(3) Kommunale indsamlinger af glas og papir, spørgeskemaundersøgelse i Frederiksborg Amt, gendan a/s, 1983.
(4) AFAV, gendan a/s, 1988. (Stencil.)
(5) Det grønne affaldssystem i Søndersø kommune beskrevet i Miljøprojekt, Miljøstyrelsen, 1988.
(6) Afrapportering af genbrugsforsøgene i Tarup og Vollsmose m.v., okt. 1984 - sept. 1985, Jysk Teknologisk, 1985.

Session 3

Biological Treatment of Solid Waste

Production and use of biogas from organic residues of agricultural farms.

Erich Dohne, KTBL, Darmstadt

The energy crisis of the 70th has been the reason to have dealings with biogas production world wide again, that means anaerobic fermentation. Also today the main task is the energy production, in spite of the favorable situation of the world petroleum market. On the other side liquid manure, the substrat of agricultural biogases plants, gets more and more a big environmental factor.

There is no registration of agricultural biogas plants. In the FRG we have between 120 and 150 plants, in Europe some 100. Because of the very cheap petroleum the first euphoria is lost. Several biogas plants are shut down by different reasons. But this only is a momental situation. For the future we see a necessity to favor the anaerobic fermentation for two reasons: use of renewable energy and help to solve the problems of residues (reduction of viscosity, to deodorize, N-content better to use for the plants). All effort should be done to make this economic too.

Principle and basis of anaerobic fermentation are well known and not necessary to explain here. Nevertheless the problems are not solved. Wants are to supply about knowledge of suitability and handling of substrat, knowledge of processhandling, development of integrated fermentation processes and digesters. Processes must operate static-free and reliable.

ISWA PROCEEDINGS
VOLUME 1 ISBN 0-12-058451-4

Extensive experiments of several test stations with liquid manure of several kinds of animal have been shown, that there are only small limits to raise the efficiency of existing agricultural biogas plants. Often the efficiency of farm plants does not reach the theoretic possible value as the datas from 50 agricultural biogas plants show (table 1). One point may be of success: to improve the reliability of the plants. The experience of the running plants - they are operated mesophylic and a few number psychrophilic - show, that in all cases the same mistakes and disturbings gives falls out or decreasing efficiency:

- leakage of fermenter
- fall out of feeding pump
- bad function of mixing system (dead volume)
- bad function of heating systems and insulation
- blockage of outlet
- an insufficient safety system or a failure of this system
- the system to measure and control is not reliable
- Corrosion.

Till today agricultural biogas plants exclusively are working with manure, feeding residues and straw. Big problems give slaughterhouse residues. Mixed with other residues and sewage they could be handled, but often this produces blockages. Several institutions try to solve these problems.

Biomass crops from several programmes to reduce surplus production may be a suitable substrat for biomethanasation like plant biomass from landscape conservation, a new business for farmers. At harvest, handling and processing of agricultural products for food or industry several by-products and residues fall up (upper parts of roots, extraction- and press residues). These materials gives problems because often they no more can be used as feedstuff or may be brought to waste dump. It's possible to increase the energy balance of the processes by making biogas out of the residues. The digester sludge is biological stabilized and may be used in several ways.

Everytime a special substrat must be seen in connection with a suitable fermenter. BAADER and his collaborators made a lot of longtime experiments with the above substrats preferable in a one-step semi-industrial loop-type fermenter. Here after sludge separation the liquid is used for fluidisation of the new substrate. First results are available with a 100 m³ fermenter of the same type. A lot of fermentation tests in small labor fermenters are done to find out the technical possible methane yield of several raw materials. Today we have datas about all kinds of liquid manure, about the most agricultural crops and residues of ethanolproduction, sugar beet extraction. vegetable processing, fruit and malt husks.It's possible too, to process highly loaded sewages of foodstuff-industry in anaerobic fermenters. Often 95 % of the organic carbon may be converted into biogas. With low loaded substrates and continuous processes special measures have to be done, to retain the slowly growing anaerobic bacterias.

In the last time, when household refuse is collected separately it is tried, to use the wet fraction for anaerobic fermentation.

In agriculture only mesophilic and psychrophilic biogas plants were used. With higher temperature (thermophilic) the degradation is quicker but the energy need is higher, the gas quality is worse and the process stability is lower.A thermophilic process nearly is the only way to get a sure hygienic sludge. This may be necessary if communal biogas plants shall be used.

Liquid manure of modern animal stables can contain several pathogenes, bacterias, parasites, virus. A sure hygienisation is not given, that means a killing of pathogenes perhaps are in the liquid manure. With communal biogas plants the problem is bigger. If of one farm the manure is infected, the whole mixed manure of the communal plant and the tank trailer must be infected. Therefore the manure of all users must be controlled periodically. Infected farms are to shut out until recovery. A two year run of a communal biogas plant did not give big problems.

For a sure hygienisation of a thermophilic biogas plant a semicontinuous feeding all two days is suggested. Looking for a stable thermophilic process is a task for future.

Tests about a possible influence of disinfectant on the gas production showed an influence only by a few agents (Decaseptol).

The tested feedstuff additives, antibiotica and sulfonamides had no influence exept of one (Monensine).It is known that thinning of manure raises the gas yield, for example a manure with 13,8 % DM brought 262 l/kg ODM, the same manure thinned to 4,84 % DM brought 423 l/kg ODM. Is the retention time the same, the fermenter volume must be higher. The higher the dry matter of a manure, the well balanced is the proportion of fermenter volume, investment and gas yield. Substrates after separation of solid matter may be treated with the new kinds of fermenters.

All high efficiency fermenters for cleaning industrial sewages which are today in work or in development are working with concentration or retaining of the active biomass in a continuous process. So contrary to an ideal mixed reactor shorter retention times and higher gas yields are possible. Retention times down to few days are possible. Methods for retaining bacterias are fermenters with contact-system, fluid-bed, fixed-bed, sludge-bed, fixed-film etc. Figure 1 shows the suitability of the processes dependend to the dry matter content. Through these enriching processes the same gas yield with about a half fermenter volume is possible.

With distiller's residue and mixed substrates tests run with a two step process. Before the biogas fermentation the substrat will be controlled microbiell hydrolized. After this the substrat is mechanically separated and with the liquid phase fed a methane fermenter. BAADER reports about his works with a mixture of potato distiller's residue, silage of sugar beet leaves and water. The first step could load up to 9 kg $ODM/m^3 \cdot d$. A gas yield of 3 m^3/m^3 has been reached. EDELMANN means, he can reduce the retention time down to 2 days, if surfaces covered with methane bacterias will moved in biogas plants pulsatory. Further on we don't have safe fermentation systems for solid manure.

Because of environmental reasons several concepts for communal biogas plants are by working. The very few running projects show the principle technological function, but they show the big problems of these kinds of plants too: only medium gas yield (< 0,4 m^3/kg ODM), very high cost to transport the manure. A plant with an average transport distance of 7 km reached transportation cost of 0,7 DM/m^3 gas or 0,11 DM/kWh gas. To that comes the cost of the plant itself and when making electricity the conversion cost.

For special regions (high animal density) it can be necessary to process the sludge after fermenting, for example phase break. If there are not enough fields to bring up the sludge, a biogas plant may be a first step of a complex manure handling plant.

Hydrogen sulfice (H_2S) is a normal component of biogas with a concentration between 0,1 and 0,5 % of the volume. An average content of 0,35 % of volume correspondents applied to the kWh nearly with a double sulfur content of heating oil and gasoline Hydrogen sulfide is toxic and corrosive. Before using the gas all should be done to reduce the sulfur content, especially when using it in engines.

There are different kinds of gas treatment. FANKHAUSER (Table 2) gave an account of these. In agricultural plants, most of them are small units. The dry cleaning method with iron hydroxyd is used because of good handling. It is looked hardly for new cheap cleaning methods. Small agricultural biogas plants now have treatment cost of about 0,06 DM/m³, 0,02 DM/m³ of this alone for the iron hydroxyd. In some projects of developing countries it has been found out that it's possible to reduce the H_2S-content of biogas down to a few ppm through giving air (1 - 2 % of volume) directly in the gasholder. It should be controlled, if this simple method is possible too under other condit_ions (reactiontime, temperature, katalyzer)

About the effect of biogas methods to environmental protection there are often wrong ideas. The flowability gets better and so clings worse to the plants. The amount of good available ammonia to the plants gets higher. It's possible to fertilize - as top dressing too - with biosludge more calculated against normal liquid manure. But biosludge gives more problems at time free of vegetation. In general the fertilize effect is not better.

The smell reduction is a function of retention time. More tests have to be done if smell molestation gets more an environmetal noxiousness.

Production and consumption of biogas from farm-owned biogas

plants are generally not congruent. A gas store may be suggestive. If a gas store is not needed it would be better (waterheating continuously or making electricity). A balance about the daily energy need, which can be replaced by biogas may be a good help to find out the right size and kind of biogas store if necessary. A balance of yearly energy need helps to find the right size of biogas plant.

One of the most expensive elements of a biogas plant is gas storage. It only should have a capacity as small as possible. One should take into account, that the energy content of 1 m^3 biogas is equivalent to 0,65 l diesel-oil. The energy content of 8 m^3 biogas is equivalent to 5 l-metal fuel container, with a price of not more than 10,-- DM. It is not possible to speak here on all kinds of storage and their pros and cons.

There are many possibilities for using biogas, but only a few are practiced in the farms of the FRG:

- house heating) atmospheric burner, fan burner
- making warm water) atmospheric burner, fan burner
- infrared radiator
- drying of grain or hay (warm water, hot air)
- kitchen appliance (stove, refrigerator, washing machine, dish washer)
- gas heat pump (absorption-compression)
- lightning
- engines (stationary, making electricity, tractor)
- giving into a gas network.

The main problems for using biogas are coming from the water vapour saturation the H_2S-component (corrosion), gasborne particulates, fluctuation of CH_4-content and gaspressure. Another problem are to small gas pipes. The implements need a gas pressure between 5 and 20 mbar. In the existing biogas plants the biogas is used preferably for house heating and making warm water.

Only in some cases it's possible to use the available net gas 100 %. Technical it's possible to give the cleaned gas into an existing gas network or to sell it to big energy user for heating. This simple using everytimes seem to be the best, because heating oil or natural gas are exchanged directly. There are no technical problems. Another possibility to use big quantities of gas is to use it as fuel for engines, for tractors or producing electricity for own need or give it into a network. When making electricity with a co-generation-set an efficiency of 80 % can be reached. But low gas cost are necessary if making electricity shall be efficient. As an average 1 m³ biogas gives 1,8 kWh_{el}. With optimal working of the sets (100% use of waste heat), whole machine costs were reached of 0,11DM/kWh_{el}, that means without the cost for gas production. So it seems impossible to produce electricity below 0,2 - 0,25 DM/kWh_{el} also when the conditions are optimal. Table 1 shows, that the average gas cost of farm biogas plants has been 0,25 DM/kWh gas.

Biogas engines require external ignition. Possible are gas-otto-engines and gas-diesel-engines. Both kinds of engines don't give problems in general. Difficulties are coming when using biogas as tractor-fuel through the necessary high pressure compression. The economy of biogas tractor looks very bad today including the necessary cleaning and compression station. A complete gas-filling station for daily 200 m³ biogas needs an investment of about 100 000 DM, the adaptation of a tractor needs about 10 000 DM more. This seems equivalent for the farmer, if he must pay about 1,5 DM/l diesel-oil.

With interest I read about new tests of the DDR with a pilot plant in which were used liquified biogas as tractor fuel. Biogas was cleaned into pure methane and then liquified through cooling down to -165,5 °C. This deepcold liqui-methane (Kryogas) is stored in a special tank with pressure between 1,5 and 5 bar. Main problem for use in motor cars is the quality of insulation of the Kryotank, recently the lossfree time before the safety valve opens. Lossfree standing between 1 and 7 days were reached. A fictive big station reached equivalent cost of 2,5 M/l diesel-oil. That's now more expensive than diesel-oil.

The main factors of influence on the efficiency of a biogas plant are:

- price of comparable energy (fuel, natural gas)
- yearly cost of the plant in % of the purchase price (depreciation, interest, repair, maintenance)
- process-energy (operating cost)
- Investment (influence of fermenter typ).
- gas yield (influence of substrat and fermenter type)

Figure 2 shows the interdependences with a static view. The graph shows, that an optimal working biogas plant must be cheaper than 1 000 DM/LU with the energy prices of today and 100 % use of net gas, otherwise it cannot be economic.

The economic marginal conditions shall change in future for the benifit of biogas plants. We mean biogas plants which have reached a high technical level today than shall be produced in series so that more and more an efficiency is given.

References

BAADER, W.; R. AHLERS: Vergärung pflanzlicher Roh- und Reststoffe zur Massenreduzierung und Biogaserzeugung. Vortrag VDI-Tagung, Braunschweig 22./23.10.1987.

BARDTKE, D., W.J. HOMANS: Bericht über den geruchsreduzierenden Effekt des Biogasverfahrens am Beispiel Schweinegülle. Inst. für Siedlungswasserbau, Stuttgart 1985

DOHNE, E.: Biogas engine, 4. Internat. Conference for anaerobic Fermentation, 11.-15. Nov. 1985, Guangzhou, China

FRANKHAUSER, J. u.a.: Erfahrungen mit Biogas als Treibstoff für Landwirtschaftstraktoren. Bericht Nr. 27 der Eidg. Forschungsanstalt für Betriebswirtschaft und Landtechnik, FAT, Tänikon, 1985

KLOSS, R.: Planung von Biogasanlagen nach technisch-wirtschaftlichen Kriterien. Verlag Oldenburg, München-Wien 1986

LINKE, B. u.a.: Zur optimalen Belastung von Biogasanlagen mit Rinder- und Schweinegülle, Agrartechnik 37 (1987), H. 10, S. 450-452

MAURER, K. u.a.: Gemeinschaftsbiogasanlage Wiesloch. Agrartechnische Berichte Nr. 20, Universtität Hohenheim 1987

PERWANGER, A.: Kurzfassung zum Schlußbericht "Untersuchung und Optimierung von Biogasanlagen in der Praxis mit technisch ökonomischer Vergleichsauswertung. Bayerische Landesanstalt für Landtechnik, Weihenstephan 1987

POHLIG-SCHMITT, M.: Seuchenhygienische Untersuchungen bei der thermophilen und mesophilen anaeroben alkalischen Faulung von kommunalem Klärschlamm. Diss. Hohenheim 1987, Univ. Fak. IV.

RIEMER, M; D. VERVIER, R. KLOSS, et al: Biogasanlagen in Europa. Ein Handbuch für die Praxis. Köln, TÜV-Rheinland, 1985

SCHEURER,E.: Untersuchungen zum anaeroben Abbau von Hühnerflüssigmist. Diss. Hohenheim 1986

STEINER, A.: Einsatzmöglichkeiten der Biogastechnologie zur Schlachthofentsorgung. GATE-Biogasinformation Nr. 24/1987, S. 30 ff.

VOLLMER, R.; B. LINKE: Zur anaeroben Fermentation von Rindergülle im thermophilen Bereich. Agrartechnik 37 (1987), H. 5, S. 208/209

WELLINGER, A.; u.a.: Biogas-Handbuch. Verlag Wirz, Aarau 1984

WINTERHALDER, K.: Untersuchungen über den Einfluß von Desinfektionsmitteln, Futterzusatzstoffen und Antibiotika auf die Biogasgewinnung aus Schweinegülle. Diss. Hohenheim 1985, Univ., Fak. IV.

-: Biogas in der Landwirtshaft. Institut f. Umweltforschung, Graz, 1986 table 1: Characteristic data out of 50 agricultural biogas plants

data		average value	limit value
added animal number	(LU)		15 - 1 000
fermenter volume	(m^3)		15 - 2 x 630
fermenter loading	(kgODM/$m^3 \cdot$d)	2,6	0,5 - 7
retention time	(d)	35	11 - 78
reduction ratio of fresh manure	(% ODM)	43	18 - 76
gross gas production	(m^3/LU·d)	0,98	0,4 - 1,45
added manure	(m^3/kgODM)	0,34	0,16 - 0,64
added manure	(m^3/m^3)	20	11 - 32
fermenter volume	(m^3/m^3)	0,71	0,2 - 1,75
methane content	(% CH_4)	65	56 - 72
sulfure content	(% H_2S)	0,35	0,11 - 0,75
process energy	(% gross energy)	38	21 - 74
net energy	(kWh/LU·d)		1,1 - 6,0
cost net energy (30 % process energy, 100 % gas use)PERWANGER	(DM/kWh)	0,25	0,03 - 0,96
capital expenditure	(DM/LU)	1 300	93 - 3 500

table 2: Treatment of biogas

A. Separation of condensate and mud	
1. Separator	- skimmer gate, rubble filter
2. Coars- and fine filter	- interchangable caramic filter elements
3. Biologic filter	- tests with compost filter
B. Extraction of gas components	
1. dry cleaning iron hydroxyd, pressure less (H_2S)	- reaction time 3-5 minutes, large surface, low gasflow velocity. Low problems with handling, less problems to put away the saturated matter
molecular sieve, medium pressure (H_2O, H_2S, CO_2)	- SiO_2- Al_2O_3 cristallin hydrous alumina silicate and silic acid. About 700 m^2 inner surface per g material. High technological expenditure.
2. wet cleaning pressure water (CO_2,H_2S)	- only together with compression. pressure water is toxic and corrosive.
Äthanolamin (H_2S, CO_2)	- solvent is toxic and corrosive
lime milk (CO_2)	- $CaCO_3$ is residue, without problems
3. membrane technology	- in development
4. gas drying	- necessary for storing in high pressure bottles, 10 °C below lowest working temperature
5. direct oxydation with air (H_2S)	- sulfur falls out directly. May be necessary higher temperature and catalyzer
C. Liquefying gas	
1. low-temperature technologie	- pressureless at -161 °C. No long-terme storage instead of high vaporation rate. Not suitable for agriculture
2. changing into methanol	- through partial oxydation a long terme storable fuel can be produced High safety requirements uneconomical for low technology plants
D. Gas compression	
1. medium pressure compression (10-20 bar)	- reduction of storage volume at 7 %. Very high safety regulations. High repair expenditure. Inspection obligation.
2. high pressure compression (> 200 bar)	- reduction of storage volume less than 1 %. Inspection obligation. High cleaning expenditure, necessary for vehicle fuel.

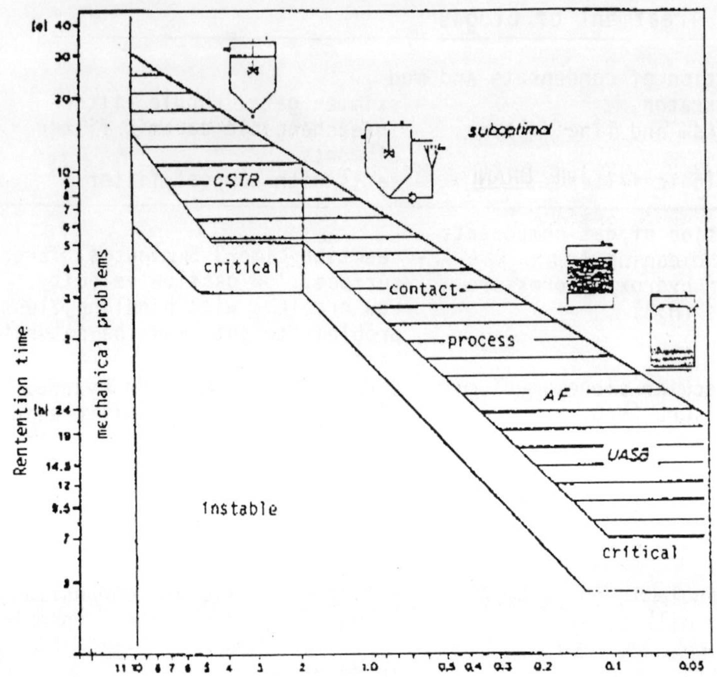

Fig.1: Optimal loading of different anaerobic processes (STEINER)

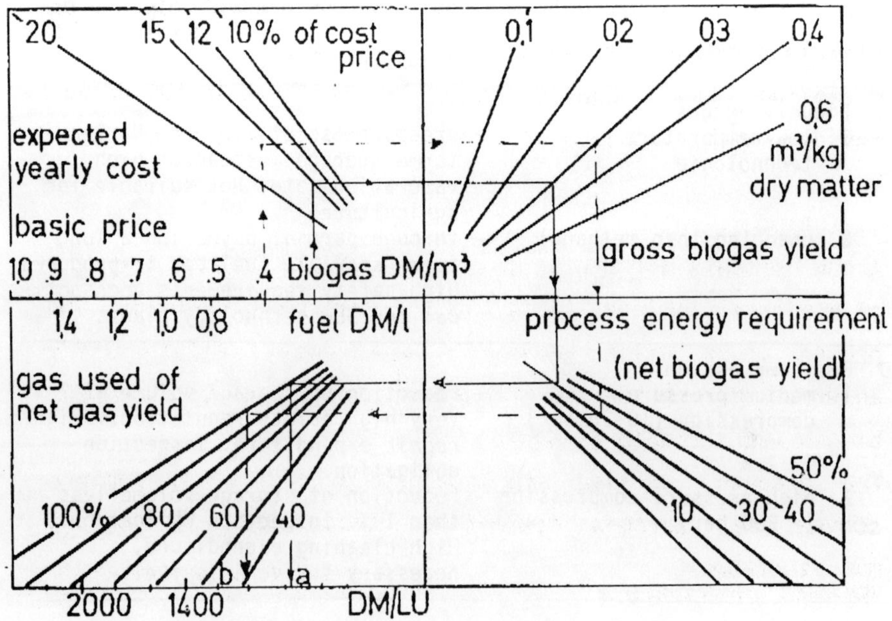

Fig.2: Admissable investments into biogas plants

DRY ANAEROBIC COMPOSTING OF MIXED AND SEPARATELY COLLECTED MSW BY MEANS OF THE DRANCO PROCESS

Six, W. and DE BAERE, L.
Organic Waste Systems, Dok Noord 7 b, B-9000 Gent, Belgium.

Summary

Pilot plant experiences are given for the dry anaerobic composting process of mixed garbage, while separately collected garbage was tried out on laboratory scale (30 l) digesters. The process works at 55°C with a total solids content of 35%. The gas production rate is 3,33 $Nm^3 CH_4/m^3$ reactor.d at the pilot plant. Separately collected garbage yielded gas production rates ranging from 1,75 to 3,85 $Nm^3 CH_4/m^3$ reactor.d depending on the manner of separation. The Duobak system which separates a recyclable fraction with 5 components from the "rest" fraction, is an efficient method of separation and results in a stable digestion. Low concentrations of heavy metals in the dried residue can be obtained through separate collection of MSW.

1. Introduction

A new process for the stabilization of the organic fraction of municipal solids waste (MSW) has been developed and proven on a pilot scale. The process utilizes a dry anaerobic fermention technique in order to produce both biogas and a stable compost.

The process, called the DRANCO process, can replace conventional aerobic composting techniques, which are

ISWA PROCEEDINGS
VOLUME 1 ISBN 0-12-058451-4

currently utilized in most solid waste recycling plants. The process functions thermophilically (50 - 55°C) and at a total solids (TS) concentration of 35%. This reduces, in comparison with the conventional anaerobic digestion which works with a slurry of 4 to 8% TS, the amount of water needed to adjust the incoming waste to the desired solids concentration. This also decreases the volume of the digesters by about a factor of five while the production rate is increased with a factor of five, because the higher TS content does not inhibit bacterial activity. Scum formation is not possible at this high solids concentration and no mixing is needed.
There is a growing interest in obtaining the recyclable fraction by separating the municipal solid waste at the source. The digestion of the organic fraction, resulting from two different collection approaches, is discussed.

2. Process description (see figure)
Organics are digested anaerobically during a period of 21 days, at an operating temperature of 55°C and a total solids concentration of 35%. The digested residue is subsequently dewatered to 65% TS by means of a hydraulic press. The dewatered cake is broken and dried to 75% TS, while the press water is used to adjust the solids content of the incoming substrate or is evaporated by means of the calories in the cooling water from the gas engines. The biogas is recovered at the top of the digester and is converted to electricity. The calories in the exhaust gases from the gas engines are used to heat up the incoming material and to dry the cake solids coming from the press.

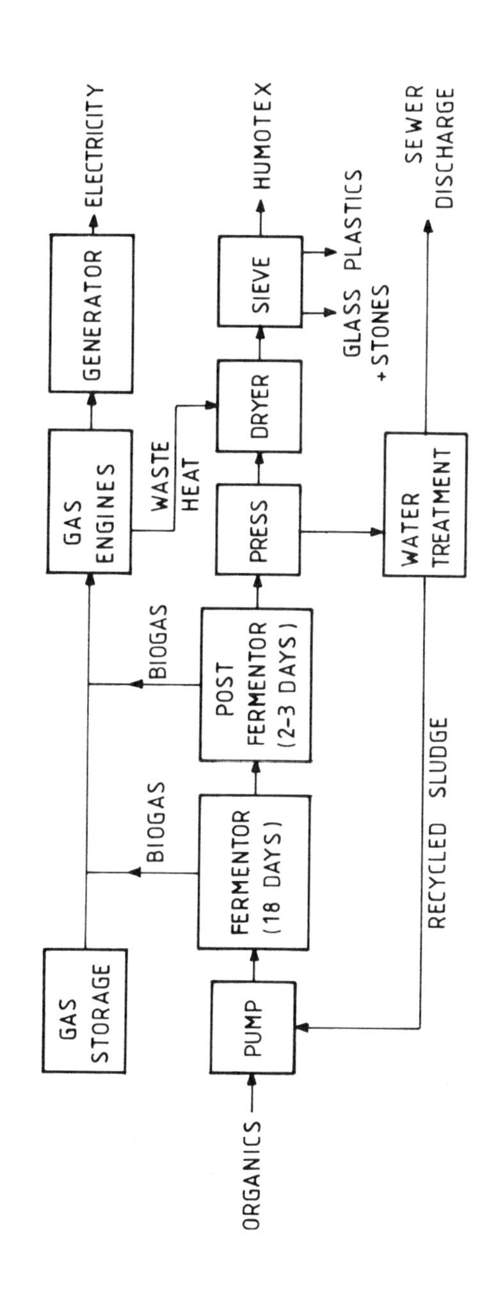
FLOW SHEET DRANCO® PROCESS
GAS STORAGE
GAS ENGINES
GENERATOR
ELECTRICITY
BIOGAS
BIOGAS
WASTE HEAT
ORGANICS
PUMP
FERMENTOR (18 DAYS)
POST FERMENTOR (2-3 DAYS)
PRESS
DRYER
SIEVE
HUMOTEX
GLASS +STONES
PLASTICS
WATER TREATMENT
SEWER DISCHARGE
RECYCLED SLUDGE

3. Dranco applied to MSW.

3.1. Mixed garbage.

Dry anaerobic fermentation of mixed garbage has been demonstrated on a pilot scale with a volume of 56 m^3 at the solid waste treatment plant in Gent, Belgium. Mixed municipal garbage is pretreated by a homogenizing and comminuting drum and sieved over a 20 mm-screen. The organic fraction (< 20 mm) had a TS content of 56% and a volatile solids (VS) content of 62% on the total solids. The organic fraction was adjusted, in order to obtain 35% total solids in the digester, and introduced at the top of the vessel. The anaerobic treatment took place in the fermentor during a period of 3 weeks. The residue was removed at the bottom of the digester by a gas-tight extraction system. At a thermophilic temperature and a volumetric loading of 17,3 kg COD/m^3 reactor.d, a 55% COD-reduction was achieved. This corresponds to a production rate of 3,33 $Nm^3 CH_4/m^3$ reactor.d. The methane content in the biogas was 55%. The weight reduction of the organic fraction to dried digested residue was 52%.

3.2. Source separated garbage.

Separately collected garbage was tried out on laboratory scale (30 l) digesters for two cities with two different collection approaches.

The inhabitants from Amersfoort (Holland) divide the garbage into a recyclable fraction with 5 components (plastics, glass, metals ,textiles, dry paper) and a "rest" fraction (mainly organics and wet paper). A mini garbage container with two compartments, called DUOBAK, is used. The trucks

which collect the garbage have also two compartments. The recyclable fraction, about 30% of the weight, is sorted.
The "rest"fraction is composted. Laboratory scale digesters were used to prove the applicability of the dry anaerobic digestion of the "rest"fraction. The "rest"fraction was homogenized and comminuted in a drum and then sieved over 20 mm. The fraction smaller than 20 mm had a TS content of 43% and a VS content of 58%. This fraction was fed to lab scale continuous digesters. At a volumetric loading of about 20 g COD/lreactor.d, a 55% COD-reduction was achieved. This corresponds to a production rate of 3,85 $Nm^3 CH_4/m^3$reactor.d. The weight reduction was 63% after a digestion time of 12 to 14 days. The paper fraction in the "rest"fraction had a favourable influence on the stability of the digestion, due to the slower conversion of paper. Eighty percent of the total amount of the "rest"fraction can be treated in the Dranco process. This means that the Duobak system results in an efficient separation of the biodegradable fraction.
A separately collected fraction in the city of De Bilt (Holland) of vegetable, fruit and garden waste (VFG-waste) yielded a 50% COD-reduction at a volumetric loading of 10 g COD/lreactor.d. This means that the production rate is 1,75 $Nm^3 CH_4/m^3$reactor.d. The retention time was 11 days. The TS content of the feed was 35% and the VS content was 55%. The weight reduction was 67%. Because of the rapid hydrolysis of this kind of feed, acidification of the reactor occurs easily at higher volumetric loading rates. The amount and composition of VFG-waste had bigger seasonable fluctuations than the "rest"fraction of the Duobak system.

3.3. Humotex quality.

After the pressing and drying of the digested residue, a dry hygienically acceptable compost, called Humotex, is produced. The product is highly stabilized. The C/N ratio is 12 and the cumulative oxygen demand during 10 days is 60 mg O_2/gOM. Humotex is virtually free of pathogens. Low concentrations of heavy metals can be obtained through separate collection of MSW (see table), due to the lack of contact between metals and the organic fraction. The concentrations in the humotex from the "rest"fraction of the Duobak system are approximately the same as the concentrations in the humotex from the VFG-waste. Heavy metals concentrations are higher in the digested residue compared with the fresh incoming waste because of the transformation of dry matter into biogas during the stabilization. The concentration factor is 100 / (100 - (% VS x % COD-red./ 100)). This factor is 1,5 for the mixed garbage and the "rest"fraction of the Duobak system, and 1,3 for VFG-waste.

Table :

Heavy metals concentrations (mg/kg DM) in Humotex (1)

	Humotex from mixed garbage after iron removal	Humotex from separately collected garbage DUOBAK Amersfoort	VFG De Bilt
Cd (1,5)	2,1	2,2	1,8
Zn (250)	1020	155	138
Cu (50)	101	48	20
Pb (150)	522	63	67
Ni (50)	46	23	25

The numbers between brackets are the proposed maximum allowed values concerning sewage sludge in the Netherlands from 1992.

(1) : State University Gent, Belgium

REFERENCES

De Baere, L., Van Meenen, P., & Verstraete, W. (1987) Anaerobic Fermentation of Refuse. Resources and Conservation, 14, p. 295-308, Elsevier Science Publishers B.V., Amsterdam.

Deboosere, S., De Baere, L., Smis, J., Six W. and Verstraete, W. (1986), Dry Anaerobic Fermentation of Concentrated Substrates, In Anaerobic treatment - a grown up technology, Conference Papers Aquatech, 15-19 Sept. 1986, Amsterdam, The Netherlands, p. 479-488.

De Baere, L. and Verdonck, O. (1986) Thermophilic Dry Anaerobic Fermentation Process for the Stabilization of Household Refuse. Proceedings E.E.C. International Symposium on Compost.

Separate collection and composting of putrescible municipal solid waste (MSW) in W. Germany

Uta Krogmann

On November 11, 1986, the 4th amendment of the West German waste law became effective. One of the essential points of the amended law refers to the avoidance and recycling of waste of all kinds. One way of utilizing the collected municipal solid waste (MSW) is the production of compost. In spite of the high concentrations of pollutants (f. e. heavy metals) in the MSW only the organic degradable household waste fraction (VFY waste ≙ vegetable-fruit-yard waste), which is the largest fraction of the MSW (about 40 percent by weight), should be composted and then be used as soil-improving material.

Due to the above mentioned reason at present, only 2 % of the MSW is processed in 17 compost facilities in the Federal Republik of Germany.

Composting of VFY waste (biowaste) has not yet been practiced on a full scale. Currently, there are about 40 projects dealing with the separate collection of biowaste in households on a small scale. The separately collected biowaste is experimentally composted in unaerated piles (windrow-system) or in MSW compost facilities. The first West German compost facility that switched over from MSW composting to biowaste composting is the facility located in Lemgo.

This paper presents a pilot project in Hamburg-Harburg called "green bin" as an example of source separation and composting of biowaste.

ISWA PROCEEDINGS
VOLUME 1 ISBN 0-12-058451-4

1. The "green bin" project in Hamburg-Harburg

1.1. Survey

The Hanseatic city of Hamburg with about 1.6 million inhabitants has no landfills on its own territory and the three MSW-incinerations have no free capacities. For this reason a part of the MSW produced in Hamburg is regulary transported over a distance of 60 km to the sanitary landfill in Schönberg (German Democratic Republic).

The "green bin" project in Hamburg-Harburg was started at the end of 1985 with the aim to contribute to a solution of this problem by reducing the amount of waste to be incinerated and/or landfilled by utilizing it. 3000 households in 5 differently structured housing estates (detached houses, terraced houses, low-storey houses, city centre houses and multistorey houses) are collecting the VFY waste separately. The biowaste is collected separately by the waste collection trucks once per week; the biowaste fraction is then composted in unaerated piles (windrows) according to the districts where it has been collected. The differentiation of the test areas according to different residential structures was based on the following consideration: Due to experiences made in other projects it could be expected that the amount and the composition as well as the degree of pollution of VFY waste varies according to the different residential structures given in the investigated areas. Different compositions of the VFY waste may also influence the composting processes.

1.2. Results

The separate collection of waste can only be realized if there is the cooperation of the residents. This fact was also one of the results of the "green bin" project; that is why an extensive information policy (home visits, information brochures, lectures, press, radio and television) are of considerable importance.
One may state that an average of about 50 % of the selected households is voluntarily participating in the project. In connection with this figure it must be taken into account that a great number of households could not be approached by direct mail information or by home visits. Only in very few cases the idea of separate collection and composting had been **criticized or objected to.**
Figure 1 shows the average annual composition of biowaste in our test area. The composition of the VFY waste of detached and terraced houses is decisively influenced by the "garden season".

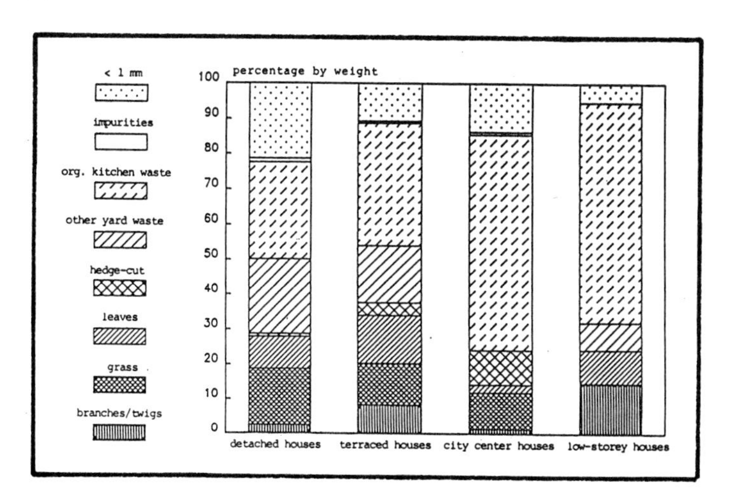

Figure 1: Average annual composition of the VFY waste in weight percent (Results of 4 sortings)

Due to the introduction of bins for biowaste, the capacity of the waste containers was increased (up to 100 %) and this probably led to the fact that garden waste in areas with detached or terraced houses became MSW as well. Accordingly, there was a considerable increase in the total waste production rate. Further investigations will have to show whether this is just a shift in the waste streams or whether it actually is an effective increase in the waste production.

Practicing the separate collection and composting of the biowaste the reduction of the MSW to be incinerated or landfilled can be estimated to 20 - 30 % in all tested areas. By 50 % voluntary participation this would correspond to a portion of 10 - 15 % if the system were introduced on a large scale.
The degree of impurities of the separately collected waste depends on the number of residents being served by a waste container, that means the smaller their number the lower the amount of impurities. Accordingly, the highest degree of impurity is found in the biowaste of multi-storey houses.

The separately collected VFY waste is transported to a compost site that was sealed by an asphalt liner. Impurities are sorted out by hand. The thus presorted biowaste is piled up to a height of about 1.3 m. As fractions of the VFY waste have a high moisture content and are poorly structured (f. e. organic kitchen waste and grass), dry and structuring materials (f. e. wood chips) are added. The piles are turned several times by a compost turner (in the first two months every 7 to 14 days then once per month) until the compost can be used for soil amelioration or as additive for substrates (after 3 - 9 months). The heavy metal concentration in the

Harburg compost is significantly lower than the respective concentration in composts originating from not separately collected waste and is even below the threshold values indicated for cultivated soils (see table 1)

	Cu	Zn	Ni	Cr	Cd	Pb	Hg	As
	mg / kg							
Harburg compost	43	235	7	29	0,8	76	0,2	7
Threshold values for cultivated soils	100	300	50	100	3,0	100	2,0	-
MSW compost	274	1570	44,9	71,4	5,5	513	2,4	5,2

Table 1: Heavy metal concentration in different composts (Krogmann 1988)

As expected, it is possible to compost biowaste according to the windrow method. However, priority should be given to a composting method for big cities like Hamburg including an intensified covered pre-composting and post-composting for the following reasons:

- minimization of the space required,
- improvement of the sieving qualities by reducing the moisture content,
- improvement of the compost exploitation,
- reduction of odor,
- influence of cold and wet weather
- reduction of bird nuisance.

The marketing of the relatively small amounts of compost does not present any problems and experience has shown that there is even an additional demand for it.

2. Comparison

Table 2 shows a comparison of 3 typical projects dealing with source separation and composting. There are variations in the amount of VFY waste obtained as well as in the rate of the reduction of the remaining waste. About 10 - 35 % by weight of the residential waste can be utilized by practicing source separation and composting of the biowaste. This is due to considerable differences in the marginal conditions related to facts like the structure of the housing estates, the question as to whether the residents' participation was on a voluntary or obligatory basis, the capacity of the container for biowaste and the importance of the residents' own composting.

Projects	Participating households (HH) resp. residents (R)	Structure of residential area	Participation (%)	Specific volume for VFY waste/ remaining waste+ ++ (l/pers·week)	Specific production of VFY waste/ remaining waste++ (kg/pers · week)	Impurities in VFY waste (weight-%)	Reduction of the remaining waste+++ (%)
Hamburg -Harburg (Krogmann 1988)	~3000 HH	detached h. (a) terraced h. (b) city center h. (c) low-storey h. (d) multi-storey h. (e)	47 (a), 50 (b) 65 (c), 49 (d) 41 (e) voluntary participation	33/39 (a,b) 7/39 (c,d,e)	4,6/4,6 (a) 3,3/3,6 (b) 1,3/2,6 (c) 1,5/3,6 (d)	0,2-0,3 (a,b) 1,6 (c) 3,0 (d) 7,0 (e)	10-15
Göttingen (NN 1987)	6031 HH 14200 R	town: mostly terraced and detached h. rural district	94 no voluntary participation	20-34/34-43	2,5/2,6	0,5-2,0	34-36
Bergischer Abfall-beseitigungs-verband (Tym 1987)	664 R	rural area	70 voluntary participation	15/70	1,8/6,0	0,3	13

\+ total waste = VFY waste + remaining waste
++ Referring to the participating households
+++ Residuals from composting were not respected

Table 2: Comparison of 3 "green bin" projects

3. Conclusion

In West Germany, the recent developments in the field of composting are going towards a source separation in the collection of waste with subsequent of the biowaste. This way of proceeding has not yet been realized on a full scale. However, the results obtained up to now are promising, especially the low content of heavy metals and the inerts content which make up the good quality of the final compost. There are still investigations necessary to optimize separate collection and composting.

References:

Krogmann, U. (1988). Kompostierung als Abfallentsorgungsverfahren von Biomüll. Be published in Wasser und Boden.
Tym, K., Doedens, H. (1987). Die braune Biotonne als dritter Abfallbehälter. Modellversuch Rösrath. Published by "Bergischer Abfallbeseitigungsverband".
NN (1987). Komposttonne Göttingen. Published by Stadtreinigungsamt Göttingen.

LOWER MOISTURE LIMIT FOR COMPOSTING

Hidehiro KANEKO and Kenji FUJITA

Department of Urban Engineering, Faculty of Engineering, The University of Tokyo, Bunkyo-ku, Tokyo, Japan

INTRODUCTION

In composting, how much the moisture content is reduced is often considered as one of important factors to evaluate the quality of produced compost. But overdrying is detrimental to microbial activity and consequently an immature compost is produced. So it is important to know the lower moisture limit for composting and maintain adequate moisture levels.

It was experimentally confirmed that the lower moisture limit depended on the nature of material when moisture content (water/solid ratio or water/total mass ratio) was used as an indicator of moisture condition (Kaneko et al. 1985).

The purpose of this paper is to determine a universal lower moisture limit criterion for composting by using water activity (a_w) as a moisture indicator of compost materials.

MATERIALS AND METHODS

The materials used in this study are shown in Table 1.

The water activity (a_w) is defined as a decimal fraction of equilibrium relative humidity (Scott, 1957), which is measured with the apparatus shown in Fig. 1.

The microbial activity is measured in terms of the oxygen consumption, using the apparatus shown in Fig. 2. The

ISWA PROCEEDINGS
VOLUME 1 ISBN 0-12-058451-4

Table 1 Materials used in the study

Sample Name	Contents
Sample S	sawdust + ground dog food (1:1)
Compost S	compost of Sample S
Sample N	shredded newsprint + ground dog food (1:1)
Sample D	ground dog food
Sample M	compost of M sewage treatment plant in Tokyo
Sample R	compost of R night soil treatment plant in Tokyo

specified amount of sample in the bottle is incubated at a constant temperature. The microbes consume oxygen and produce carbon dioxide stoichiometrically. As the carbon dioxide produced is absorbed by the potassium hydroxide solution, a change in pressure in the manometer occurs. Thus, the pressure change is related to the oxygen consumption.

RESULTS AND DISCUSSION

Relationship between moisture and water activity

The definition of a_w and Raoult's law gives the following equation :

$$a_w = n_2/(n_1+n_2) \qquad (1)$$

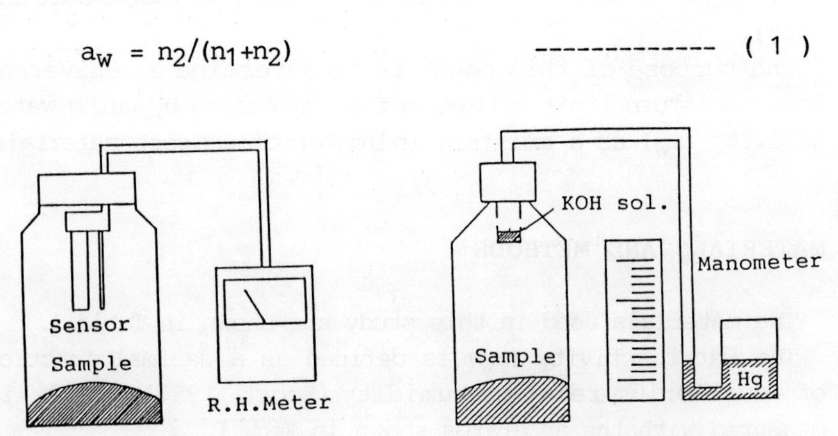

Fig. 1 Measurement of water activity.

Fig. 2 Measurement of oxygen consumption.

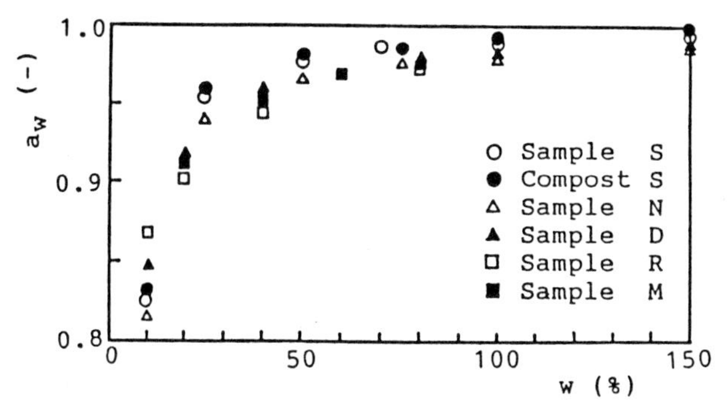

Fig. 3 Relationship between water/solid ratio and water activity

where n_1 and n_2 are the number of moles of solute and solvent.

Referring this equation, the following equation was assumed to relate water/solid ratio (w) and a_w :

$$a_w = w / (K + w) \qquad (2)$$

or

$$w/a_w = w + K \qquad (3)$$

where K is a constant.

Fig. 3 shows the relationship between w and a_w measured, and Fig. 4 illustrates the plots of w versus w/a_w. Fig. 4 proves the validity of equation (2) and the value of K is given by the intercept of the straight line as a result of regression.

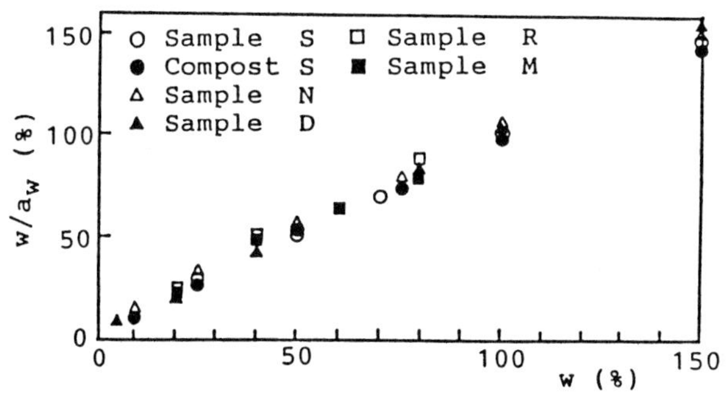

Fig. 4 Confirmation of equation (3).

Table 2 Value of K for each sample

Sample	K(%)	Sample	K(%)	Sample	K(%)
Sample S	2.0	Sample N	3.6	Sample M	4.0
Compost S	2.2	Sample D	4.1	Sample R	5.0

Table 2 shows the value of K for each sample. The values of K for Sample S and Compost S, the latter being the composted product of the former, indicate that the K does not change appreciably during the composting process.

Relationship between a_w and microbial activity

To examine the relationship between moisture content and microbial activity, oxygen consumption of each of the samples was measured under various moisture conditions. Fig. 5 shows a typical time course of cumulative oxygen consumption of a sample under various moisture conditions. It is seen that the sample with less moisture content consumes less oxygen.

The oxygen consumption rate of each sample under the various moisture conditions was calculated as the slope of straight part of the cumulative oxygen consumption curve. a_w was estimated using the K values in Table 2 along with equation (2). Fig. 6 illustrates the plots of a_w versus oxygen consumption rate. It is evident that the oxygen consumption rate drops as a_w decreases, and samples with a_w

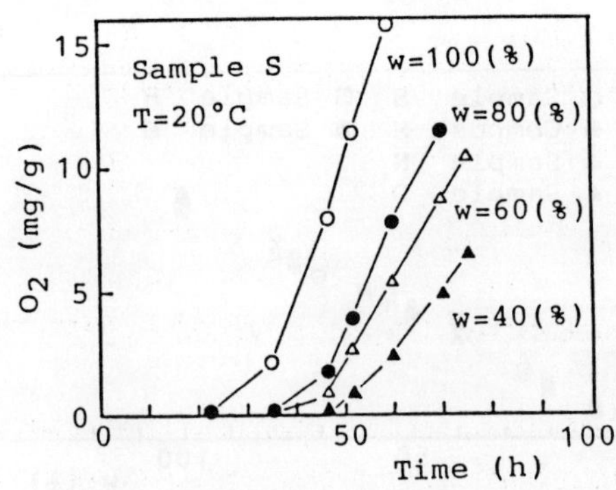

Fig. 5 Time course of oxygen consumption.

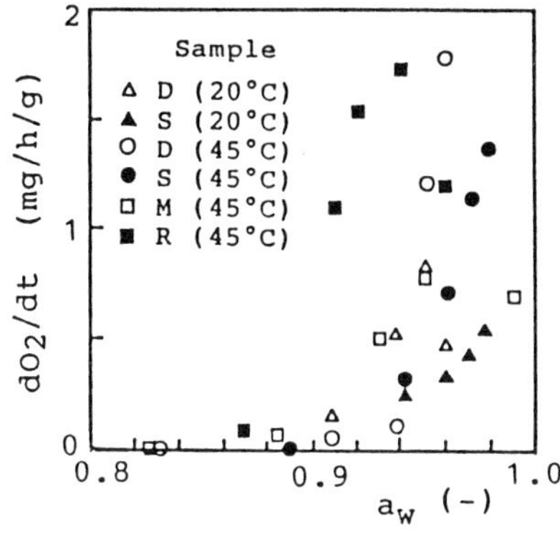

Fig. 6 Effect of water activity on oxygen consumption rate.

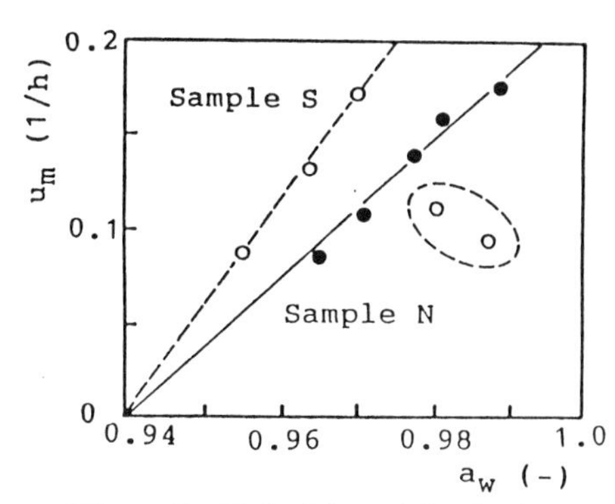

Fig. 7 Relationship between water activity and specific growth rate.

less than 0.9 show no microbial activity.

Effect of a_w on composting process

Kinetic analysis of laboratory scale composting with Sample S and Sample N had been done and the maximum specific growth rate (u_m) of microbes in compost on each moisture condition had been estimated (Kaneko et al. 1985).

Fig. 7 illustrates the relationship between a_w and u_m for two different samples. This shows that u_m varies linearly with a_w for the experimental conditions of this study except for the data bounded by a broken line where the oxygen supply may probably have been insufficient because of high moisture and low porosity.

Assuming thus that u_m is linear to a_w, the following equation is obtained by fitting a straight line to the data for each sample in Fig. 7 :

$$u_m = u_a (a_w - a_{wo}) \qquad (4)$$

where u_a is a constant and a_{wo} is water activity which gives u_m=0.

a_{wo} means the minimum a_w at which composting is possible, i.e. the lower moisture limit for composting. The values of a_{wo} are about 0.94 and almost same for both samples. But the values of w which are equivalent to a_w=0.94 are quite different between these two samples (w=32% for Sample S and w=57% for Sample N). This implies that a_w is a better moisture indicator than w because of its universality.

CONCLUSIONS

To determine a universal lower moisture limit criterion for composting, a_w is used to describe the state of moisture in compost. The relationship between a_w and microbial activity reflected by the oxygen consumption rate was experimentally determined and following is concluded :

1) a_w and w can be related by equation (2).
2) The microbial activity is influenced by the moisture condition and samples with a_w<0.9 consume no oxygen.
3) Equation (4) is proposed to relate a_w and u_m.
4) a_w>0.94 is essential for composting.

In conclusion, it is stressed that, compared with moisture content, a_w is a better indicator of lower moisture limit criterion for composting because of its universality and ease of measurement.

REFERENCES

Kaneko, H. and Fujita, K. (1985). Effect of moisture on decomposition rate in composting, Proc. of Environ. & Sani. Eng. Res., Vol. 21, 115-121.

Scott, W. J. (1957). Water relations of food spoilage microorganisms, Adv. Food Res., Vol. 7, 83-127.

Anaerobic digestion of the organic fraction of municipal solid waste in the BIOCEL-process

I.W. Koster[1], E. ten Brummeler[1], J.A. Zeevalkink[2], R.O. Visser[2]
1: Agricultural University, De Dreijen 12, 6703 BC Wageningen, The Netherlands; 2: Heidemij Consultancy, P.O. Box 264, 6800 AG Arnhem, The Netherlands

Introduction

The BIOCEL-process is a batch process for the dry anaerobic digestion of organic solid waste. The BIOCEL-process yields biogas and a compost-like residue. At the time the development of the BIOCEL-process started, already some other systems for biogas production from solid waste existed on laboratory or semi-technical scale. However, in a comparative study it was concluded that these systems were either too expensive because of a 'high-tech' set-up in order to allow for continuous operation or produced enormous amounts of wastewater because they require slurrying of the waste before digestion [1]. Since BIOCEL-process is operated batch-wise its design can be relatively simple, which implies relatively low construction and operation costs. In this paper the results of research concerning the operation of the BIOCEL-process is described, and a process lay-out is given together with an estimation of the costs for treatment of source separated municipal solid waste.

Operation of the BIOCEL-dry anaerobic digestion

In the anaerobic digestion of organic material several microbial conversion processes take place at the same time. These conversion

ISWA PROCEEDINGS
VOLUME 1 ISBN 0-12-058451-4

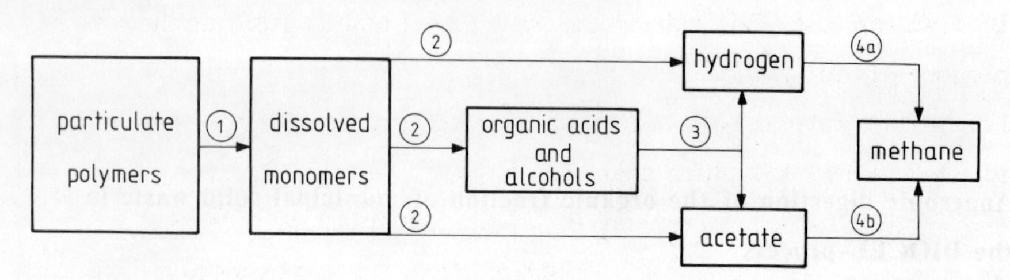

fig.1 The four metabolic stages of anaerobic digestion of complex waste (1 = hydrolysis, 2 = acidogenesis, 3 = acetogenesis, 4 = methanogenesis)

processes can be divided into four groups [2] which are shown in figure 1. Each of these groups has its own specific environmental requirements. Of most engineering significance is the fact that the pH tolerance of the methane forming bacteria is much narrower than that of the other bacteria involved in anaerobic digestion, especially the acid producing ones. Therefore in batch processes such as the BIOCEL-process there should be enough buffering capacity present to allow for the temporary build-up of organic acids which takes place in the first part of the digestion process [3]. If the buffering capacity of the material to be digested is not adequate, the build-up of organic acids might cause a pH-drop below 5.5, from which a cure is not possible because the methanogenic bacteria which would have to consume the acids will cease activity at these circumstances. Our own laboratory scale research indicated that for adequate buffering in the dry anaerobic digestion of the organic fraction of hammer-milled municipal solid waste such amounts of sodium bicarbonate were required that the digetic process became inhibited by the sodium ions. Therfore partial composting as a pretreatment step has also been investigated. By means of composting the easily degradable organic material (from which during anaerobic digestion acids can be formed at such a rate that a build-up of acids occurs) is degraded. It appeared that at least 15 % of the volatile solids have to be degraded in the composting process to provide a noticable stimulation of the subsequent anaerobic digestion in the

BIOCEL-process. This reduces the maximum amount of biogas that can be produced to 75%.

For obvious reasons both buffer addition and pre-composting were rejected as methods which could be applied in practice for prevention of a fatal pH-drop. Instead, mixing of the raw substrate with inert material was chosen as the most appropriate in practical conditions [4]. The various methods of prevention of a fatal pH-drop during batch digestion of the organic fraction of municipal solid waste are summarized in figure 2. At the first start-up of a BIOCEL-process compost should be used as inert material to dilute the substrate with. In subsequent digestions instead of compost the digested material from previous digestions can be applied.

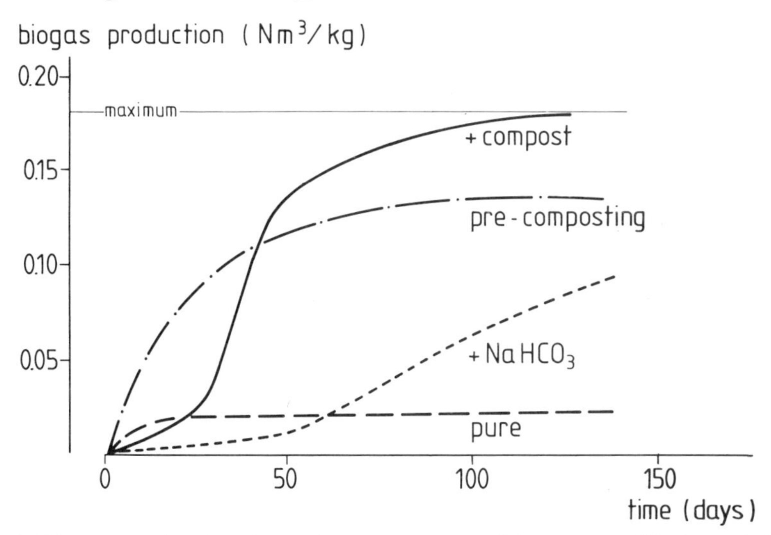

fig.2 Biogas production from factory separated hammer-milled municipal solid waste at various BIOCEL start-up methods

Pilot plant results

In figure 3 some results are shown of a BIOCEL-reactor treating 5 m^3 of source separated municipal solid waste (the so-called vegetables, fruit and garden waste) at 35 ^{0}C and 30 % total solids. These results indicate that at these conditions the digestion time can be circa 50 days. In that

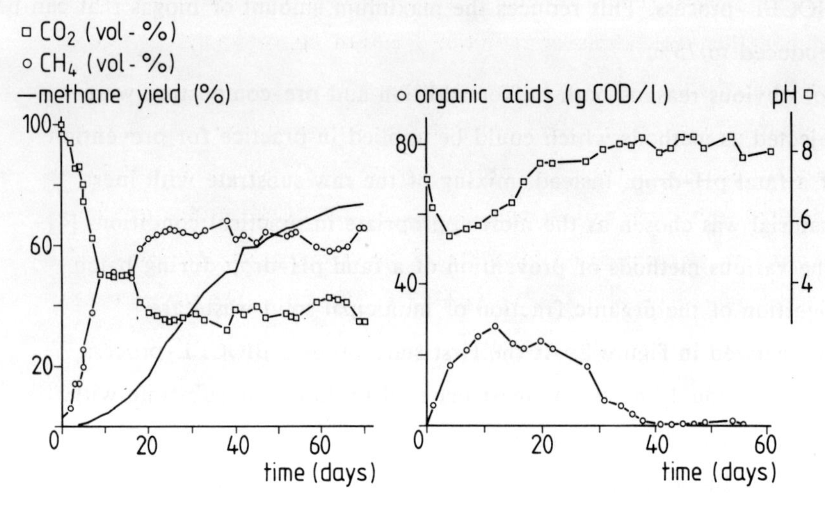

fig.3 Results of a BIOCEL pilot plant treating vegetable, fruit and garden waste at 30 % TS and 35 ^{o}C.

time all organic acids have been removed from the digesting mass while more than 65 % of the maximum biogas yield has been produced. When the organic acids have been converted, the obnoxious organic compounds present in the waste and/or formed as intermediates during digestion have also been converted. The more than 65 % completion of the digestion indicates that the residue is more or less stabilized. This was confirmed by the fact that it proved to be impossible to start a composting process with the BIOCEL-residue.

Process lay-out and economics

Based on laboratory scale research a flow sheet for treatment of source separated municipal solid waste was developed (figure 4). With this flow sheet an economic evaluation of the BIOCEL-process has been made. The gross exploitation costs (including investment, operation and maintenance) were found to be Dfl. 29.50 per ton of source separated municipal solid waste, in case of treatment of 34.000 tonnes/year. This is a rather conservative estimate of the gross exploitation costs, since it is based on the assumption that the retention time of a batch load of waste in a

BIOCEL reactor should be nearly 3 months. The pilot plant research indicates that much shorter retention times of approximately 7 weeks can be applied.

The net costs of the BIOCEL-process will be much lower, since the biogas represents a value of approximately Dfl. 15.-- per ton of source separated municipal solid waste. Moreover, the dried residue can be sold as compost. Depending on the state of the compost market, the dried residue could represent a value as high as Dfl. 20.-- per ton of source separated municipal solid waste.

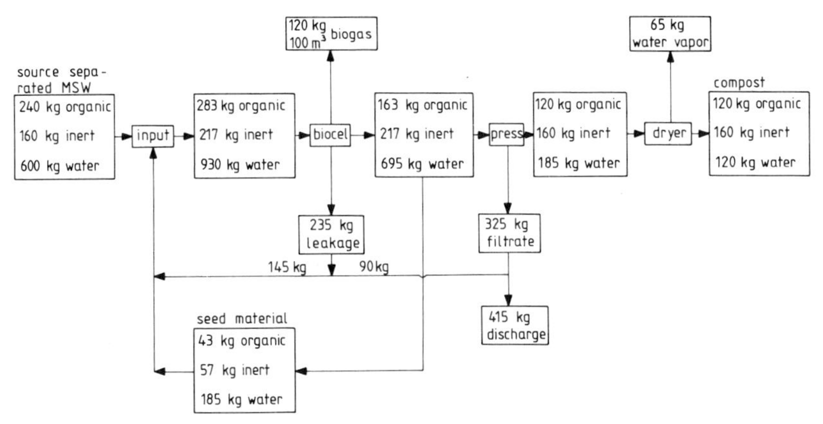

fig.4 Flow sheet of a treatment operation for source separated municipal solid waste, incorporating the BIOCEL dry anaerobic digestion.

Acknowledgement

This work is financially supported by the Project bureau for Energ Research (PEO) within the framework of the National Research Program for Recycling of Waste (NOH).

References

1) Ten Brummeler, E., Koster, I.W. and Zeevalkink, J.A. (1986). in: Materials and Energy from Refuse (eds. A. Buekens and M. Tels), 6.49-6.58, KVIV, Antwerp, Belgium
2) Koster, I.W. (1988). in: Biotreatment Systems (ed. D.L. Wise) vol. 1, CRC Press, Inc., Boca Raton, U.S.A. (in press)

3) Buivid, M.G. et al. (1981) Fuel gas enhancement by controlled landfilling of municipal solid waste, Resources and Conservation 6, 3-20

4) Ten Brummeler, E., Koster, I.W. and Zeevalkink, J.A. (1988) Dry digestion of the organic fraction of municipal solid waste in a batch process, J. Adv. Water Pollut. Control. 5, 37-46

QUALITY AND MARKETING
OF MSW COMPOSTS

by

Régis de Lauzanne

French Agency for Waste Recovery and Disposal
Les Transformeurs

The Situation Today

In France, as in other European countries, there are two types of composts:

1) Urban compost, or those made from household refuse,
2) Composts of relatively non-synthetic origin, made from cattle manure, mixtures of sawdust with poultry droppings or of liquid pig manure with bark, residue from grapes pressed in wine-making (rapes), sludge from sewage treatment plants, slaughterhouse wastes, etc...

French urban compost production came to 650,000 metric tonnes in 1984, of which 550,000 metric tonnes were sold at an average price of 25 French Francs a tonne, with pick-up at the plant.

Production of other composts is more scattered and more difficult to estimate. However, a production of 80,000 to 100,000 metric tonnes would be a reasonnable estimate, including 10,000 to 12,000 metric tonnes made from sewage sludge from treatment plant.

We'll not come back here to "Why One Should Compost" these by-products, since much has already been said on that topic. For our purposes, it would suffice to recall that composting is a treatment process which converts basic material and its components into a product marketable for agricultural use - and yet it can even convert it into a non-marketable product.

Moreover, the interest of this means of processing is that it allows the compost producer - if not actually to make a profit - at least to optimize the costs of a public service. Composting is the best way of processing the by-products coming from that public service.

This treatment process is but one of many other refuse treatment systems, such as dumping, or incineration. Before choosing composting for treating refuse, a community must first conduct an in-depth study outlining the relative advantages and drawbacks of each of the various treatment

ISWA PROCEEDINGS
VOLUME 1 ISBN 0-12-058451-4

processes. In particular, this study should address the financial consideration relevant both to the initial investment and the operating costs.

The choice of composting offers quite a few advantages from the investor's point of view:

1) a reduction of the volume of material in dumps which become necessary mainly for bulky or non-recyclable products;

2) limited risk of pollution;

3) enhancement of the compost producer's public image;

4) reduction of removal costs, not to mention the possibility of financial assistance or lowered pollution charges.

Choosing composting usually implies a commercial task. In other words, it will not be enough simply to produce compost, you will also have to sell it. There is a specific regulatory framework which allows composts to be placed in the category of organic soil-improvers, or when used in association with other products - in the category of soil for crops. In spite of that, it will nevertheless be necessary to define the compost produced in terms of a commercial product and thus to draw up a real commercial strategy.

This strategy must be founded on three basic tenets:

A) Quality

B) Suitability to crops and demand

C) Creation of appropriate commercial structures

<u>A - QUALITY</u>

Compost is used as an organic soil-improver, and even as soil for crops when used in association with other products. Thus its place is on the market for fertilizing substances.

This means that compost quality must be attractive enough to the potential user so as to atone for what could be composts' "original sin" in his opinion: the fact that compost is produced from wastes. In order to achieve this, phyto-toxicity tests alone will obviously not be sufficient.

The accent should be on quality, as much for the outward appearance of the compost as for its agronomic value:

▪ Its organic matter content, which improves the physical and microbiological properties of the soil; this content can be assessed from the isothermal coefficient as well as from the C to N ratio.

▪ Its content in fertilizing elements:

nitrogen, phosphorus and potassium, but also lime, trace elements, etc...

For these two criteria, concrete reference systems will be necessary, based on laboratory tests as well as trials conducted by specialized testing bodies (such as the INRA or the CTIFL, for example).

Concerning the chemical composition, it might be added that it could prove worthwhile to add nitrogenous phosphoric or potassic nutritive elements to compost in order to bring together the organic and fertilizing aspects of compost. A well-conceived chemical composition can in fact be an advantage, commercially. The association of several products (sludge and urban composts), potentially offering a significant technical or commercial advantage becomes a possibility to consider.

A sanitary compost. A compost product should contain neither infectious forms nor parasites. A particular effort to this end must be undertaken during the supervision of composting.

Outward appearance of the product. Urban compost contains a certain amount of undesirable elements (plastics and glass) which can give it an unappealing appearance. This is not the case with other composts.

It will be necessary to do the utmost in order to give the consumer real visual satisfaction by carefully looking after the various physical factors: granulometry, colour, odours, moisture and density.

This was the reason the Agency, in 1986, created the NF Urban Compost Mark of Quality in order to provide the consumer with a guarantee of quality going far beyond the requirements spelt out in French Standard NFU 44051. The NF Mark of Quality is duly certified by testing body which is independent of the producer and maintains a constant and unflagging standard of excellence.

Administered by the Agency, in accordance with the French Standards Association (AFNOR), this Mark sets "NF" quality composts apart from those of only average or mediocre quality found on the market.

The Mark should enable producers to reach larger and more lucrative markets and, above all, to raise their prices which are currently ridiculously low in comparison to the true worth of the product.

It should also protect and give new vigour to composting as a treatment option, even if it requires investment expenditures to improve compost quality. Composting is of particular interest to medium-size communities (20,000 to 80,000 inhabitants), both technically and economically.

Although today it is still too soon to draw any concrete conclusions from the Mark experiment (5 communities out of 90 have subscribed to the Mark), it could certainly already be said that the Mark has had a beneficial influence on compost producers and contractors who take the quality requirements into account on their specification sheets.

A public awareness campaign to inform compost users is currently being carried out by certain compost producers with the active support of the Agency.

B - SUITABILITY TO CROPS AND DEMAND

This point, too, seems essential. There are composts produced that do not correspond to the demand or that are ill-suited for the various types of agriculture in the area.

In such a situation, either the compost user will continue to accept the product free of charge without much use for it or he will simply refuse to be community's "dump". Farmers' ways of thinking are changing, too, and compost producers must bear that in mind.

In particular it is becoming ever more important to supply the compost user with a product that incorporates the physical, chemical and biochemical properties that respond to his specific needs, both technically and economically.

For example, the main use of peat was initially as a fuel; today its use in horticulture has become a reality.

Why? Simply because its drawbacks, such as its lightness and moistness, have been overcome and because new peat products are being made: pressed peat planting cups, mixtures, etc...

Your approach must be firmly rooted in market realities.

The specific use made of organic soil-improvers varies depending on whether one is addressing the needs of large-scale farming, vegetable farming, orchards, crops grown above soil, mushroom cultivation, viticulture, reforestation of eroded soils, tree farming, etc...

The approach, the characteristics expected of the products, the amounts and dates of usage are all different depending on the specific and differing needs of each one of these agricultural activities.

This means that the compost producer must identify a readily available and locally predominant target group of consumers and aim production towards meeting the particular needs of those consumers. To this end, the Agency, in cooperation with the Assembly of Chambers of Agriculture, in 1987, published a guide to using organic soil-improvers, made up of data sheets addressing, as comprehensively as possible, questions concerning the use of organic soil-improvers. It addressed these questions crop by crop. These sheets are supplemented by appendices that present, according to themes, useful - and indeed necessary - information, such as regulations, analysis methods, etc...

ANALYSIS OF DEMAND

Finally, we must be careful not to forget that organic soil-improvers are used in large quantities, potentially ranging from 20 to 100 metric tonnes per hectare (1 hectare = 2,5 acres], for one application every two or three years.

Producing 1500 to 2000 metric tonnes - which is common - supplies enough compost for only 50 to 100 hectares.

The chart in Appendix 1 outlines the present French needs in this area by type of agriculture, expressed in equivalence to tonnes of manure. These needs have increased notably in the last few years due to modern agricultural methods.

It will be noted that large-scale agriculture could be a market of considerable size. However, the product must be sold at a low price because the crops produced have a low profit margin and the demand for these crops is not, in fact, related to supply. (There are problems of transportation and initial investment costs, with a more or less long term effect that has often been badly estimated by the farmer.)

On the contrary, special crops (market gardening, horticulture, nurseries, etc...) are major users of organic matter and demand products of very high quality. Prices are therefore much higher.

As concerns amateur gardening, its clientele is widely scattered, sales channels are long and sales costs are high.

CONCLUSION

Composts have the advantage of being lowest-costing means of providing organic matter. Some of its characteristics bring it very close to those of soils for certain crops. It is clear that the origin of the product, its image and its appearance are all important factors affecting the selling price of compost, whence the need for a veritable commercial strategy.

C - THE CREATION OF APPROPRIATE COMMERCIAL STRUCTURES

On several occasions, the Agency has encountered cases of certain household refuse composting plants that, although they produced high-quality compost, proved unable to sell what they had produced. This serves to prove that in this field commercial considerations are at least equal in importance to production.

A market study is therefore necessary in every case and must answer the following questions: Is there a market for the compost produced by the plant? If so, how much? What quality of compost? For which agricultural activities?

And also:

What market strategy would reach the largest number of compost users? What promotional approach should be used? What distribution network? What packaging?

This comprehensive approach could be broken down in the following manner:

1) A preliminary general study outlining the potential sales opportunities, the market competition and a rough idea

of possible prices. The conclusions drawn from this first study serve to eliminate certain market sectors where the competition is too intense or where the sales opportunities are not guaranteed and to target potential sales opportunities if there are, in fact, any (for example: market gardening, viti-culture, parks and lawns...).

2) A more in-depth study, both quantitatively and qualitatively, of each promising sales opportunity indicated in the preliminary study. This would allow the compost producer precisely to define the following: the sales structure and means to be set up, market shares, prices, distribution networks and the quantities to be produced, etc...

Market Strategy

Once the results of the study have reached the compost producer, it remains for him to define his market strategy which will include:

1) The setting of a price
2) The choice of a distribution network
3) The choice of promotional means

1) The Price

It must be recognized that those involved in refuse processing think in terms of treatment and disposal costs. Incorporating an idea of profit doesn't seem to be an easy task for them. If it is true that the treatment process chosen can allow the producer to optimize his costs, it is no less true that it is possible to evaluate the market value of processed by-products. The producer can, by contacting a representative of a service specialized in this area, attempt to introduce an idea of profit, or more precisely, an idea of a sales margin which covers distribution costs and, consequently, a purchase price for the producer.

It is the value of this purchase price which allows the producer to establish the operation or not.

2) The Distribution Policy

The choice of a distribution network is of great importance.

Distribution can either be done directly, the producer thus having his own commercial organization, or it can be done indirectly through cooperatives, distributors or other industries. Although direct distribution has the advantage of absolute control of the product and prices, it is not at all the more common method, since communities are ill-equipped to act in this area. It would seem, in fact, a business better left in the hands of professionals, such as cooperatives or distributors. Choosing from among them, of course, requires reflection and a choice should be made considering their efficiency, their drive and ambition, and their impact in the region...

3) The Choice of Promotional Means

It must be remembered that compost users don't spontaneously come to the producer in order to get information or to buy a product. It is therefore up to the compost producer to make himself known and thereby create demand.

A variety of means are available in order to do this:

a) Publicity:

It can take several forms: written documents, information meetings, such as "Open Houses", sample trial packages with information cards, published articles...

The choice of one of these forms or several of them combined is to be made in direct relationship to the objectives to be achieved.

Furthermore, it is clear that launching a new product is a job for professionals. A community or a compost plant owner will usually have to seek help in order to make wise choices concerning the production of an appropriate product, the choice of a clientele and type of agriculture and the appropriate means needed to reach these goals.

b) The Sales Arguments

Finally, whatever the method decided upon, it is important to have a text listing the sales arguments and product's selling points, available for those selling the compost. Arguments of this sort are a tool which will serve to give conviction and inspiration to sales representatives. It is, furthermore, a means of dialogue between the producer and the compost user and buyer. These arguments must be forceful, thorough and relevant to all situations.

CONCLUSION

The compost producer, accustomed to viewing compost with a certain fatalism, must place himself in a new framework today if he wants to optimize the costs of the public service for which he is responsible.

He must try to make the most financially of his product and in order to do this, he must learn to think of himself as the purveyor of a product that is of value to agriculture. He must bear in mind of the needs of agriculture in terms of quality and of the product's ability to respond to specific needs.

And in this, although the problems he must overcome are numerous, there is an exhilarating challenge as well as a real need for action.

Appendix 1

Theoretical Organic Matter Needs of Cultivated Land

REGION	Manure Needs in T	Arable Lands		Vineyards - Orchards		Market Gardening - Special Crops Flowers	
		in T	% of Total	in T	% of Total	in T	% of Total
ALSACE	1534400	1260000	82,12	244800	15,95	29600	1,93
AQUITAINE	2982900	5413800	67,82	2458800	30,80	110300	1,38
AUVERGNE	3200600	3078600	96,19	98000	3,06	24000	0,75
BASSE NORMANDIE	3101000	3058200	98,62	13700	0,44	29100	0,94
BOURGOGNE	6570200	6081600	92,56	420300	6,40	68300	1,04
BRETAGNE	9582300	9447600	98,59	55000	0,57	79700	0,83
CENTRE	13210700	12494400	94,58	610000	4,62	106300	0,80
CHAMPAGNE ARDENNE	2344300	6942600	94,53	384200	5,23	17500	0,24
CORSE	518000	63000	12,16	416000	80,31	39000	7,53
FRANCHE COMTE	1473900	1428000	96,89	39700	2,69	6200	0,42
HAUTE NORMANDIE	3143400	3076800	97,88	19600	0,62	47000	1,50
ILE DE FRANCE	3638900	3415800	93,87	54800	1,51	168300	4,63
LANGUEDOC ROUSSILL.	8331700	1504800	18,06	6552000	78,64	274900	3,30
LIMOUSIN	1801600	1719000	95,42	73300	4,02	9300	0,52
LORRAINE	3380000	3313200	98,02	41400	1,22	25400	0,75
MIDI PYRENEES	10951500	9530400	87,02	1336900	12,21	84200	0,77
NORD PAS DE CALAIS	4044100	3912000	96,73	8300	0,21	123800	3,06
PAYS DE LA LOIRE	10163400	8947200	88,03	909900	8,95	306300	3,01
PICARDIE	6830000	6744000	98,74	58500	0,86	27500	0,40
POITOU CHARENTES	9547400	7838400	82,10	1649700	17,28	59300	0,62
PROVENCE ALPES C.A.	5023700	1504800	29,95	2670500	53,16	848400	16,89
RHONE ALPES	6057100	4220400	69,68	1728400	28,54	108300	1,79
TOTAL FRANCE	127431100	104994600	82,39	19843800	15,57	2592700	2,03

Session 4

Sanitary Landfills

Sanitary Landfill Technology has to be improved!!

by Ing. D. Louwman, The Netherlands

The working group Sanitary Landfilling, one of the working groups of the Scientific and Technical Committee of ISWA, started her activities at the end of 1986.
Besides the organization of specialized seminars (1987 in Cagliary, Italy and this year in Amsterdam as a post-seminar of this ISWA-congress) the working group is also occupied with the collection and exchange of data from sanitary landfill sites and their technology, the research programmes, regulations and legislation in several ISWA-countries.
As a result we will publish in 1989 a `State of the Art of Sanitary Landfilling in the several ISWA-countries`.
As chairman of the working group I will now present here an interim report.

At the end of the seventies, begin eighties, regulations and legislation to protect the environment were promulgated in most of the Western countries and in North-America. With respect to our field of interest this legislation focussed on protection of soil and groundwater.
The disposal of waste is, therefore, restricted by rules in all countries. Moreover, at a further increase of population density and welfare, society will demand a better protection of soil and groundwater.
Therefore, in future the disposal of waste will be an ever increasing complicated soil project where only high tech is required based on high standards.

In the field of waste disposal many research programmes

ISWA PROCEEDINGS
VOLUME 1 ISBN 0-12-058451-4

have been carried out in the last decades in the mentioned countries.
These studies were aimed to determine the kind and volume of waste with the result amongst other things that the quantity of waste to treat turned out to be larger than thought before.
Domestic waste, bulky domestic waste, construction and demolition waste, sewage sludge, agriculture waste, hospital waste, waste gypsum, dredging spoil, combustion residues and hazardous waste are kinds of waste which were not separated from each other for twenty years. They were treated together in those days by disposal or by combustion.
These days it is obligatory to collect them separately and/or to treat them separately as well. The more complicated the collection and treatment, the more complicated will be the last phase for all waste-handling: sanitary landfill.
The availability of landfill sites is a relentless demand for the continuity in every well-functioning waste-disposal system.
Because of the soil protection and zero-emissions to soil, air and groundwater technical precaution measures have to be taken within the legislation and requirements of landscape, climatic circumstances and hydrogeological aspects.
Highlights in future Sanitary Landfill management will be - within reasonable costs and without any indication that the system is affecting human health or the quality of the environment -:

A. A technical and scientific approach to the landfill locations, design and construction.

Expertise from different disciplines has to improve the necessary disposal methods of the several kinds of waste.
Much research is done at the development of new sites, geophysical methods are applied to study the hydrogeological conditions at the potential landfill sites and its surrounding.
It is necessary to carry out location research systematically because a landfill site will - during his using phase as well as after closing - have a large impact on the area where it is situated.
Previous Environmental Impact Research is already obligatory in many countries in order to know and to weigh the several impacts, so the licence conditions can remove or reduce possible negative effects.
Design and construction of sites have to fulfil now and in the future the following premises:

- the way of disposal of waste on or in the ground has to be controlled and regulated now and in the future;
- direct contact between the waste and bottom-, ground- and open surface water has to be avoided;
- dispersion in the ground of leachate from sites is often forbidden;
- regular checks on the sites themselves and on the equipment have to take place and precautions must be taken;
- the processes in the disposed waste have to be known and, if necessary to be influenced and

supervised.

Because of the above mentioned premises the lay-out of sites must foresee in compartments so that the total site surface needed is spread out over years.

In every compartment construction, exploitation, final phase and after care will succeed each other.

Because of the partition in compartments several activities from the above mentioned phases overlap each other.

Normally the after-care is not carried out per compartment while it is necessary for the total surface of the site.

To prevent contact of groundwater with leachate from disposed waste it is necessary to provide for isolation. The natural hydrological situation (percolation water or infiltration) must not influence the hydrologic balance in the dump itself.

Therefore, covering of the bottom is a must.

Artificial sealing (e.g. flexible membrane liners such as HDPE bottom liners) or natural sealing (e.g. clay, loam and bentonite) are possible applications.

Double precautions (a first and a second bottomliner) are applied at the dumping of hazardous waste and/or contaminated soil from reconstruction areas out of the past.

For protection of the bottomliners a 25-30 cm thick layer of sand or gravel can be used, which is also useful to drain the leachate.

The protection has to extend to the top of the compartment.

In this way the necessary measures can be taken per compartment to collect and to treat the leachate and

to utilize landfill gas.
Recirculation of leachate back to the site is also possible. By doing this the tip can be brought into the methanogenic phase quicker than normally.
The leachate from a site in the methanogenic phase is easier to treat and the landfill gas production will be enhanced.
The dump can also be used as a buffer for water in times of very heavy rainfall or a temporary breakdown of the leachate treatment plant.
When an infiltration system is installed at the dump it has to be located beneath the upper level, so there will be no odour emissions or damaging of the vegetation.
There will be a growing number of gas utilization systems on municipal solid waste sites for environmental and economical reasons.
After the tip is closed or even immediately at the beginning of the exploitation horizontal or vertical tubes are installed to win the gas. Research has shown that the permeability factor of the landfill gas through the bulk of the solid waste is 7 times larger at horizontal winning than at its vertical counterpart. For this reason horizontal winning tubes seem to be more interesting.
However, the vertical pipe lay-out is superior with respect to maintenance costs, accessibility for repair, flexibility in building up the site and the opportunity to monitor the composition of the gas and differences in gas pressure.
Purification and further quality enhancement are necessary for domestic or industrial use of the

landfill gas. Several installations are in operation in Europe and North America with an environmental and economical positive result. It is true that this is a very attractive result of advanced waste disposal technology.
The disposal of organic municipal waste will not be acceptable anymore in the future in the developed countries, but surely it will continue to be the solution in the third world for a long time.
An important problem in the future is the question whether municipal waste and `normal` industrial waste can be disposed of together with hazardous waste.
The research into the advantages and disadvantages of co-disposal is still in its infancy. Much work has to be carried out in this field.
In Germany and Holland legislating authorities have the opinion that co-disposal sites are unacceptable; this is contrary in e.g. the UK and Italy.
Further research has to answer this question.
Pre-treatment of hazardous waste can possibly become a solution to this problem.

B. Professional management and use of the right equipment

Professional management and qualified personnel are necessary for an constant alertness in acceptating waste and an adequate operation of the site during the construction, the operational and the after-care period. Highly qualified personnel is needed to assess dilemmas and to introduce advanced

technologies in practice.
Training programmes for operators have to be developed. The knowledge transferred by adequate management training programmes will enable implementation of a correct working plan divided in phases which extend over several years and will ensure a proper preparation, operation, restoration and after-care.
The psychological impact on the inhabitants living in the surrounding area can not be neglected. Periodical consultation of the neighbours is a good way to solve problems that may arise on both sides. Also attention has to be paid to the movements of the car/trucks from and to the site.
Reducing the movements of the traffic or intelligent construction of (new) roads can be beneficial. Reduction of the noise from the activities on the tip (compactors, bulldozers and so on) can be realized by zone location and/or by acoustic baffles.
Annoyance arising from blow-over of sand, papers or plastics to the surrounding fields can be prevented by planting trees and bushes or by disposal of the waste in nets. Also blow-over of seeds to the farmland is mostly forbidden by law.
Bird damage to the farmland can not easily be proven, but from a psychological point of view it is advisable to hunt or shoot the several kinds of birds. However, this action has only a temporary result. Experiments with natural hunting with hawk or falcon are going on at this moment. The first results seem positive.

The right equipment has to be selected to ensure a proper operation of the landfill. Every landfill site needs machines to prepare the site, to handle and compact the waste, to excavate and transport cover material and to spread and compact the final cover.

C. Scientific research into all aspects of a sanitary landfill site remains necessary

Research into the several aspects of a sanitary landfill site is of great importance for future design, construction and operational management. Attention has to be paid to the following issues:

- location investigation, together with Environmental Impact Research. After physical planning the location must be investigated on various aspects.
- Geohydrological research has to be carried out so the water balance both in the compartments and in the site as a whole can be controlled.
  Also the depressions in the underground have to be calculated.
- Research into bottomsealings (HDPE, LDPE, clay-loam and bentonite) with respect to permeability, solvability, strength of the connections have to answer the question on what kind of sealings can be used in a specific situation.
  Protocols have to be made to ensure a right construction and great strength over a long life.
- Research on quantity and quality of the leachate.

Prevention of leachate by topliners (surface capping), by sidelong surface draining of the rainwater or installation of bottomliners to avoid percolating water entering the dump seem meaningful but under these circumstances the deposited waste is not leached. Future disturbance of the precaution measures can always lead to contamination of the surrounding groundwaters with leachate.

Possibly it is better to leach the deposited waste as quickly as possible. Answers to this problem have not yet been given.

Treatment of the leachate is necessary anyhow.

At every specific situation one has to look at the quality and to assess the opportunities to:

- discharge directly to a sewage treatment plant;
- discharge to a sewage treatment plant after pre-treatment of the leachate (for example, only reducing the heavy metals by physical-chemical treatment);
- treatment of the leachate at the site location enabling a direct effluent to the surface water without environmental damage.

Decisions about the right method can be taken - in phases and in time - based on measurements of the leachate quality and the restriction on the discharge to the surface water.

Mobile installations can be used for temporary solutions. Without constant monitoring of the quality and quantity of the leachate neither the right treatment method at the right time nor the right capacity can be found. Per situation the method of treatment has to be found by research and the advantages and disadvantages have to be weighed in environmental and economic senses. The final choice out of the distinguished treatment methods (table 1, showing different purification projects) depends on the input quality of the leachate. Further refinement/optimalization of the chosen treatment methods may not be neglected.

Research into the effects of co-disposal (disposal of hazardous waste together with municipal waste and "normal" industrial waste) is a demand for the future, in order to enable a more rational discussion on this topic.
Must hazardous waste be pretreated? What kind of neutralization processes are going on? What kind of leachate is coming out? What kind of precautions are desired or necessary. A field of research that has to be studied in depth.

With respect to research in general one can say that mostly applied and fundamental scientific research have been carried out in laboratories of universities and consultants.
We are pleading for a good cooperation between the above mentioned institutes and the management of landfill sites on behalf of upscaling from pilot plants to operational plants.

The working group Sanitary Landfill of ISWA had the intention to show you that landfill technology at this moment is still growing to a higher level and has not yet reached the end of the development, but mostly is determined by the specific circumstances in a country or a continent.
The disposal technology is present and will be improved by research.
The application will be determined by nature and composition of the site and of the waste, by climatic circumstances, by the underground, the level above groundwater and sea, the legislation and the social acceptance and last but not least the economic welfare in the country involved.

We will go on to inform you about our work and - in that way - about the developments in landfill technology.
Permanent study, research and application of the results are high topics.

TABLE 1-1

RESULTS IN PRACTICE WITH DIFFERENT TREATMENT METHODS

Technology	Purification efficiency ACID LEACHATE			Methanogenic leachate		
Aerob treatment with nitrification and denitrification	CZV	50-95	(90)	CZV	20	
	BZV	> 95	(95)	BZV	45-85	(80)
	N-Kj	[60]		N-Kj	75-90	(80)
	Me	[70]		Me	15-50	(50)
Aerob treatment	CZV	50-85	(80)	n.a.		
	BZV	[95]				
	N-Kj	< 10	(5)			
	Me	< 80	(70)			
Hyperfiltration - 2 sections (reverse osmosis)	n.a.			CZV	> 99	
				BZV	> 99	
				N-Kj	> 99	
				Me	> 99	
Hyperfiltration - 1 section (reverse osmosis)	n.a.			CZV	98	
				BZV	97	
				N-Kj	75	
				Me	> 98	
Evaporation	CZV	moderate	[60]	CZV	good	[95]
	N-Kj	moderate	[60]	N-Kj	moderate	[60]
	Me	good	[95]	Me	good	[95]

.../see next page

TABLE 1-2

Technology	Purification efficiency ACID LEACHATE			Methanogenic leachate		
Flocculation, precipitation	CZV	below	[10]	CZV	below	[10]
	N-Kj	below	[10]	N-Kj	below	[10]
	Me	good	[90]	Me	moderate	[50]
Recirculation into methanogenic site	CZV	good	[91]	n.a.		
	BZV	good	[99]			
	N-Kj	0				
	Me	good				

Me = (heavy) metals

Between round brackets the efficiency is shown as used in the calculations.

Between square brackets an estimated percentage has been given.

ASSIMILATIVE CAPACITY OF LANDFILLS FOR SOLID AND HAZARDOUS WASTES

F.G. Pohland, W.H. Cross, J.P. Gould and D.R. Reinhart
School of Civil Engineering
Georgia Institute of Technology
Atlanta, GA 30332, USA

Introduction

The threat of adverse environmental impacts from landfill disposal of solid and hazardous wastes has become a focus of international concern. Much of this concern is accountable to the problems consequenced by past disposal practices and a tendency to perpetuate traditional approaches to waste management. As a result, regulatory activity has flourished, as evidenced in the U.S.A., by the enactment of the Comprehensive Environmental Response, Compensation and Liability Act (CERCLA) for old sites and the Resource Conservation and Recovery Act (RCRA) for existing and proposed new sites. Indeed, Subtitles C and D of RCRA deal specifically with hazardous and solid wastes, respectively, with the latter receiving most recent attention by the U.S. Environmental Protection Agency in its development of regulations for municipal landfills.

Preliminary Considerations

Over the past several decades, considerable emphasis has been placed on the characterization of landfill behavior in terms of those processes responsible for the release and migration of leachate and gas. A critical review of published information (Pohland and Harper, 1985) and associated research investigations (Pohland, 1975; Pohland, 1980; Pohland, et al., 1983; Pohland, et al., 1985; Pohland and Gould, 1986; Pohland and Harper, 1987) have led to a better

ISWA PROCEEDINGS
VOLUME 1 ISBN 0-12-058451-4

understanding of this behavior, including the consequences of inadequate landfill management practices and the requirements for more efficient operation and control. Crucial to this understanding is a recognition that, regardless of environmental setting, most landfills exist as dynamic, microbially mediated, physical-chemical processes that tend to progress through a series of stabilization events often reflected by parameters descriptive of the magnitude and intensity of leachate and gas generation. Accordingly, landfill stabilization results either randomly in time and space, or in a more temporally and spatially uniform and predictable fashion, depending on circumstances and technology applied. These include availability of waste substrate, nutrients and moisture; reaction opportunity between the waste matrix and major transport phases (gas and leachate); presence of inhibitors, catalysts, and reaction mediators; and, nature and viability of the microbial consortia in mediating the *in situ* conversion processes, particularly during the acid fermentation and methane formation phases of landfill stabilization.

Attenuation and Assimilation

In considering the origin and fate of landfill leachate and gas phase constituents, it is necessary to make a distinction between attenuation, the lessening of amounts and their consequences, and assimilation, the incorporation or conversion within the process. The former may be consequenced by washout and/or dilution and also by assimilative action, whereas the latter includes all biochemical and physical-chemical alterations leading to a particular condition. Hence, organic and inorganic constituents may be susceptible to not only washout, but to an array of biotic and abiotic conversions, including, but not limited to, acid-base, oxidation-reduction and precipitation/complexation reactions as well as sorption, filtration and ion exchange.

Based upon an understanding of conditions prevailing during the phases of landfill stabilization, it is possible to predict the behavior and fate of many waste constituents as they are released to the leachate and gas phases. Many of these can be analyzed in terms of either gross or specific parameters. For example, leachate chemical oxygen demand (COD) or total volatile acids (TVA), pH or alkalinity, and conductivity are gross parameters indicative of organic strength, acid-base condition, and ionic strength-activity, respectively, whereas analyses for acetic acid, sulfide, and zinc are representative of specific parameters indicative of ingredients constituting these gross analyses. Likewise, gas quantity is a gross analysis, whereas gas components (CH_4, CO_2, H_2, N_2, volatiles) are specific constituents of the gross analysis. Moreover, when considered in an integrated manner with knowledge of the prevailing microbial processes, predictability of dominant chemical species becomes possible and more meaningful. Hence, a "young" landfill undergoing acid fermentation will produce a leachate of low pH and total alkalinity, high COD and TVA, an abundance of leached ions (high ionic strength), and high migration and environmental impact potential. In contrast, leachate from an "older" landfill will usually exhibit a near neutral or higher pH, higher total alkalinity, lower COD and a near absence of TVA, and a lower migration and environmental impact potential. Gas production and quality would similarly reflect the microbial nature (aerobic/anaerobic; acid fermentation/methane fermentation) of the respective phases, with methane generation a significant contributor when active methane fermentation is in progress.

Within a conventional landfill setting, the transition between the various stabilization phases may not be so discrete, since each operating section of the landfill serves as an individual reaction system with a characteristic "age". When the leachate (or gas) from these systems merge, the constituents detected reflect a mixture of contributions from each component section and may increase or decrease accordingly. Nevertheless, as time passes, all parametric indices tend to be moderated by internal attenuation and assimilative mechanisms.

When organic and inorganic hazardous constituents are present, attenuation may be halted or interrupted due to retardation of the microbially mediated processes of waste stabilization. For instance, concentrations of hazardous constituents at levels inhibitory to the microbial populations present will eliminate or suppress their contribution to effective assimilation. Therefore, the "loading" of such constituents into a landfill can impede and extend the normal progress of waste stabilization, and exacerbate the potential for adverse environmental impact. In terms of the acid fermentation and methane formation phases, the latter phase is more vulnerable to such inhibition due to the greater sensitivity of the methane formers, thereby creating a condition analogous to an acid-stuck anaerobic sludge digester. Both organic and inorganic compounds can exert such an influence if a toxic limit or threshold is exceeded. Fortunately, however, both types of compounds may be eventually assimilated due to microbial adaptation and reactivity within the waste matrix.

There is evidence that both organic and inorganic hazardous and toxic compounds can be assimilated when present below a threshold loading limit within the landfill environment. Results of studies with selected classes of organic compounds (phenols, pesticides, monocyclic and polycyclic aromatics, halogenated aliphatics, and phthalate esters) and inorganic heavy metals (Cd, Cr, Ni, Pb, Hg, Zn) codisposed with municipal refuse have corroborated this fact and have suggested some of the potential mechanisms involved. The data on assimilation of heavy metals are more compelling, primarily because the reaction mechanisms are more predictable and generally involve reduction and precipitation or encapsulation as sulfides or hydroxides. The data for the organic compounds are less convincing, primarily because of the difficulty of distinguishing between biodegradative and volatilization, complexation and sorptive phenomena which are more subtle, long term and often influenced by competition. Nevertheless, the probable assimilative mechanism for these leachate (and gas) constituents is beginning to emerge; confirmation of these trends will eventually provide a sounder basis for operational and regulatory initiative. Moreover, although literature accounts indicate the susceptibility of many of the organic compounds to biodegradation under anaerobic conditions (Vogel, et al., 1984), most of these efforts have not considered the heterogeneity and complexity of the landfill environment. However, such studies provide presumptive evidence which may eventually lead to better experimental design and evaluation of either laboratory or field-scale results.

Operational Significance

As already suggested, operational control over the leachate and gas transport phases as well as the waste materials to be subjected to landfill stabilization is paramount, not only to proper understanding of landfill behavior and its potential environmental impacts, but to enhancement of predictability and reliability of process performance. Shortening the time necessary for landfill stabilization by optimizing environmental conditions and associated microbially mediated conversion processes concomitantly reduces operational and maintenance costs and beneficiates gas production and recovery as an energy source. Operational control requires containment and retrieval of both gas and leachate phases. As demonstrated at many landfill gas recovery installations (Anon., 1988), this facet of the strategy is a proven technology which is already linked to leachate (and condensate) management. Effective leachate management can be accomplished by restricting the amount accumulated to that necessary for maintaining accelerated stabilization or rapid biological stabilization (RBS) through leachate collection and recycle, thereby converting the landfill (or sections thereof) into a controlled anaerobic biological reactor system. When stabilization is complete, the leachate may be removed for ultimate disposal (spray irrigation, sewer discharge or evaporation ponds) with or without pretreatment. Final leachate removal deprives the landfill of this crucial transport phase and dramatically reduces the potential for further microbially mediated conversion and release of more complex compounds such as the humic-like substances and fractionation and desorption of various other waste, intermediate, and complexation products. The tenure of recycle operations would also be a function of intent, i.e., the ultimate requirements for encouraging adaptation to permit biodegradation and detoxification of hazardous compounds.

With regard to codisposal issues, loading criteria would establish the limits of such inputs to a given landfill situation, both in terms of avoidance of the inhibitory effects and an associated assimilative capacity which cannot be exceeded. As with many natural environmental systems, this *in situ* assimilative capacity should be used to advantage since most municipal landfills probably receive some amounts of hazardous and toxic wastes. Again, control and management of the principal transport phases, particularly the leachate phase, are crucial to routine operations as well as the implementation of corrective procedures when adjustments are required.

Concluding Remarks

Although a unified approach to leachate and gas management at landfill disposal sites is not yet available, increased understanding of the factors influencing leachate and gas formation and its integration into landfill design, operation and maintenance is becoming more prevalent. The advantages of accelerated stabilization and its control in terms of promoting more effective utilization of the assimilative capacity of landfills for both solid and hazardous wastes are becoming clearer and are being used to develop a new generation of landfills. Effective gas production, collection and utilization are integral to this development, and the innovation of *in situ* leachate management by containment and recycle is being recognized as the principal operational control element capable of converting the landfill into a controlled treatment process. Collectively, this approach promotes greater process predictability and operational control, and provides better assurance of protection from adverse environmental impacts.

References

Anon. (1988). Landfill gas survey update. Waste Age, March, 167-172.

Pohland, F. G. (1975). Accelerated solid waste stabilization and leachate treatment by leachate recycle through sanitary landfills. Prog. Wat. Tech., 7, 3/4, 753-765.

Pohland, F. G. (1980). Leachate recycle as landfill management option. J. Env. Engr. Div., ASCE, 106, EE6, 1057-1069.

Pohland, F. G., Dertien, J. T. and Ghosh, S. B. (1983). Leachate and gas quality changes during landfill stabilization of municipal refuse. Proc. 3rd Intl. Symp. Anaerob. Dig., Cambridge, MA, 185-201.

Pohland, F. G. and Harper, S. R. (1985). Critical review and summary of leachate and gas production from landfills. PB 86-240 181/AS, NTIS, Springfield, VA 22161.

Pohland, F. G., Gould, J. P. and Ghosh, S. B. (1985). Management of hazardous waste by codisposal with municipal refuse. Haz. Waste J., 1, 2, 143-158.

Pohland, F. G. and Gould, F. G. (1986). Co-disposal of municipal refuse and industrial waste sludge in landfills. Wat. Sci. Tech., 18, 12, 177-192.

Pohland, F. G. and Harper, S. R. (1987). Retrospective evaluation of the effects of selected industrial wastes on municipal solid waste stabilization in simulated landfills. PB 87-198 701/AS, NTIS, Springfield, VA 22161.

Vogel, T. M., Criddle, C. S. and McCarty, P. L. (1984). Transformations of halogenated aliphatic compounds. Env. Sci. Tech., 21, 722-736.

HEAT FLUX FROM A SANITARY LANDFILL

Matti O. Ettala, Paavo Ristola Ltd Consulting Engineers, Hollola, Finland

INTRODUCTION

Aerobic degradation can raise the temperature of the refuse up to 70°C. Much less heat is generated by anaerobic than by aerobic decomposition (e.g. Rees 1980). During the anaerobic stage, the temperature in sanitary landfills decreases to 25-40°C in a couple of years (e.g. Valdmaa 1981), and much lower temperatures have been measured in Finland, 9-22°C (Ettala 1986). An increase in the temperature would improve evapotranspiration (e.g. Ettala 1987) and promote degradation (e.g. Willumsen 1987), but high temperatures may interfere with revegetation (e.g. Flower et al. 1978). This study of the heat flux from a sanitary landfill also deals with the factors affecting heat generation and the results of this process.

MATERIAL AND METHODS

The heat flux through the upper layer of the landfill was investigated in 1986-1987 at Lahti (60°57'N, 25°45'E) and Hollola (60°57'N, 25°30'E). The temperature of the covering soil and refuse were measured with an accuracy of 0.1°C every two weeks at 11 points at depths of 10, 20 and 50 cm. The thermal conductivity of the substrate was estimated at each point by analysing samples of the material. The volume of the samples was measured by water volumetry. Snow depth was measured with measuring stakes and frost depth with a methylene blue tube (National Board of Waters 1984). The points are divided into four groups differing in disposal technology and landfill characteristics (Table 1). The snow water equivalent was measured three times in winter at each of 20 points at three landfills with the Korhonen-Melander snow sampler (National Board of Waters 1984). The temperatures deeper in the refuse were measured with an accuracy of 0.5°C at seven landfills, mostly in October, in boreholes or pits. To eliminate the influence of the age of the refuse, temperatures were measured once a week in June-December 1987 at 12 points at a depth of 0.8 m in 0-6-month-old refuse.

ISWA PROCEEDINGS
VOLUME 1 ISBN 0-12-058451-4

Table 1. Observation groups

Group	Number of points	Age of refuse (a)	Volume percentage of mineral soil(X_m)	organic soil(X_o)	water (X_w)	air (X_a)
1	3	2	15.6	10.2	34.9	39.3
2	2	5	29.9	7.6	15.5	47.1
3	3	5	44.3	6.2	23.7	25.7
4	3	>5	40.1	10.3	44.4	5.2

RESULTS

The measurements (Fig.1) show that the temperature is decreased by compacting the refuse strongly and covering it with soil (Groups 3-4). This agrees with Valdmaa's (1981) observation that the temperature is higher at the edge of the site (Group 1) than in the middle. Short-rotation plantations irrigated with leachate (Group 4) keep the substrate temperature lower in summer but higher in winter. Degradation of refuse decreases the frost and snow depths (Fig.2) and the snow and frost melt earlier in spring than in natural soils. As in winter 1984-1985 (Ettala 1986), the snow water equivalent was smaller at new than at old parts of the site in winter 1985-1986, too. Winter 1986-1987 was especially cold and no differences could be noted in the snow water equivalent or snow depth.

The heat flux from the refuse (Table 2.) was calculated for the period without frost and snow with the heat conduction equation (1), which includes a rough estimate of thermal conductivity obtained from equation (2) (Van Wijk 1963).

$$H = \lambda \frac{dT}{dz} \quad \text{in which} \qquad (1)$$

H = heat flux from the refuse ($W\,m^{-2}$)
λ = thermal conductivity ($J\;C^{-1}\;m^{-1}\;s^{-1}$)
T = landfill temperature (^oC)
z = vertical distance (m)

$$\lambda = X_m \lambda_m + X_o \lambda_o + X_w \lambda_w + X_a \lambda_a \qquad (2)$$

The temperatures (Table 3.) at depths of 1.0-9.0 m correspond to the results for small Danish landfills (Willumsen 1987), but are rather low compared with most other results (e.g. Valdmaa 1981). This can be explained by the low moisture content of the refuse samples (n=97, $\bar{x}$= 24 % w.w, s=13 % w.w) and their low loss on ignition (n=90, 12 % d.w., s=16 % d.w.) (Willumsen 1987, Rees 1980). The refuse age

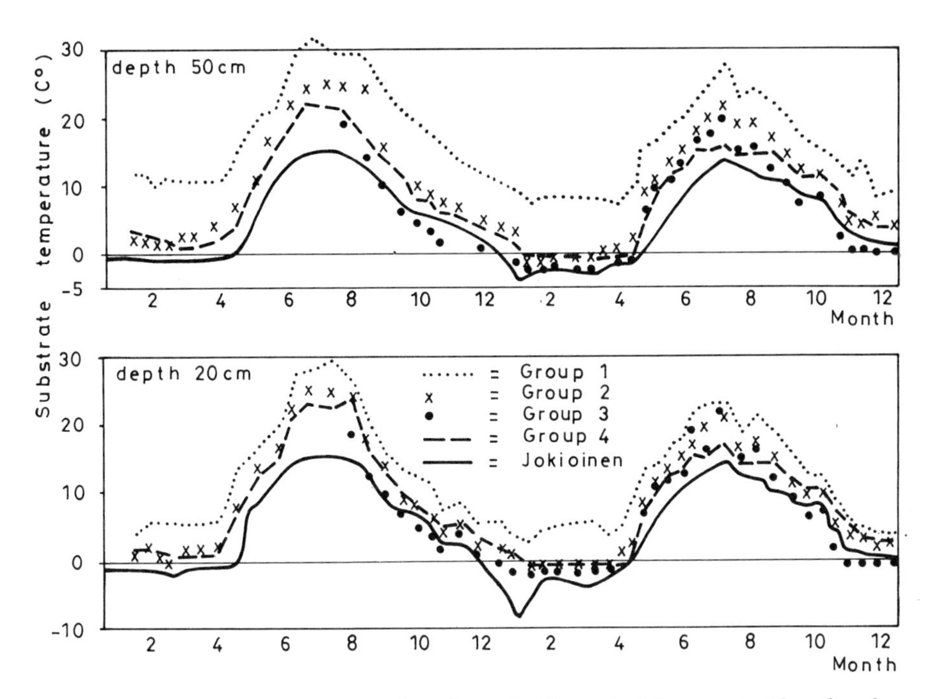

Fig.1. Temperatures at depths of 20 and 50 cm at the landfills studied (key in Table 1.) and at the Jokioinen observation station in 1986-1987.

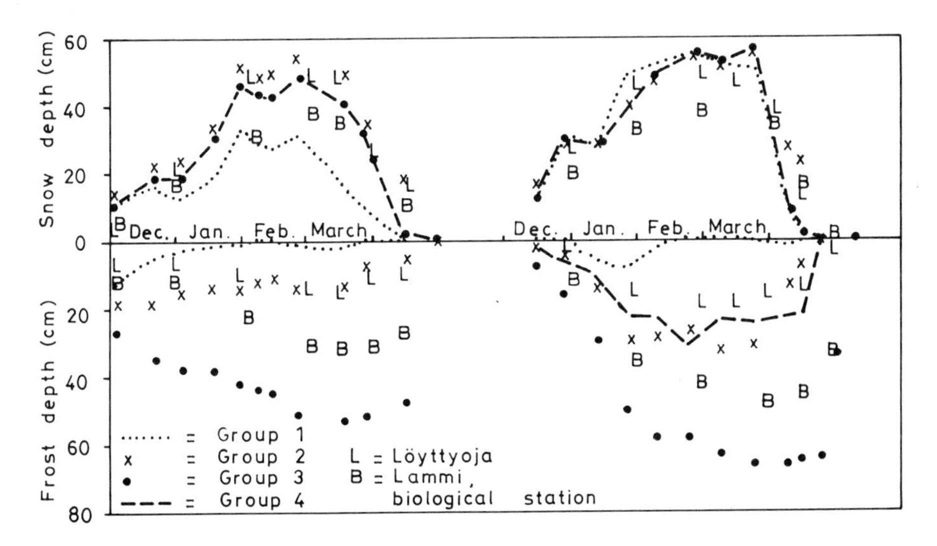

Fig.2. Depth of frost and snow at the landfills studied (key in Table 1.) and at the Lammi biological station and Löyttyoja station in Dec. 1985-April 1987.

is not the only reason for the low values, because the temperature of fresh refuse was also low: with strong compaction n=150, $\bar{x}$= 12.6°C, s=3.9°C and with slight compaction n=141, $\bar{x}$=12.2°C, s=4.2°C. Water temperature in the upper part of saturated refuse at four of the studied landfills averaged 12.0°C (n=13, s=2.1°C).

Table 2. Thermal conductivity (λ) of the refuse and heat flux from the refuse at the landfills of Lahti and Hollola in May-October 1986 and 1987 (key in Table 1).

Group	λ J °C^{-1} m^{-1} s^{-1}	Heat flux from the refuse (W m^{-2}) n	$\bar{x}$	s	range
1	1.1	24	4.1	2.6	0.23...9.0
2	1.9	24	1.4	2.1	-4.1...4.8
3	2.7	19	-0.41	2.7	-5.8...4.7
4	2.6	24	-0.92	2.5	-7.9...2.1

Table 3. Temperature measurements at seven sanitary landfills at depths of 1.0-9.0 m.

Landfill no.	Established in (year)	Refuse temperature (°C) at depth of 1.0-2.9 m n	$\bar{x}$	s	3.0-5.9 m n	$\bar{x}$	s	> 6.0 m n	$\bar{x}$	s
1	1966	1	17	-	1	26	-	1	12	-
2	1955	4	5.5	1.0	5	7.1	0.8	1	6.7	-
3	1957	-	-	-	2	14	-	1	21	-
4	1955	5	13	4.5	8	12	2.7	2	10	-
5	1960	1	25	-	3	13	3.9	1	18	-
6	1967	6	17	2.9	9	14	2.6	12	11	2.5
7	1964	1	20	-	1	13	-	1	24	-

DISCUSSION

The temperature at 20 cm depth was low enough for plant survival (Ettala 1987), but higher than in natural soils, being favourable for plants (e.g. Kramer 1983). The short-rotation plantations had sufficient nitrogen and lost their leaves as late as November (Ettala 1987), but the low winter temperatures did not cause more growth disturbance at the landfills than in other experiments (Hytönen 1988, pers. comm.).

Because of strong daily variation in the temperature at 10 cm depth (Van Wijk 1963), the temperatures at 20 and 50 cm were used in the heat flux calculations. The accuracy of the difference in the temperatures (dT) is ± 3 % in groups 1-2 and ± 15 % in groups 3-4. The distance between measuring points (dz) is estimated to be 30 cm ± 3.0 cm. The moisture content in group 4, irrigated with leachate, is fairly stable, but in the other groups the seasonal variation is larger. An accuracy of ± 50 % in the moisture volume percentage means an accuracy of -7.9...+8.8 % in the thermal conductivity in groups 1-2 and -4.7...+1.6 % in groups 3-4. The calculations of heat flux are most sensitive for the volume proportion of mineral soil, which results in a total accuracy of ± 50 % in groups 1-2 and ± 30 % in groups 3-4. However, the influence of the disposal technology and vegetation on the temperature and its seasonal variation can be clearly seen.

When account is taken of studies on the annual fluctuation of temperature at various depths in the soil (Van Wijk 1963) and in refuse (Valdmaa 1981), and of the time of the measurement, the temperatures at depths greater than 3.0 m can be considered to represent the average value with an accuracy of ± 2°C, except for landfill no. 2, which was flat and measured in winter. The temperatures are mostly in the range 10 - 15°C, which is unfavourable for methanogenic degradation. The moisture content of less than 45 % limits the possibility of raising the refuse temperature (e.g. Rees 1980). The abundant use of covering soil between refuse layers lowers the moisture content and the loss on ignition below the optimal values for the methanogenic phase. The wide variation in the results can be ascribed to differences in the water level and age and quality of the refuse, though no correlation could be noted between those parameters and the refuse temperature. Willumsen (1987) has reported the highest temperatures for deep landfills.

CONCLUSIONS

The heat flux from the sanitary landfills was low and conditions unfavourable for aerobic degradation and the methanogenic phase. The frost and snow depths measured indicate that the leachate discharge is smaller and steadier than the run-off in natural areas, which should be considered in dimensioning the leachate basin. Removal of snow from the site to reduce infiltration would also decrease cooling of the refuse. The low temperatures measured indicate that the results of laboratory tests using temperatures of 20-30°C cannot be applied directly.

Refuse temperatures should be raised in order to achieve methanogenic degradation and increase evapotranspiration. This could be done by having higher landfills, limiting mineral soil disposal, controlling the moisture content and promoting aerobic degradation at the beginning of disposal.

REFERENCES

Ettala, M. (1986), Snow cover and maximum leachate discharge of a sanitary landfill. Aqua Fennica 16,2:187-202.

Ettala, M. (1987), Influence of irrigation with leachate on biomass production and evapotranspiration on a sanitary landfill. Aqua Fennica 17,1:69-86.

Flower, F.B., Leone, I.A., Gilman, E.F. & Arthur, J.J. (1978), A study of Vegetation Problems Associated with Refuse Landfills. Cincinnati, EPA-600/2-78-094. 130 p.

Hytönen, J.(1988). The Finnish Forest Research Institute. Personal communication.

Kramer, P.J. (1983), Water Relations of Plants. Academic Press Inc. New York. 415 p.

National Board of Waters. (1984), Hydrologiset havainto- ja mittausmenetelmät (Methods of hydrological observations and measurements). Publications of the National Board of Waters 47. Helsinki. 88 p.

Rees, J.F. (1980), Optimisation of methane production and refuse decomposition in landfills by temperature control. Journal of Chem. Techn. Biotechnology 30: 458-465.

Valdmaa, K. (1981), Avfallsupplagets inre processer (Landfill processes). In Lectures: Helsinki waste symposium 81:310-323.

Van Wijk, W.R. (ed.) (1963), Physics of plant environment. North-Holland Publishing Co. Amsterdam. 382 p.

Willumsen, H.C. (1987), Landfill gas utilization, especially optimalization at small landfills. Proceedings of ISWA International symposium on Process, Technology and Environmental Impact of Sanitary Landfill. Gagliari, Sardinia, 19th-23rd October 1987. Vol I. 14 p.

Cancelli, A. - Dept. of Earth Science-Univ. of Milan, Italy
Cossu, R.- Dept. of Hydraulic - Univ. of Cagliari, Italy
Malpei, F - Civil Engineer, Milan, Italy
Pessina, D. - Civil Engineer, Milan, Italy

PERMEABILITY OF DIFFERENT MATERIALS TO LANDFILL LEACHATE.

1. INTRODUCTION

In the design of modern sanitary landfills many materials, performing different mechanical and hydraulic functions, are used. With regards to the lining and drainage functions, the performance of materials is normally tested with water (tap or distilled). In most cases, nevertheless, leachate comes into contact with the materials. Interactions between leachate and materials can include both physical (hydraulic) and chemical phenomena which could severely affect the performance of the materials.
In this paper the results of a research study on the influence of landfill leachate on the properties of sand - bentonite mixtures and geotextiles are reported. Hydraulic conductivity was tested for both materials.

2. MATERIALS TESTED AND PERMEANT LIQUIDS

2.1 Sand-bentonite mixture The basic component of the mixture tested is a siliceous sand, having the characteristics reported in Tab. 1. The bentonite (treated with Na salts) is commercially available in form of slightly moist powder. Its general properties are reported in Tab. 2. The mixture was formed of 95% of sand and 5% of bentonite.
The grain size curves of each component and the result of Proctor compaction test (Standard AASHO) are published in [1]. Standard AASHO was selected for compaction of the mixture and the preparation of specimens.
2.2 Geotextiles The complete list of tested geotextiles is reported in table 3, together with the most significant physical properties. The list includes only nonwoven geotextiles: mostly monofilament, both needled and thermobonded, and some staple-type Italian products. Where possible, data concerning the opening size O95 were inserted into the table, as drawn from literature [2].

3. PROPERTIES OF FLUIDS USED FOR PERMEABILITY TESTS

3.1 Water For permeability tests to water, the selection

ISWA PROCEEDINGS
VOLUME 1 ISBN 0-12-058451-4

Tab.1 - Geotechnical characteristic of the siliceous sand.

- Unit weight of solid particles	2.68 g/cm3
- Fraction passing through 2.0 mm sieve	100 %
- Fraction passing through 0.074 mm sieve	1.5 %
- Effective diam. D10	0.185 mm
- Coefficient of uniformity CU	2.22

Tab.2 - Properties of the bentonite

Mineralogical composition	
Smectite	80 %
Plagioclase	10%
K-feldspar	4-5 %
Quartz	2-3 %
Calcite	2-4 %
Dolomite	traces
Amphiboles	traces
Cation exchange capacity	90-100 meq/100 g
Swelling	15-20 cm3
Eslin's Index: after 2 h.	300%
after 8 h.	500%

between distilled and non-distilled water is still a controversial matter [3]. In this research study all samples are moulded and compacted with common tap water, in order to avoid expansion of the adsorbed layer and leaching of salts and to be more adherent to field compaction procedures. The same tap water, previously de-aired in order to enhance as much as possible the saturation of testing

Tab.3 - General characteristic and type of permittivity test developed on the considered geotextiles.

Commercial name and manufactures	Polymer and process	Type	μ (g/m²)	Tg (mm)	O95 (μm)	Permettivity test Water	Leachate
BIDIM Rhone-Poulenc (F)	PES NP CF	U 34	290	2.9	120	2	A
DREFON Man. Fontana (I)	PES NP SF	S-45	200	2.2	48	-	B
TECNOFELT Tecnofibra (I)	PES NP SF	FAG	300	3.4	-	-	C
GEODREN Edilfloor (I)	PES NP SF	PE/S	300	2.8	-	-	D
POLYFELT Chemie Linz (A)	PP NP CF	TS 750	370	3.0	125	6	E
DREFON Man. Fontana (I)	PP NP SF	SIA200	200	2.9	-	-	F
		SIA400	400	5.2	-	-	I
STRATUM Viganò Pav. (I)	PP NP SF		450	5.4	-	-	L
TYPAR Du Pont (LUX)	PP TB CF	3807	280	0.7	37	8	H
TERRAM I.C.I. (U.K)	PP-PE TB CF	1000	140	0.8	100	-	G

PES: polyester; PP: polypropylene; PE: polyetilene; NP: needle-punched
TB: thermo-bonded; CF: continuos-filament; SF: staple-fibre
μ: mass per unit area; O95: opening size; Tg: thickness of the geotextiles.

specimens, is used for all permeability tests and is taken as a reference fluid.

3.2 Geotextiles Four different kinds of leachate were used for the tests:

L1 - Leachate sampled in a young MSW landfill (Mozzate).

L2 - Leachate sampled in an old MSW landfill (Imola).

L3 - Leachate from a young MSW landfill (as L1 but sampled

in a different period).
L4 - Leachate from a hazardous industrial waste landfill.
Chemical analyses of the leachates are reported in Tab. 4.
Raw (unfiltered) leachates L1 and L4, and filtered leachates L1 and L2 were used for permeability tests on sand - bentonite mixtures. Raw L1 and L3 has been used for the tests on the geotextiles.

4. TESTING EQUIPMENT AND METHODOLOGY

4.1 Test on sand-bentonite layer A flexible wall permeameter was selected for tests. The main advantages of this solution are:

- a flexible wall inhibits undesired flow along side walls;
- the application of a back - pressure to the hydraulic circuit allows an easier, and more controllable, preliminary saturation of the test specimens.

Constant hydraulic head tests were carried out; special care was taken to use a low hydraulic gradient, in order to avoid particle migration. The testing conditions are characterized by the following values: all around cell pressure = 397 kPa; hydraulic head = 265 kPa; back-pressure = 245 kPa.

Tab.4- Chemical analysis of the tested leachates.

PARAMETER (*)	L1	L2	L3	L4
pH	6.0	8.5	6.3	7.3
COD	38,520	7,750	28,060	1,924
BOD	3,000	2,125	10,400	1,230
Volatile Fatty Acids (C)	1,574	n.a.	435	n.a.
Organic Nitrogen (N-NH4)	60	125	554	173
Ammonia (N-NH4)	1,293	1,040	1,203	n.a.
Phospate (P-PO4)	n.a.	2.3	n.a.	2.5(*)
Alkalinity (CaCO3)	5,125	8,250	4,250	n.a.
Chlorine (Cl)	2,231	3,650	1,868	n.a.
Sulfate (SO4)	1,600	219	1,860	n.a.
Sulphides (SO3)	n.a.	10.5	n.a.	n.a.
Calcium (Ca)	175	n.a.	n.a.	n.a.
Sodium (Na)	1,400	n.a.	1,300	n.a.
Potassium (K)	1,200	n.a.	1,200	n.a.
Magnesium (Mg)	1,469	n.a.	827	n.a.
Iron (Fe)	47	n.a.	330	0.3
Manganese (Mn)	42	n.a.	27	n.a.
Zinc (Zn)	7	n.a.	5	0.1
Lead (Pb)	n.a.	n.a.	n.a.	0.23
Copper (Cu)	n.a.	n.a.	n.a.	0.07
Cadmium (Cd)	n.a.	n.a.	n.a.	0.04
Nichel (Ni)	n.a.	n.a.	n.a.	0.58
Alluminium (Al)	n.a.	n.a.	n.a.	0.02
Surfactants (MBAS)	n.a.	4.2	n.a.	6.3
Phenols	n.a.	n.a.	n.a.	6.3

(*) All values in milligrams per liter, except pH.
(*) This value is not referred to phosphate, but to the total present phosphorus.

The hydraulic conductivity values are computed by Darcy's Law:

$$k = V/A \cdot t \cdot i$$

where V is the volume outflowing during time t, A is the section area of the soil specimen (38,47 cm2); t is the elapsed time between subsequent measures and i is the hydraulic gradient (13.3)
Leachates were invariably introduced into the hydraulic circuit after a preliminary specimens saturation with de-aired water; subsequently measurements of permeability to water were performed on the same specimens. The duration of the tests ranged from 25 to 30 days.
4.2 Tests on geotextiles All tests were carried out by means of a permeameter that was purpose-designed for testing the coefficient of normal permeability (kN) of geotextiles.
The hydraulic scheme of the apparatus is shown in Fig. 1.
The permeameter is formed of 3 coaxial plexiglass cylinders (respectively Ø = 284,124 and 51 mm) and the geotextiles are placed at the base of the internal cylinder. Constant head were performed. To limit the values of the apparent fluid velocity (Vmax = 0,035 m/s), in view of the validity of the Darcy-Ritter's law [4],[5], a minimum value of the specimen

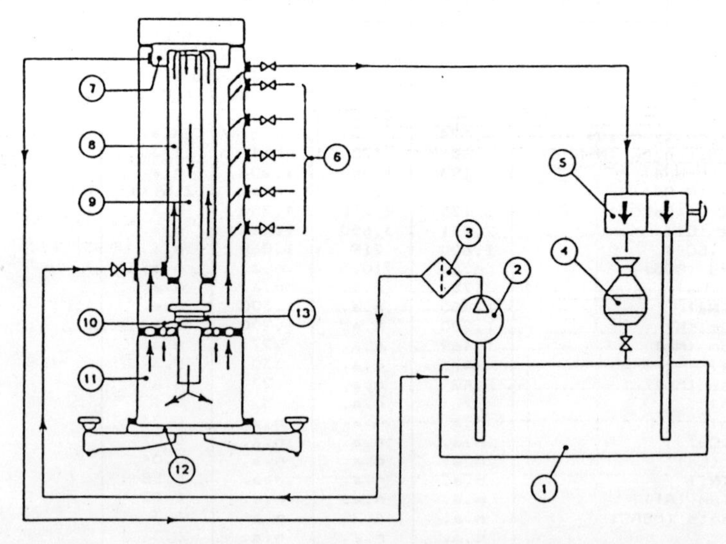

Fig. 1 Hydraulic scheme of the closed circuit, constant head permeameter for permittivity tests: (1) backwater tank; (2) feed pump; (3) filter; (4) calibrated container; (5) two-way valve; (6) outlets (6 positions for 6 different levels); (7) overflow system; (8) constant level cylinder; (9) feed cylinder; (10) specimen bearing; (11) external cylinder; (12) base; (13) geotextile specimen

thickness H, function of the hydrostatic load, was used. The coefficient of normal permeability kN is computed by:

kN = V • H/(A • h • t)

where: V is the calibrated volume (1450 •10-6 m3), A is the specimen surface (2043 •10-6 m2), H is the specimen thickness, h is the hydrostatic load and t is the required time to fill calibrated volume during a single test.

The permeameter as decribed above permits testing in absence of normal stesses on the specimen surface.

The duration of the test, for most materials, ranged from 50 to 100 minutes.

5. RESULTS AND DISCUSSION

5.1 Permeability test on sand-bentonite mixture The results of all tests, expressed by k vs. time plots, are reported in Fig. 2. Curve A refers to de-aired tap water. The final value of hydraulic conductivity is about 2,3 x 10-5 cm/s.

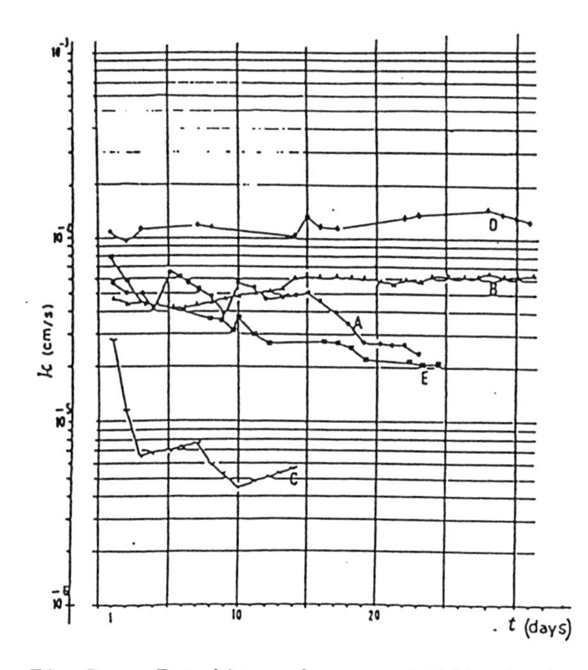

Fig.2 - Results of permeability tests: A=de-aired water; B=filtered L1; C=unfiltered L1; D=filtered L2; E = L4.

Filtered leachate L1 was used for the second test, represented in the figure by curve B. The hydraulic conductivity gradually increased and reached a value of 6.3 x 10-5 cm/s after two weeks. When the filtered leachate L1 was replaced by the unfiltered leachate L1, the hydraulic conductivity showed a sharp decrease; the final value was 5.6 x 10-6 cm/s (curve C).

Curve D refers to the hydraulic conductivity tests with filtered leachate L2. This specimen behaved a little different, because even the initial permeability to water was larger than the previous one; when water was replaced by leachate L2 the permeability showed a light decrease and then a moderate increase.

The last test was performed with leachate L4. The general trend of this test is very similar to that obtained with the first permeability test to water; the hydraulic conductivity moderately decreased with time, reaching a final value of 2.1 x 10-5 cm/s (curve E).

These results suggest that two phenomenon may occur when a leachate comes in contact with a sand-bentonite liner: the physico-chemical characteristics of leachates affect the hydraulic conductivity of the mixture, causing an increase of its values with respect to those obtained with water; on the other hand the presence of suspended particles in the leachates causes pore clogging and, consequently decrease of permeability.

Further investigation of the effect of different leachates on bentonite, Atterberg limits were determined on bentonite remoulded with filtered leachates L1 and L2, leachate L4 and tap water. The values thus obtained are reported in the plasticity chart (Fig. 3). The decrease of the plasticity and liquid limits of the bentonite remoulded with leachates is particularly evident. As a consequence of reduced plasticity, reduction in the qualities of bentonite as sealing and lining agent should be expected.

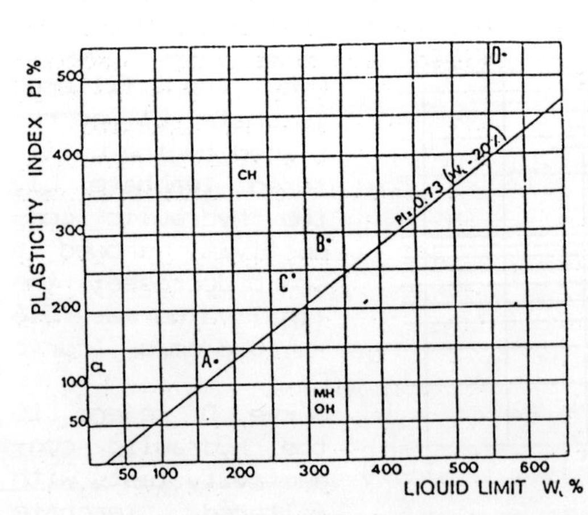

Fig.3 Atterberg's limits: A=filtered L1; B=filtered L2; C=L4; D= tap water.

5.2 Permeability test on geotextiles In Fig. 4 the coefficient of normal permeability (kN) to water of some of the geotextiles tested with the leachates is reported. These results are from [4] and were performed under normal stresses σ different from zero.

The results of permittivity tests carried out with the leachates are reported in Fig. 5. Leachate L3 was used for testing BIDIM (plot A) and POLYFELT (plot E), while L1 was used

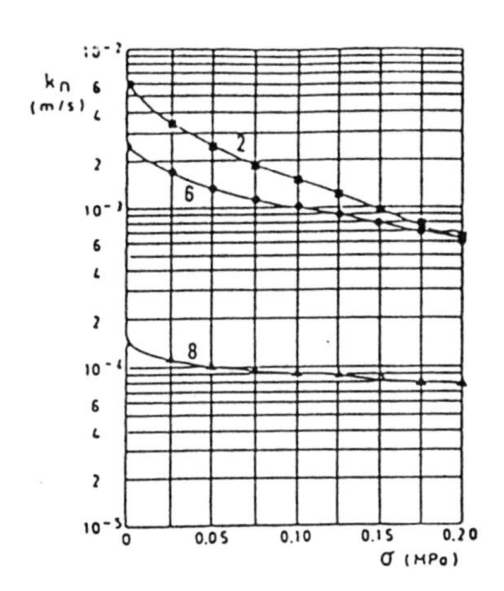

Fig. 4. Results of permittivity tests to water

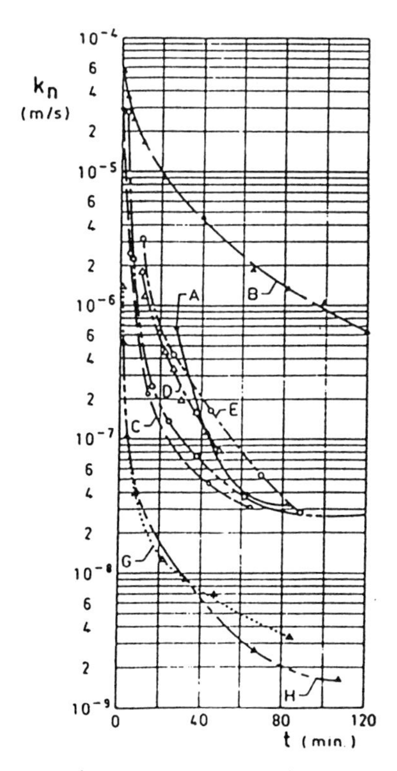

Fig. 5 Results of permittivity tests to leachate

for all other geotextiles.

Owing to the (feared) high clogging capacity of leachate, testing specimens formed by a single geotextile layer were adopted. Consequent to the changing of the fluid in the closed hydraulic circuit every two geotextiles, only some materials could be tested with fresh leachate, while the remaining ones were tested with filtered leachate. Therefore:

- the results obtained for the former group (BIDIM U34, GEODREN PE/S, TECNOFELT FAG, DREFON S1A-200, and TERRAM 1000) can be considered valid and comparable to the present ones;
- for the latter group of geotextiles (POLYFELT TS/750, STRATUM, DREFON S1A-400, DREFON S45-200, TYPAR 3807), the measured values are only indicative, and should be considered with extreme caution (it has to be remembered that, in field applications, geotextiles do not come into contact with pre-filtered leachate).

On the basis of all results, the following considerations can be made:

a) all tested materials show a marked decrease of kN according to time elapsed from start of tests; the ratio between initial and final permittivity is 100; b) the decrease should be ascribed essentially to the deposition of solid substances suspended in leachates, prevalently on the geotextile surface, and partly also in the pores of the geotextile itself;

c) for those materials where

comparison is possible (i.e. BIDIM, POLYFELT and TYPAR) it can be observed that leachate is able to reduce permability to less than 1/100.000 of the values referred to water.

ACKNOWLEDGEMENTS

Laboratory tests were carried out partly at the Geotechnical laboratory of the Politecnico of Milan and partly at the ENEL-CRIS (Hydraulic Research Centre of the National Board for Electric Energy). The valid cooperation of all technicians is kindly **acknowledged.** Leachate were supplied by I.G.M. (Milan) and Ecodeco (Pavia). Bentonite was supplied by Laviosa S.p.A., which provided also chemical analyses of it.

REFERENCES

[1] Cancelli, A.; Cossu, R. & Malpei, F.;(1987)"Laboratory Investigation on Bentonite as Sealing Agent in Sanitary Landfill. Proc. International Sanitary Landfill Symposium, Cagliari, 19-23/10/1987, XXXI.

[2] Fayoux, D.; Cazzuffi, D. & Faure, Y.,(1984) "La détermination des caractéristiques de filtration des géotextiles: comparison des resultats de différent laboratoires", Proc. Materials for Dams 84, Monte Carlo, 1984.

[3] Olson, R.E. & Daniel, D.E. "Measurements of the Hydraulic Conductivity of Fine grained Soils". Permeability and Groundwater Contaminant Transport, ASTM STP 746, pp.18-44.

[4] Francia, L.(1983) "Studi sperimentali sui geotessili per applicazioni in campo geotecnico ed idraulico", unpublished thesis, Faculty of Engineering, University of Bologna, 1983.

[5] Gourc, J.P.; Telliez, Ch.; Sotton, M. & Leclerq, B. (1981) "Perméabilité des géotextiles et perméamètres", Matériaux et Construction, No.82, 1981.

COMBINED BIOLOGICAL AND PHYSICAL/CHEMICAL TREATMENT OF SANITARY LANDFILL LEACHATE

Henning Albers/Gerd Krückeberg

1. Introduction

In the past first preference of sanitary landfill leachate treatment in Germany was the removal of oxygen-consuming substances, i.e. BOD_5 and ammonia-nitrogen. The new water legislation acts of 1986/87 brought stricter requirements for leachate treatment: besides oxygen-consuming also "hazardous" substances have to be removed to acceptable limits by using means of prior art. For leachate treatment especially COD, heavy metal and AOX (= Adsorbable Organic Halogen) effluent levels need further reductions.

In many cases it is expected that these values cannot be met by biological treatment methods alone. So the landfills discharging to publicly owned treatment works are searching for suitable pre-treatment steps, whereas for separate treatment post-treatment processes have to be developed.

Two options for the latter case are combinations of biological treatment with reverse osmosis on the one hand and with coagulation and adsorption on the other hand.

2. Process Combination at Heisterholz

A process scheme of the plant at Minden-Heisterholz, located west of Hannover, is given in Figure 1. The chemical/physical part has operated since 1977. In 1982 a first biological plant was placed ahead as a result of a research project (Ehrig, 1983). This year the activated sludge plant was completed by a denitrification tank and a fixed-bed reactor for nitrification. These provisions will help to overcome problems associated with hydraulic overloading and insufficient nitrogen

ISWA PROCEEDINGS
VOLUME 1 ISBN 0-12-058451-4

removal. Besides a belt filter press was installed to improve the sludges`mechanical and leaching characteristics before disposal to the landfill.

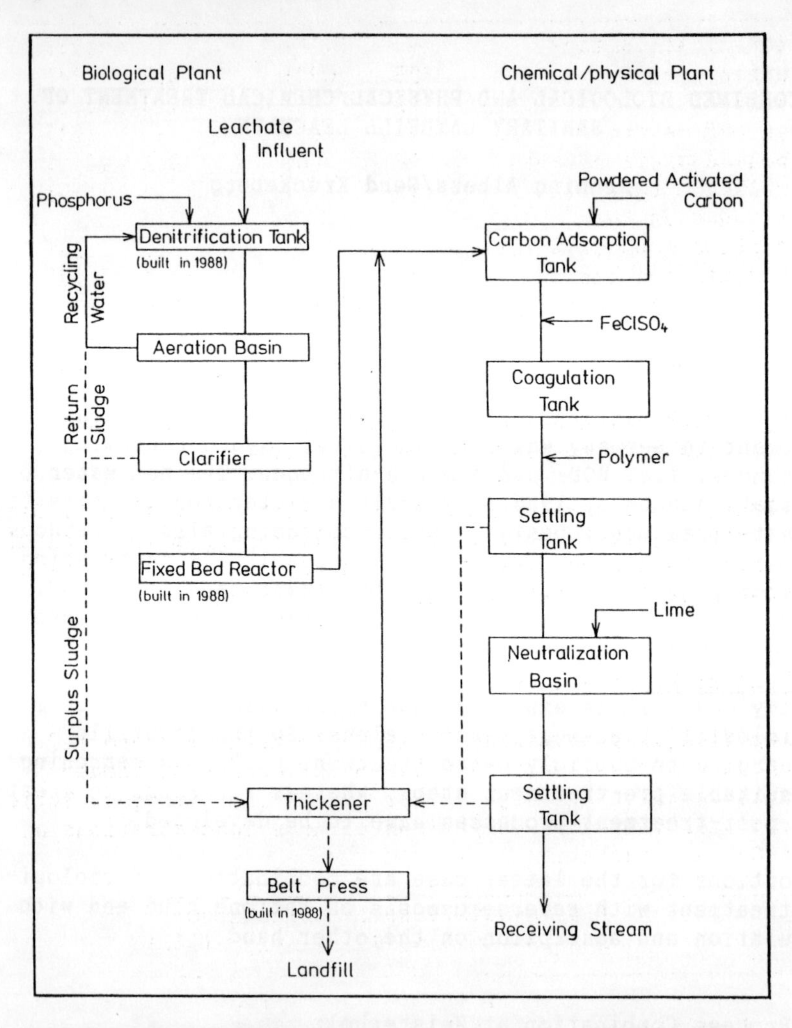

Fig. 1: Process scheme of the sanitary landfill leachate treatment plant at Minden-Heisterholz

	Tank volume $[m^3]$		Mean retention time [h]	
Denitrification tank	-	(417)	-	(21)
Aeration basin	725	(1670)	72.5	(84)
Clarifier	19	(146)	1.9	(7.3)
Fixed bed reactor	-	(460)	-	(23)
Carbon adsorption tank	36	(80)	3.6	(4.0)
Coagulation tank	36	(30)	3.6	(1.5)
Settling tank for adsorption + coagulation	60	(277)	6.0	(14)
Neutralization basin	131	(48)	13.1	(2.4)
Settling tank neutral.	60	(filter)	6.0	-
Total	1067	(3128)	100.7	(157)

Table 1: Tank volumes and retention times at mean flow (values from 1988 on in brackets)

The plant is designed now for a water flow of 20 $m^3 \cdot h^{-1}$. Mean value from August 1985 to January 1987, the time interval of which results are presented here, was 10 $m^3 \cdot h^{-1}$. Tank volumes and mean retention times are given in Table 1.

3. Removal Efficiencies

The landfill is progressing to a stable methane phase and therefore produces leachate with small amounts of biodegradable organics but high levels of ammonia-nitrogen. Table 2 gives mean values in comparison to mean concentrations of other German leachates.

		Minden-Heisterholz Aug.`85 to Jan.`87	Leachates (Ehrig, 1986)
pH	[-]	7.1	8.0
COD	$[mg \cdot l^{-1}]$	1,400	3,000
BOD_5	$[mg \cdot l^{-1}]$	343	180
NH_4-N	$[mg \cdot l^{-1}]$	493	750
Cl	$[mg \cdot l^{-1}]$	1,130	2,100
AOX*	$[\mu g \cdot l^{-1}]$	742	2,000

Table 2: Leachate concentrations for Minden-Heisterholz and other German landfills
*: only few data points

Because of special characteristics of landfill shape and former use as a clay pit this leachate is more diluted as usually found in Germany. This fact has to be taken into account when looking at removal efficiencies and effluent values in Table 3.

	BOD_5		COD		AOX	
	Values [$mg \cdot l^{-1}$]	Removal [%]	Values [$mg \cdot l^{-1}$]	Removal [%]	Values [$\mu g \cdot l^{-1}$]	Removal [%]
Influent	343	-	1400	-	742	-
Effluents						
Activated Sludge	51	85	742	47	605	18
Powdered Act. Carbon	20	9	420	23	343	36
Coagulation	7	4	112	22	122	30
Lime Neutralization	7	0	110	0	122	0
Total Removal	-	98	-	92	-	84

Table 3: BOD_5, COD and AOX levels and removal efficiencies (mean values)

About 50% of Total-COD was biodegradable matter and mainly removed in the activated sludge plant. But a significant portion (analyzed as BOD_5) remained for the chemical/physical plant. Almost every time the oxygen level in the aeration basin was near zero, indicating that the aeration capacity was not sufficient for the arising leachate pollution any more. Because of the same reason a nitrification process could not be established.

The non-biodegradable COD was removed to nearly the same extent by the adsorption and the coagulation step, respectively. No further reductions were observed in the lime neutralization stage. Only a slight AOX removal was found in the activated sludge plant, mainly due to volatilization and probably to adsorption onto biomass. Main portions of AOX in sanitary landfill leachate are of non-volatile nature and can effectively be removed by both adsorption and coagulation processes.

For COD as an example the statistical approach in Figure 2 shows large fluctuations of influent concentrations. But in nearly all cases these peak levels could already be compensated in the activated sludge plant so that effluent concentrations were held at constant levels.

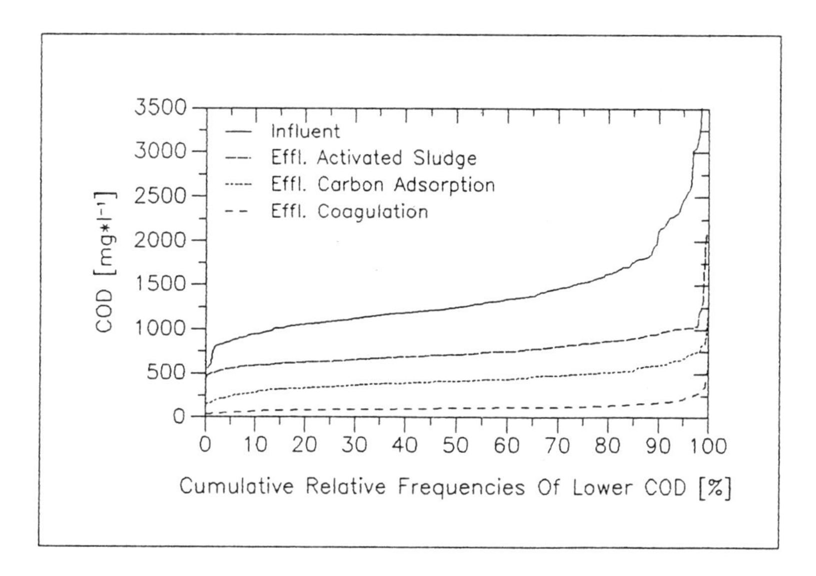

Fig. 2: Cumulative relative frequencies of lower COD values for influent and effluents of the various stages

The given levels for COD and AOX effluents were achieved with carbon doses of 800 to 1,200 $g \cdot m^{-3}$ leachate and chemical coagulant ($FeClSO_4$) doses of 600 to 1,200 g $Fe^{3+} \cdot m^{-3}$ leachate. Costs were about DM 2.- to DM 3.- for carbon and DM 1.10 to DM 2.20 for coagulant per cubicmetre leachate, respectively.

4. Special Aspects of Physical/Chemical Treatment

In the adsorption stage a linear relationship was found between carbon dose and COD removal: with a 800 $g \cdot m^{-3}$ dose about 250 $mg \cdot l^{-1}$ of COD were removed, at 1,200 $g \cdot m^{-3}$ applied the COD removal reached nearly 400 $mg \cdot l^{-1}$. This fact is somewhat surprising because theoretically a decrease in adsorptive capacity with lower COD equilibrium concentration has to be expected. But on the contrary, Figure 3 shows a constant capacity over the whole COD range.

Since isotherm theories could not describe this reaction by using the parameter COD a super- position of competing and reversible adsorption processes might have happened. Biochemical processes could also have caused interferences because the reactor was stirred by air injection.

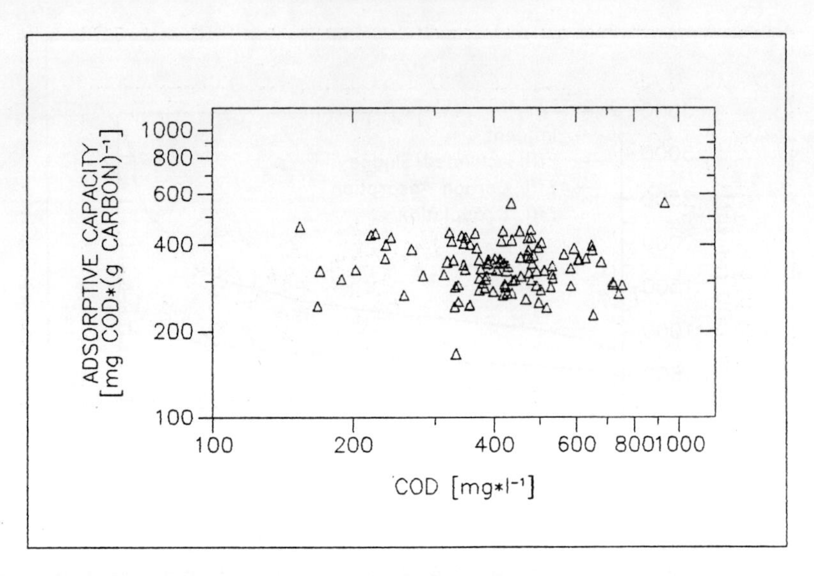

Fig. 3: Adsorptive capacity versus COD equilibrium concentration in the reactor

COD removal efficiency in the coagulation stage was mainly influenced by pH. Optimum values for leachate coagulation with ironsalts have been found at pH levels between 4 and 5 (Ehrig, 1983). So the amount of iron required depends on the leachate's alkalinity after biological treatment. This relation can be described by the following equation:

$$Fe^{3+}\ [g \cdot m^{-3}] = 0.373 \cdot CaCO_3\ [mg \cdot l^{-1}]$$

The data points derived from Heisterholz fit well to this equation (here not shown in detail). Iron salt addition was controlled by pH. The set value was between 4.8 and 5.0.

Of special interest were interactions between the adsorption and the coagulation step. Figure 5 shows COD reductions by coagulation versus corresponding values for adsorption and indicates a better removal by coagulation when, at the same time, COD concentrations after adsorption remain high. This fact is also valid for the case vice versa, so that about 500 to 600 $mg \cdot l^{-1}$ COD were removed totally, independent from removal rates in the individual steps. To draw a definite distinction between adsorbable organic matter of the leachate and constituents which are only removable by coagulation is not possible.

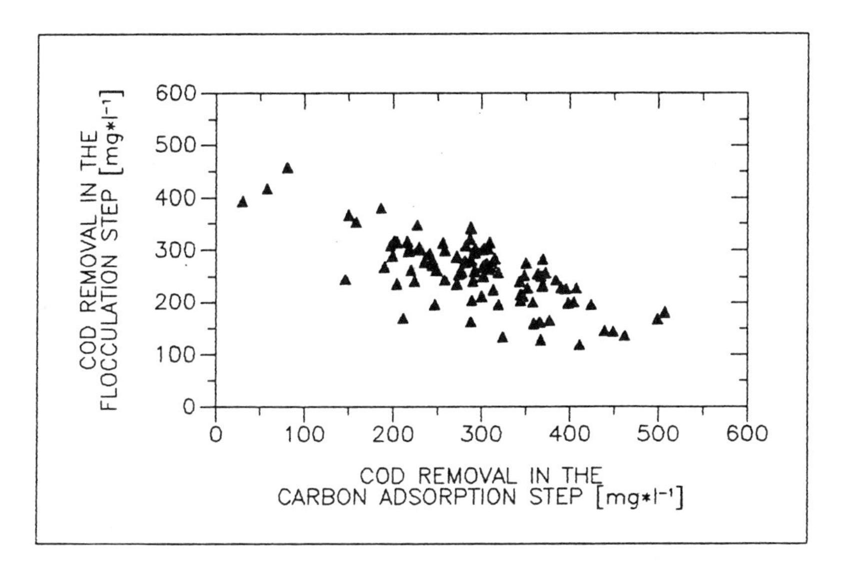

Fig. 4: COD reduction by coagulation versus COD removal by adsorption

5. Process Modifications and Future Options

Main improvements have been achieved this year by establishing a nitrification process in the activated sludge plant. To meet the effluent limits of 3 mg NH_4-N · l^{-1} a fixed bed reactor with packed plastic material was installed. The pre-denitrification step was introduced to improve total nitrogen removal and process stability and especially to avoid a possible lack of alkalinity. The higher alkalinity consumption due to nitrification in the biological plant will reduce the amount of iron salt needed for coagulation. It is therefore expected that chemical sludge volumes will decrease.

If a more extensive nitrogen removal by denitrification is required, this will be possible with this plant in principle. But it has to be kept in mind that these leachates from older landfills do not contain enough biodegradable organic matter (BOD) for complete denitrification (Mennerich/ Albers, 1987). One possibility then is adding organic concentrates like methanol or acetic acid.

6. Summary and Conclusions

BOD_5, COD and AOX removal efficiencies and influencing factors were investigated during a 18 month period at the biological and chemical/physical leachate treatment plant of Minden-Heisterholz. These full-scale results show that the main BOD_5 is removed in the biological plant. Further improvements in BOD_5 removal and full nitrification are expected from this year on when the new biological plant will be completed.

Biological pre-treatment makes an efficient chemical/physical treatment using carbon adsorption and coagulation with iron salts possible. This combination removes non-biodegradable organic matter (COD) and halogenated hydrocarbons (AOX) to low levels which will meet the new German effluent standards for leachate treatment. The three individual treatment methods are able to react flexible so that a high process stability as a whole is achieved.

7. References

Ehrig, H.-J. (1983). Großtechnische Verfahrensentwicklung zur chemisch/physikalischen Behandlung von Deponiesickerwässern (Development of a Full-Scale Chemical/Physical Treatment for Landfill Leachates), Research Project Report, TU Braunschweig, unpublished.

Ehrig, H.-J. (1986). Was ist Deponiesickerwasser? - Mengen und Inhaltsstoffe - (What is Landfill Leachate - Quantities and Constituents -), ATV-Dokumentation, Number 4, 19-36.

Mennerich, A./Albers H. (1988). Nitrification/ Denitrification of Landfill Leachates, Wat. Supply, Vol. 6, 157-166.

USING BIOASSAYS TO EVALUATE THE TOXIC AND MUTAGENIC PROPERTIES OF LANDFILL LEACHATE

Mezzanotte,V.,Sora,S.,Viganò,L.,Vismara,R., Politecnico di Milano, Italy

The Italian law (1,2) allows the disposal of wastes by different kinds of landfill according to the hazard they pose , not so much for themselves as for the leachate they could produce. That's why wastes are classified by lists of industrial processes, or of toxic components (for which concentration limits are indicated). Such classification, however, is used to define the appropriate kind of landfill for a given waste, but can say nothing about the true risk that will arise from the leaching once that waste is disposed of with others. No further indication is specified about the monitoring of leachate or of groundwater , assuming the lining of landfill could absolutely prevent groundwater pollution, or of the receptor, in case leachate is disposed after treatment in surface water .
With respect to both disposal according to the law indications and accidental events, the Italian law shows, in our opinion, two major lackings: the first one concerns the monitoring of the combined effects which the great number of pollutants in leachate may have on the organisms in the receiving water body and on man itself, and the second one is about the possible presence of mutagens in landfill leachate.

PROPOSAL OF BIOASSAYS FOR MONITORING

For the above mentioned reasons, an analytical evaluation of waste, even if accurate, seems to be inexhaustive for certain aspects. In particular, the individuation of various organic compounds (pesticides, solvents, polycyclic aromatic hydrocarbons, etc.) presents serious analytical problems, which, moreover, affect significantly the global cost of the analysis. At the usual concentrations detectable in the various environmental compartments, such compounds are held to be hazardous not as much for their short term toxicity as for their genotoxic potential.
That's why two kinds of bioassay are taken into account to integrate the answers of the chemical analysis to be used both for an a priori evaluation of the waste to be disposed (performing the bioassay on the liquid obtained by artificial leaching of the waste) and for the in situ monitoring

ISWA PROCEEDINGS
VOLUME 1 ISBN 0-12-058451-4

of landfill leachate.
The crustacean *Daphnia magna* is presently proposed as the best substitute to rainbow trout for acute toxicity bioassays. *Daphnia* shows in fact a sensitivity equal or higher than *Salmo gairdneri* and its use presents significant practical advantages:few liters of water are in fact enough for a massive growing of such organisms and only some milliliters are necessary for the bioassay (5).
The mutagenesis bioassay, on the contrary, aims at a qualitative evaluation of the mutagenic properties of the tested sample.The short term mutagenesis assays are performed at the highest possible concentration of the tested material. The applicability of the datum obtained on microorganisms to man is affected by various limitations, but important informations can be drawn about the mutagenic potential of the tested material without difficult and expensive analytical determinations, which is finally the goal of a certain kind of environmental evaluations.

MATERIALS AND METHODS

The toxicity and the mutagenicity bioassays were performed on leachates from municipal solid waste landfill and on the liquids obtained by leaching, at the laboratory, ashes and slags from municipal solid waste incinerators. Ash and slag leachates were obtained by artificial rain leaching in plastic lysimeters (6). The chemical composition of landfill leachates is reported in Table 1, while that of ash and slag leachates is reported in Table 2.
The toxicity test was carried out according to the method described by OECD (5) and aimed at defining the levels corresponding to acute toxicity (24 hours exposure) on the 50% of the tested organisms (EC 50).
The effect was observed after 24 hours and measured in terms of immobilization of the tested organisms, as specified by the above mentioned OECD method (5).
The mutagenicity assay (Ames test), whose details are described in the works of the author (4), is based on the detection of the increased number of bacteria able to grow without istidine in a population needing istidine, exposed to the tested substance. Since it was necessary to test sterile material, the samples were homogenized and centrifuged, and the supernatants,filtered on a 0.45 um membrane, were directly used for the test. To obtain exhaustive informations, the particulate fraction was tested too, after extracting the organic component in dichloromethane.

Table 1 – Chemical characterization of the tested landfill leachates (all concentrations expressed as mg/l)

Parameters	L1	L2	L3
pH	7.85	7.80	6.00
COD	64,022	56,909	38,523
BOD	7,000	6,400	12,000
N org.	1,703	1,526	60.1
N-NH3	2,478	3,040	1,293
Chlorides	4,411	4,235	2,232
Sulphides	2.2	1.8	n.d.
Fe	48	60	47
Cr tot.	0.1	0.1	<0.2
Mn	6.8	7.4	42
Ni	3.1	3.5	1.6
Cu	0.10	0.13	<0.20
Zn	1.08	1.06	6.80
Cd	0.10	0.11	0.08
Pb	1.40	1.30	<1.00

Table 2 – Chemical characterization of the tested ash and slag artificial leachates (all concentrations expressed as mg/l)

Parameters	Ash leachates			Slag leachates		
	A1	A2	A3	S1	S2	S3
pH	7.1	7.1	11.4	6.4	6.9	6.8
Chlorides	252,593	79,711	5,442	3,110	1,542	1,413
Fe	4.00	1.65	0.17	0.32	<0.44	<0.40
Zn	13,875	382	0.18	3.62	1.66	1.00
Pb	52	<7.50	0.18	<1.00	<1.00	<1.00
Ni	8.00	3.00	<1.00	0.30	<0.21	<0.20
Cr	2.00	<1.48	0.49	0.65	0.41	0.38
Mn	10.40	<0.85	3.00	<0.30	<0.30	<0.30
Cu	1.50	<0.46	0.15	0.25	<0.16	<0.14
Cd	1,632	112	0.07	0.44	<0.18	<0.04

RESULTS AND DISCUSSION

The results obtained by Daphnia bioassay are shown in Table 3.
While the toxicity test showed no effect only for the samples S2 and S3, Ames test showed no mutagenicity in most

cases. As reported in Table 4 , only samples A1, A2, S1 induced significant increases in the spontaneous frequency of mutation .

Table 3 – Dilutions of the tested materials corresponding to acute toxicity in 50% of the exposed organisms in 24 hours (EC 50 = Effective Concentration)

Sample	EC 50 (24 h)
L 1	35%
L 2	2.3%
L 3	7%
A 1	0.022%
A 2	0.9%
A 3	11%
S 1	88%
S 2	>100%
S 3	>100%

Table 4 – Results obtained with Ames test on 5 strains of Salmonella typhimurium: + indicates significant increases in the spontaneous frequency of mutation.

Sample	Strains				
	TA 98	TA 100	TA 1535	TA 1537	TA 1538
A1	+	+	–	+	–
A2	+	+	–	+	+
S1	–	+	–	–	–

As previously specified, the two bioassays lead to different kinds of answers; however, the first evidence of the work carried out is that the most toxic samples (A1 and A2) also show the highest rate of induced mutation with respect to the 5 tested strains of Salmonella.
In Table 5 are reported the metal concentrations corresponding to the EC 50. Such elements have been considered as a reference, due to the availability of great amounts of data concerning their toxicity on Daphnia, whose threshold values are also reported in the table, and to their being routine analysis parameters. The last column shows the limits set by the Italian law (Table A of the Law n.319, 10-5-76) for the discharge in surface water (3). Table 5 does not

include data about the samples S2 and S3, showing no toxicity.

Table 5 - Metal concentrations in leachates corresponding to acute toxic effects on *Daphnia*, toxicity thresholds (T.T.)for the single metals (7), and relevant limits set by the Italian law (Table A)(mg/l)(3).

	T.T.	L1	L2	L3	D1	D2	D3	S1	Table A
Fe	9.6	16.8	1.4	3.3	0.0009	0.01	0.02	0.28	2.0
Zn	3.0	0.4	0.02	0.5	3.10	3.40	0.02	3.2	0.5
Pb	1.815	0.5	0.03	<0.07	0.01	<0.07	0.02	<0.88	0.2
Ni	8.2	1.1	0.08	0.1	0.002	0.03	<0.11	0.26	2.0
Cr(*)	0.6	0.04	0.002	<0.01	0.0004	<0.01	0.05	0.57	0.2-2
Mn	9.8	2.4	0.20	2.9	0.002	<0.008	0.30	<0.26	2.0
Cu	0.15	0.04	0.003	<0.01	0.0003	<0.004	0.02	0.22	0.1
Cd	0.203	0.04	0.003	0.006	0.40	1.00	0.008	0.39	0.02

(*) Concentrations expressed as total chromium, except for the Table A column, where the first value refers to Cr(VI) while the second one refers to Cr(III)

The toxic effect shown by L1 (MSW leachate) can depend on the high value of iron concentration, overcoming the generally accepted threshold value and, among other things, also the limits set by Table A. Other metals are likely to contribute to the toxic effect, three of which (Pb,Mn and Cd) show values higher than the limits. For L2 and L3 no metal shows values high enough to be specifically responsible for the observed toxic effect. With respect to the limits set by law, then, the concentrations of the eight considered elements appear acceptable for L2, while L3 is characterized by iron, zinc and manganese concentrations higher than the limit values. Again metals can be responsible for the toxicity of A1, A2 (in particular zinc and cadmium) and S1 (zinc, copper and cadmium). The more represented elements, which can thus be involved in the observed toxicity, also overcome, for A1, A2 and S1, the law limits. A different case is represented by A3, somehow comparable to L2 and L3, for which toxicity can be attributed either to the combined effect of various low level components or to the presence of other, unidentified, pollutants, difficultly detectable by routine monitoring.
Obviously, the present work does not aim at concluding about the general toxicity or mutagenicity of landfill leachate,

or of ash or slag leachate. An interesting result, however, is concerning the applicability of the two kinds of bioassay to leachates of various origin and the complementarity of the answers they allow to obtain. Actually, with both Ames test and Daphnia test, we have been able to detect effects unidentifiable by routine analysis, supplying at least preliminary information about two aspects of major concern, i.e. short-term effects and mutagenic potential. In general, the kind of answer obtained by tests performed on leachate itself can then suggest the opportunity of performing the tests on the receiving water bodies, and/or on the surrounding ones, and on groundwaters as routine monitoring or as periodical survey, or of adopting a different classification for the hazards posed by the wastes to be disposed of.

REFERENCES

(1) DPR 10.09.82,n.915 : Attuazione delle Direttive CEE n.73/442 relativa ai rifiuti,n.76/403 relativa allo smaltimento dei policlorobifenili e dei policlorotrifenili e n.78/79 relativa ai rifiuti tossici e nocivi.

(2) Comitato Interministeriale di cui all´art.5 del DPR 10.09.82,n.915:Disposizioni per la prima applicazione dell´art.4 del DPR 10.09.82,n.915, concernente lo smaltimento dei rifiuti (Deliberazione del 27.07.84).

(3) Legge 10.05.76 , n.319: Norme per la tutela delle acque dall´inquinamento.

(4) Ames, B.N.(1981): The detection of chemical mutagens with enteric bacteria.Chemical mutagens: principle and methods for their detection. A.Hollander ,Ed., Plenum Press, New York

(5) OECD (1984): Effects on biotic systems. In:Guidelines for testing of chemicals, 202.

(6) Giugliano, M., Cernuschi,S.,Marforio,R. (1986): I rilasci a breve termine dalle scorie e dalle ceneri volanti prodotte dall´incenerimento dei rifiuti solidi urbani. Recycling International, Berlin, 29-30 October 1986.

(7) Viganò, L.(1987): Metalli e non metalli tossici totali: esame del limite della legge 319/76 con Daphnia magna. Ingegneria Ambientale,16,341-345.

MODELLING LANDFILL GAS PROCESSES

P E Rushbrook, J E Pearson
Environmental Safety Centre, Harwell Laboratory
Didcot, Oxfordshire, UK

Introduction

Throughout the world most household solid wastes are disposed of to landfill with little or no pre-treatment. Only relatively small quantities, except perhaps by one or two Western nations, are incinerated or composted. In those countries where modern controlled landfill techniques have been adopted higher waste emplacement densities result and the environmental conditions governing microbial decomposition of organic materials are anaerobic. Where conditions are fully anaerobic, generally regarded as characterised by an Eh below -200mV, the generation of methane in the landfill gas is inevitable. Under these conditions the landfill gas (approximately 65% methane and 35% carbon dioxide) can be viewed as either a naturally-occurring energy source or an environmental nuisance depending on the possibilities for utilisation and off-site migration at each landfill. Therefore, it is essential that the management of landfill gas is considered at all stages of landfilling, ie. design, operation and after-care.

The prevailing trend in several Western nations is towards fewer, but larger, landfills serving a wider catchment and hence containing substantially more waste than former town dump sites. Consequently, the potential quantities of landfill gas that could be generated are very large. Properly planned and executed gas management at these landfills is therefore increasing in importance.

At some sites this has led to landfill gas being used constructively as a fuel supplement, for space heating and for electricity production (1). Elsewhere there are several well documented cases of methane gas explosions arising from uncontrolled or poorly controlled gas migration (2-4).

Over the last decade many research teams have studied in the laboratory and at field sites the processes involved in waste decomposition. Gradually, a sizeable body of scientific data has become available from which empirical interpretations of the processes responsible for the generation of landfill gas, its production rate and migration characteristics have been proposed (Figure 1). The knowledge gained from this research activity has certainly led to a better understanding of landfill processes, but does not provide a means to relate together, quantitatively, the complex interactions which are believed to relate the decomposition processes to the physical environment. Ultimately, the best approach is to model all the processes and their interactions which influence gas generation and migration. Such a model would be a "framework" upon which research results from one study can be mounted and their interactions with other decomposition processes assessed. Such a model will also have the potential to allow the results from research into one landfill environment to be extrapolated to others. Such a model, alas, currently does not exist.

However, as the body of knowledge in landfill science grows and the nature of the interactions between processes becomes better understood, then the use of models as descriptive, investigational and predictive tools by researchers and site operators will increase.

The work described in this paper will ultimately provide landfill designers and operators with the same computational facilities as their counterparts in wastewater treatment. The aim of the three-year project is to develop a computer model which draws together the current information on waste decomposition and landfill gas production to enable practical assessments of different landfill management gas collection and migration control schemes. The model will not be absolute; it will "evolve" and be refined in the light of future research work by landfill scientists. When finished it will represent a flexible aid for the investigation and prediction of landfill gas generation and migration within different landfill environments and surrounding geologies.

Approach

The development of a comprehensive landfill gas model involves two elements. First, the modelling of gas "generation" (ie. the complex physical, chemical and biochemical interactions leading to the production of landfill gas) to produce the "source term" (rate of production, gas composition etc). Second, the modelling of gas "migration" through the body of waste and surrounding strata. The modelling and validation of gas generation equations is technically more complex than gas migration. Some authors have attempted to model gas generation but efforts so far have been relatively crude (6). Therefore, in the initial stages of the Harwell project effort has concentrated upon fundamentally rethinking the existing theories postulated to explain the gas migration mechanisms that prevail within landfills. Once a properly validated gas migration model has been completed, attention at Harwell will switch to gas generation. This approach is considered to be the most practical since the immediate concerns in the UK over landfill gas are related to migration ie. i) problems arising from uncontrolled gas migration off-site; and ii) commercial incentives to improve gas recovery from gas collection wells.

Previous Gas Migration Models

Rather than attempting to simulate all aspects of gas recovery and hazard abatement earlier models tended to address particular problems. Usually they only considered transport in the gaseous phase. However, gas migration in the aqueous phase could be of significance in the case of some trace gases. Moreover, the transport properties of the gas within the landfill will vary with the degree of saturation.

The simplest models (7, 8, 9) are concerned with one-dimensional movement of gases through cover material. One study (7) assessed the ability of different covers to enhance methane recovery by reducing the amount lost to the atmosphere, reducing oxygen ingress during extraction and increasing the sphere of the influence around wells. Their model included a threshold concentration of oxygen above which there was no methane production. The other two (8, 9) investigated the enhancing effect of cyclical barometric pressure variations on gas flux. This is also related to the permeability of the cover to gas, which is itself a function of water saturation. Another model (10), considering one-dimensional simulation of gas flow both into and out of a landfill, included a time varying source term in the form of a first-order, substrate-limited kinetic equation.

Lateral migration away from a landfill, under different natural environmental conditions and with or without operation of abatement or recovery strategies has also been considered in up to three dimensions (6, 11, 12, 13).

We believe that at least two dimensions are required in order to model the effects of landfill geometry and extraction equipment layouts, and to follow pressure and concentration profiles due to migration laterally and through the cover. To represent irregular geometries and inhomogeneities the use of a non-uniform spatial grid is considered to be necessary. The landfill system to be modelled should include the waste, its final cover and the surrounding strata. This has not been the case in the earlier models.

A detailed review and critique of earlier gas migration models is being prepared and will be published later this year.

Work at Harwell on Gas Migration Modelling

We believe that previous models have not adequately considered all relevant gas migration mechanisms and have been relatively inflexible in their application to variations in landfill conditions (eg. the presence of perched water tables and layers of intermediate cover material). In order to simulate gas movement under the influence of total pressure and partial pressure (or concentration) gradients, convection and two diffusion mechanisms (Knudsen and molecular*) are being considered. The significant step taken beyond the approaches of earlier studies is that we intend

* molecular diffusion is due to molecule-molecule collisions, Knudsen diffusion is due to molecule-wall collisions.

also to consider "slip flow", (ie. the transition regime between pressure-driven flow and Knudsen diffusion) in order to accommodate the range of pressure gradients and pore sizes possibly existing in a landfill. A summary of the development plan being followed is presented in Figure 2.

The initial stages of model development consider convection and diffusion as separate sub-models. First, transport of a single component (ie. methane) is modelled in two dimensions (vertical and horizontal) with varying boundary conditions, transport coefficients and physical properties. Additional components are then added so that as a minimum the four key constituent gases in landfills - methane, carbon dioxide, and atmospheric nitrogen and oxygen - are included in the final sub-model. In the landfill the proportion of atmospheric nitrogen involved in decomposition processes is negligible and therefore as a "conservative" parameter it provides a useful indication of the extent of air infiltration. Modelling the presence of oxygen is equally important since its presence in landfills inhibits methanogenic bacteria. Predicting the concentration of this gas also provides one of the fundamental links with modelling gas generation.

Modelling convection and diffusion separately enables more rigorous attention to be paid in the sub-models to the various landfill processes known to be influencing each mechanism. However, once the algorithms for each sub-model have been prepared, and tested in field trials planned for Summer 1988, they will be combined. These field trials will also provide data for transport coefficients. The first "combined" model represents four-component convection and independent molecular diffusion of each for as wide a range of site geometries, cover materials, boundary conditions, atmospheric pressure gradients and well pumping rates for which data are available. This combined model, in turn, will be broadened further to include multi-component molecular diffusion with Knudsen diffusion and slip flow. Once this final model development is complete its expected predictive capabilities will be tested against documented field data from the UK, and continental European sites, before its ultimate use as a design and predictive aid in its own right.

In 1987 only flow within the landfill was considered and the two mechanisms, diffusion and convection, were studied separately. No flow boundary conditions were applied at the sides and bottom of the fill. At the surface one of two boundary conditions was imposed (clay cover or a more permeable cover). The external gas concentration was fixed when diffusive flow was considered and external pressure fixed when convective flow was considered.

The Diffusion Model

The initial diffusion model was greatly simplified by modelling only one component (ie. methane). Thus the driving force for flow was the total pressure gradient, which in this case was the same as the concentration gradient. In order to represent a wide range of pore sizes both molecular and Knudsen diffusion are to be included. However, for a one component system a single coefficient was used since in this case there were no compositional effects on diffusivity. Two series of simulations were performed using this simple diffusion model:

i) simple gas generation and gas extraction via a vertical well. These runs illustrated the effects of a "pump" mechanism in reducing gas pressure in the landfill around a well. Variations in the size of this 'region of influence' under varying operating conditions were also explored; and
ii) exchange of gases between the atmosphere and the landfill, and the influence of a clay cap. Again this feature of the model will be developed further in 1988 to enable landfill operators to predict and monitor the direction, volume and distance of landfill gas migration off-site under varying conditions of hydrogeology, landfill operation and restoration.

The Convection Model

The convection model was run for one component (ie. methane gas) in 1987 but in this case extension to several constituent gases is relatively straightforward. A Darcy expression was used to predict gas flowrates and the ideal gas law was used to relate gas density to pressure and temperature. The situations simulated using this model involved variation of atmospheric pressure relative to the initial pressure existing in the landfill. A case in which the atmospheric pressure varied sinusoidally with time (simulating daily or seasonal fluctuations) was also examined. The resultant 'barometric pumping' was found to be important with regard to predicting landfill gas generation rates (eg. from oxygen ingress inhibiting methanogenesis) and methane egress rates through the surface of the landfill. These results were concordant with gas migration findings from a recent field study in Illinois, USA (14).

Conclusion

In 1987 the emphasis was to establish a practical and theoretical foundation for the modelling of gas migration. This built upon the work of earlier researchers. The provisions in the new model reported descriptively in this paper include the initial treatment of convection and diffusion as separate sub-models, the consideration of more than one component gas, and the incorporation of molecular and Knudsen diffusion terms and slip flow. These reflect the dependence of gas migration and release at the surface on cyclical climatic conditions, site geometry, physical characteristics of deposited wastes and boundary geology. The final migration model is scheduled for completion in 1988 with field testing in early 1989. So far we have identified a serious lack of properly validated field data on gas flow transport characteristics of waste. Such data are essential and their absence could hamper the completion of a working model. Therefore, a separate project is to be established in 1988 to evaluate gas transport coefficients in the field. Results from both the gas modelling and field data collection projects will be reported in future papers.

Acknowledgements

The assistance of Dr P Bell, University of Queensland, Australia and Dr A Krol at Harwell in the project work reported here is gratefully acknowledged.

References

1 Richards K (1987). "Landfill Gas. A global review". 7th Int. Biodeterioration Symposium. Emmanuel College, Cambridge University, UK. 8 September 1987.
2 Parker A (1981). "Landfill gas problems - case histories". Proc. Harwell Landfill Gas Symposium. 6 May 1981. Harwell Laboratory, UK.
3 Parker A (1986). "Landfill gas - a potential environmental hazard". Disasters. 10(1) pp65-69.
4 Derbyshire County Council (UK) (1988). "Report of the non-statutory public inquiry into the gas explosion at Loscoe, Derbyshire, 24 March 1986". Derbyshire County Council, Matlock, UK.
5 Rees J F (1980). "The fate of carbon compounds in the landfill disposal of organic matter". J Chem. Tech. Biotechnol. 30 pp 161-175.
6 Massman J W, Moore C A, Sykes R M (1981). "Development of computer simulations for landfill methane recovery". Report ANL/CNSV-26. Argonne National Laboratory, USA.
7 Massman J W, Moore C A (1982). "Computer optimisation of landfill cover design". Argonne National Laboratory, USA. Municipal Waste Program. ANL/CNSV-TM-109. December 1982.

8 Lu A H, Matuszek J M (1978). "Transport through a trench cover of gaseous tritiated compounds from buried radioactive wastes". IAEA-SM-232/60, pp 655-670.
9 Thibodeaux L J, Springer C, Riley L M (1982). "Models of mechanisms for the vapour phase emission of hazardous chemicals from landfills". J Hazardous Materials. 7 pp 63-74.
10 Findikakis A N, Leckie J O (1979). "Numerical simulation of gas flow in sanitary landfills". J Environmental Engineering Division. ASCE 105 (EE5) pp 927-945.
11 Mohsen M F N, Farquhar G J, Kouwen N (1978). "Modelling methane migration in soil". Appl. Math Modelling. 2 pp 294-301.
12 Metcalfe D E, Farquhar G J (1982). "Modelling gas transport from waste disposal sites". MSc Thesis University of Waterloo, Canada.
13 Metcalfe D E, Farquhar G J (1986). "Modelling gas migration through unsaturated soils from waste disposal sites". Water, Air and Soil Pollution. 32 pp 247-259.
14 Bogner J E (1986). "Understanding natural and induced gas migration through landfill cover materials - The basis for improved landfill gas recovery". Proc. 21st Intersociety Energy Conversion Engineering Conference. San Diego, California, USA. 25-29 August 1986.

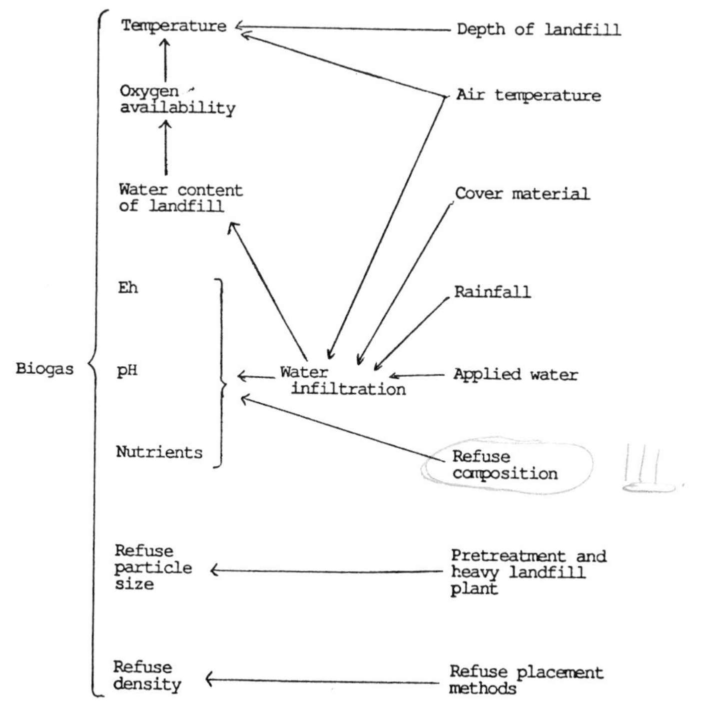

FIGURE 1 Factors affecting gas production in landfills (5)

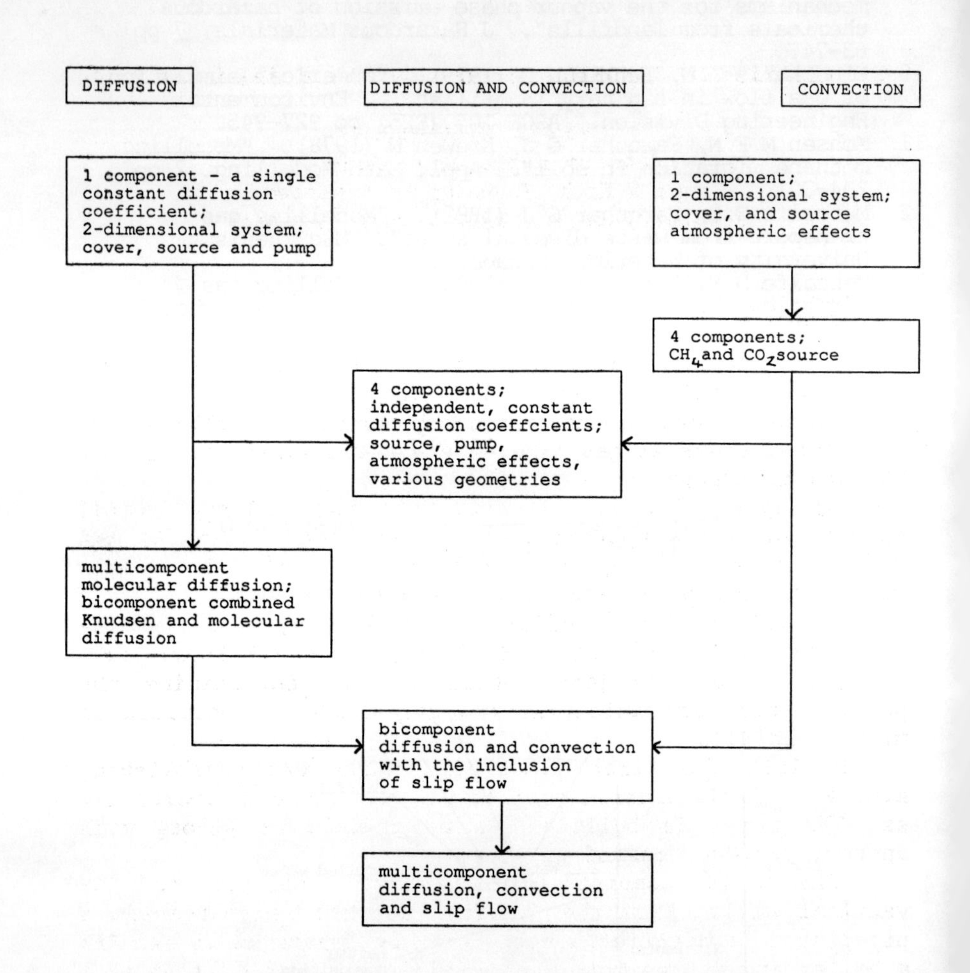

FIGURE 2 Harwell Landfill Gas Migration Model Development Plan

OPTIMIZATION OF LANDFILL GAS RECOVERY FROM SMALL LANDFILL

Hans C. WILLUMSEN
Head of Department, B.Sc.
Crone & Koch, Consulting Engineers
Jernbanegade 22, 8800 Viborg, Denmark

1. INTRODUCTION

Exploitation of biogas from landfills was first attempted in the USA around 1975. Since then there has been a pronounced development in this field and, notably around 1980, many landfill gas plants were established both in the USA and in Europe.

In April 1987 we collected data from 120 operational plants, of which approximately 40 are located in the USA and the rest in Europe with the exception of a few plants which are located in Brazil and Canada.

In Denmark the landfills are relatively small, wherefore the Danish Ministry of Energy since 1981 has subsidized various projects which aim at determining the possibilities of exploiting the gas which is generated in minor landfills.

In 1985 the first plant in Denmark was established, a.o. as a demonstration project for the European Communities. The plant is built up on a landfill at Viborg with approx. 325,000 tons of waste.

The plant consists of a recovery section with 32 vertical wells. From the landfill the gas is routed via a pipeline to a district heating station, where it is used in a boiler and in two motor/generator units. One of them is a Stirling machine, which can run on different types of fuel and with a fluctuating calorific value.

In order to optimize small plants comprehensive measuring and adjusting equipment has been established which, among other things, by measuring the gas quality automatically and continuously, ensures that the gas flow from the single wells is adjusted so that stable and optimum gas quality is obtained.

The measuring and adjusting system has shown an effective function and proves that the utilization can be optimized and improved with the system. Further it has

ISWA PROCEEDINGS
VOLUME 1 ISBN 0-12-058451-4

shown to be a profitable solution establishing plants for utilization of landfill gas, Danish conditions taken in consideration.

2. GAS PRODUCTION

In most landfills refuse consisting of a mixture of household, industrial and gardening waste is deposed. The refuse has a great content of organic matter.

Immediately after the refuse has been dumped in the landfill an aerob decomposition of the organic waste using the oxygen sets in. Once the oxygen has been exhausted an anaerob decomposition will begin whereby biogas is produced, with a methane content of approx. 50% methane (CH_4) which can be used for energy purpose. The gas will diffuse up through the top soil of the landfill or the surrounding areas.

Roughly, the gas will be generated as shown in figure 1, and the duration of the first three phases may vary from a few months up to about four years.

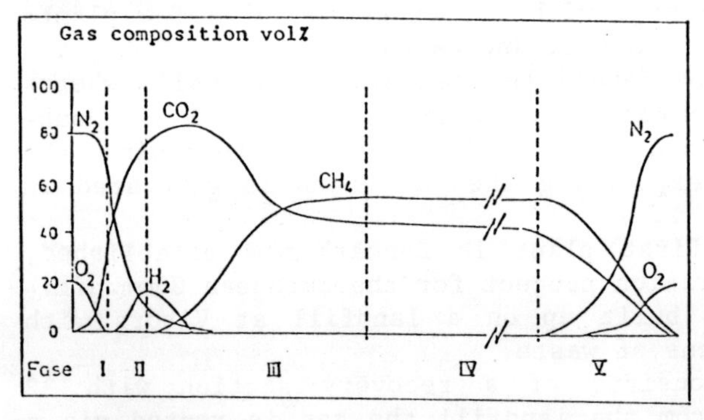

I. Aerobic
II. Anaerobic, Non-Methanogenic
III. Anaerobic, Methanogenic, Unsteady
IV. Anaerobic, Methanogenic, Steady

3 GAS RECOVERY

A landfill gas plant consists of a recovery system and a production system. A recovery system can e.g. consist of vertical perforated pipe wells, horizontal perforated pipes or ditches, or a membrane covers serve to collect the generated gas.

The landfill gas is sucked out of the landfill by a pump or a compressor, which at the same time presses the gas to the production plant. The gas can be used in a gas boiler for the production of hot water for heating or process heat. Very often the landfill gas is used as fuel in a gas engine, which drives a power generator. Under normal conditions it will not be necessary to leach the gas, except for the removal of impurities (particles) if the gas is used in a gas boiler or a gas engine.

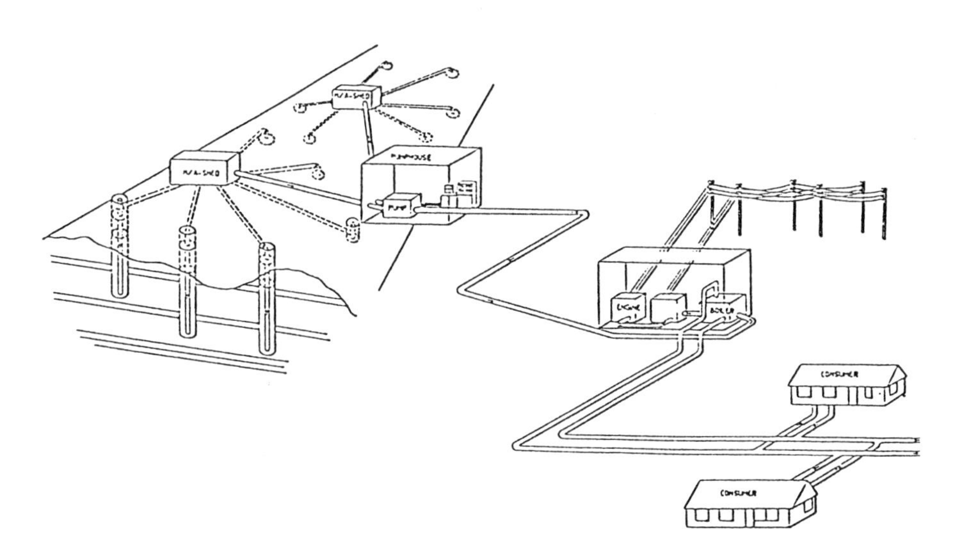

Figure 2: The Principle of a landfill gas plant

In April 1987 we have found data for about 120 full scala projects. Fig. 3 shows the gas production rate in landfills in relation to the age, from some of the 120 landfill gas plants, where it has been possible to get exact informations.

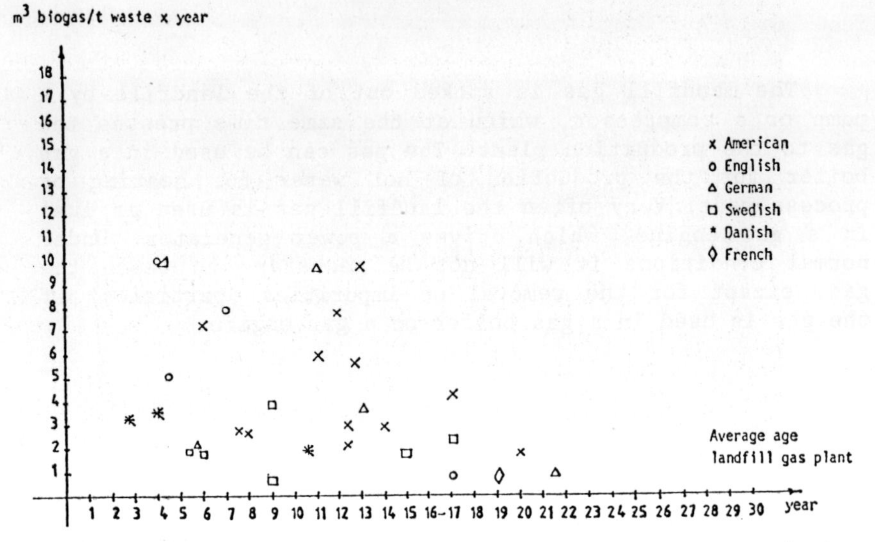

Fig. 3: Gas Production Rate depending on the Age of the Landfill.

The results do not show the very picture of whether it is possibly to utilize bigger amount of gas per ton waste from bigger landfills than from smaller. Though the recovering system must be better optimized at the small landfills to obtain the same gas production rate than at the bigger.

4. OPTIMIZATION, CONTROL AND ADJUSTMENT IN SMALL LANDFILLS

To find the most optimal gas utilization from minor landfills the Danish Ministry of Energy and the EEC has financially supported a demonstration project, which is more detailed described in (1).

The project shall show whether it is technically possible to optimize the gas utilization from smaller landfills and whether it is financially profitably.

In connection with the EEC demonstration project, an automatic measuring and adjusting system was introduced at the landfill site at Viborg.

The gas is sucked out of the waste through vertical perforated wells. From each well a horizontal pipe leads to a measuring and adjustment shed.

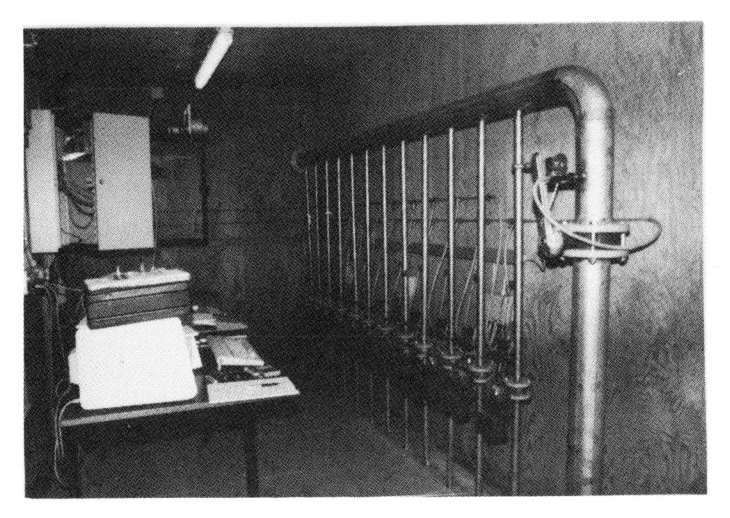

Measuring and adjusting system.

Where the gas pipes from the indivudal wells enter the measuring and adjustment sheds, samples are taken automatically. After a predetermined scanning procedure the samples are pumped to the pump house for analysis. In case of fluctuations of the calorific value, the gas flow from each well is adjusted automatically by means of a motorised valve; the purpose of this adjustment being to allow a swift change in conditions, as landfills with low filling height and slight top soil are easily affected by a change of pressure, weather, etc., and penetration by ambient air soon reduces gas production.

Figure 4 shows how one of the valves is automatically adjusted every 4 hours, and that the CH_4 value gradually stabilizes and remains at a constant value.

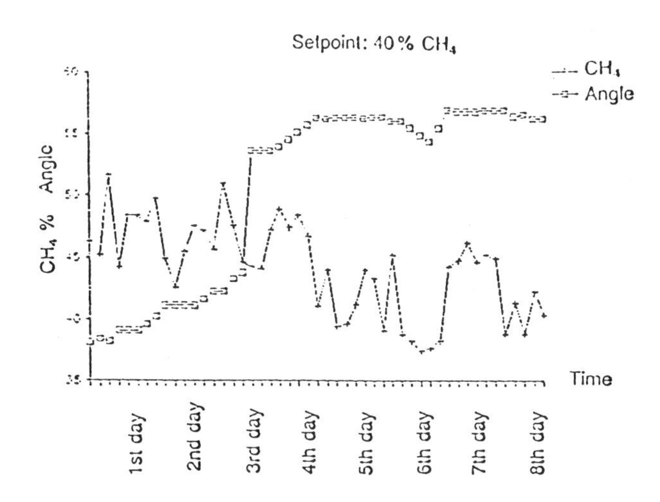

Figure 4: Flow Adjustment as a result of CH_2%.

5. PROBLEMS

At the two minor Danish plants which were mentioned earlier, problems have been encountered with the secondary water surface in the vertical wells. The water rose in these wells and closed off the holes (filter surface) through which the gas was supposed to be pumped. Therefore, pumping had to be increased through the remaining holes in order to get the same gas volume out, which resulted in additional atmospheric air permeating the site.

The problem was solved by mounting pumps which suck up the water from the wells.

At the Grindsted plant major problems were encountered at the beginning, related to the O_2 content of the gas, which was higher than the limit that had been set to protect the plant against explosions (3.5%).

It was found that the plant only tolerated very low suction pressures of 100-200 mm water column, and that the plant is extremely sensitive to adjustment of flow and pressure. Fig. 5 shows the sensitivity at varying gas flow.

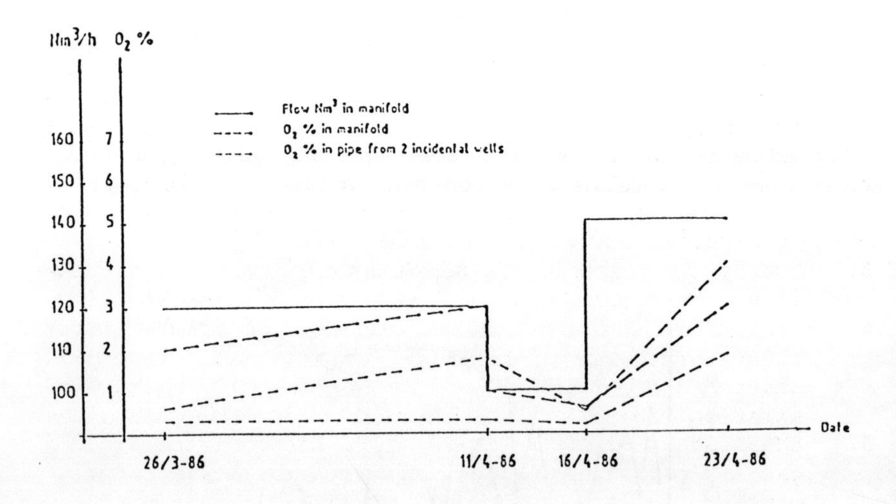

Fig. 5: Oxygen Penetration in Wells at Grindsted

6. CONCLUSIONS

The very first experience that was gained in American landfills date from around 1980 shows an average gas production rate of approx. 8 m^3/biogas/ton waste x year.

As it will appear from the preceding, the results from most minor landfills show a gas production rate of 2-5 m^3 biogas/ton waste x year.

Contrary to our original assumptions, i.e. that it would take 3 to 4 years before utilization of the gas would be profitable, it now seems that it will not take more than 1 year. On the other hand, the results seem to indicate that it will not be profitable to utilize the gas over as long a period of time as was first assumed; a more likely time horizon would be 15 to 20 years.

As can be seen from the cost tables in para 10, small landfills have the poorest economy, but a comparison with plants for other alternative energy sources shows that an investment in landfill gas plants in minor landfills is still a relatively good investment.

REFERENCES

1. RECOVERY AND USE OF LANDFILL GAS, European Community Demonstration project for energy saving and alternative energy sources, Project no. BM/741/83-DK, Flag brochure no. 58.
2. REES, J.F. (1980a): The Fate of Carbon Compounds in the Landfill Disposal of Organic Matter. Journal of Chemical Technology & Biotechnology, 30, 161-175.
3. HOEKS, J. Significance of Biogas Production in Waste Tips. Waste Management & Research, vol. 1 no. 4, 1983.
4. WILLUMSEN, H.C. and BURIAN-HANSEN, P. Crone & Koch (1986): Recovery of Landfill Gas from small Landfill. Anaerobic Digestion. Results of Research and Demonstration Projects. Elsevier applied Science. 1986.
5. EMBERTON, J.R and EMBERTON, R.F. Energy from Landfill gas - Proceedings of a conference in Solihull, Birmingham, England. 28-31 October 1986.

BUILDING REDEVELOPMENT ON DISUSED LANDFILL SITES - OVERCOMING THE LANDFILL GAS PROBLEM?

R. J. Carpenter, CChem, MRSC, M.Inst.WM, M.Inst.E
London Scientific Services, London, England

ABSTRACT

There is a growing pressure in the area within a twenty mile radius of London for old landfill sites to be redeveloped by building either industrial warehousing or residential houses. Although the problems to be overcome to ensure the health and safety of future site occupiers are numerous, the most critical factor is probably the extent of the presence of landfill gas within the site. Obviously, where there are large quantities of gas, redevelopment for building should not be considered and equally, where an investigation shows that all the landfill gas has dissipated from the site, it can be recommended that building redevelopment may take place given that suitable remedial measures are taken to overcome any other problems. The difficulty in the decision making process occurs when gas quantities and concentrations fall between these two extremes.
Site investigation strategy and the interpretation of gas monitoring results are discussed. The use of gas membranes and underfloor ventilation are discussed in relation to experimental work carried out on a full scale house foundation test rig. This experimental work showed the efficiency of natural underfloor ventilation in preventing gas build-up in confined spaces.
Case histories (from the experience of London Scientific Services) are briefly discussed. In two cases, housing redevelopment is taking place on former landfill sites and another landfill is being redeveloped for industrial warehousing.

1. INTRODUCTION

No one of sound mind would consider building on land that had very recently been used for the disposal of household waste. Similarly, no one would be likely to doubt that safe building redevelopment could be carried out on a thirty-year old site that had been filled with inert waste. Between these two extremes, however, there exists a potential minefield of decision making with respect to the safety of building development on disused landfill sites.
London Scientific Services (formerly the Scientific Services Branch of the Greater London Council) have been investigating the problems caused by the generation of landfill gas (produced by the anaerobic decomposition of organic matter) since the early 1970s[1].
Methane gas mixtures in air between 5-15% methane gas v.v. are explosive and migrating gas can provide a significant threat to the lives of occupants of buildings if it becomes entrapped in enclosed spaces. Over the past few years several explosions have occurred in the U.K. as a result of gas migrating from landfill sites into adjacent properties.

ISWA PROCEEDINGS
VOLUME 1 ISBN 0-12-058451-4

2. THE REDEVELOPMENT OF LONDON DOCKLANDS

This major redevelopment, which began in 1978, was put in jeopardy when a minor explosion during the drilling of a geotechnical borehole led to the discovery that methane gas was present within the ground. Although the old docks had been backfilled with inert fill (construction rubble), they had been filled wet because the dock walls would have collapsed if the water had been pumped out. It was not therefore possible to remove all the river silt from the dock bottoms and it is from this silt that the methane gas was generated. The gas was odourless and not detected until the minor explosion in the borehole. The extent of methane contamination was not just confined to the refilled dock areas since these areas had been dynamically consolidated squeezing the methane gas into adjacent land. Before any remedial measures could be proposed it was necessary to know the extent of the problem. This led to gas boreholes being sunk throughout the site and the monitoring of methane and carbon dioxide concentrations and gas emission rates from these boreholes. There are problems of data interpretation with most portable gas detection equipment when monitoring landfill gas[2] and gas samples need to be taken from the boreholes and analysed by gas chromatography in the laboratory to check on-site measurements. Gas emission rates were measured by flushing the borehole with nitrogen gas and then monitoring the build up of methane gas concentration with time once the flushing had ceased. The flushing technique was used to measure gas flow rates below 0.01 m/s in the 50mm diameter borehole (this equates to flows of 1.2 litres/minute) and flow rates in excess of 0.01 m/s were measured using a hot wire anamometer. Although methane gas concentrations were often in the range of 20%-30% gas v.v. the majority of gas emission rates were below 0.01m/sec and it was decided that it should be possible to overcome the problem through careful building design. The three prime objectives must be:-

(i) landfill gas being emitted from the ground beneath the building must be prevented from entering the building through the floor slab.

(ii) landfill gas must not be allowed to build up in the area beneath the floor slab.

(iii) service routes into the building must not be allowed to provide easy routes for landfill gas to enter the building.

To achieve these objectives the design solution shown in Figure 1 was suggested. Services were brought up outside the building into a ventilated chamber, from which they are taken into the building horizontally above ground level. The floor slab was designed to be as impermeable to gas as possible being in-situ concrete with a gas membrane and a further protective screed. Any drainage pipes that had to penetrate the floor slab were puddle-flanged and sealed into the surrounding floor with dense concrete. The floor slab was raised above ground level to create a 225mm void beneath it that could be vented directly to atmosphere. The target figure was for 50% of the

FLOOR SLABS AND METHANE EXCLUSION

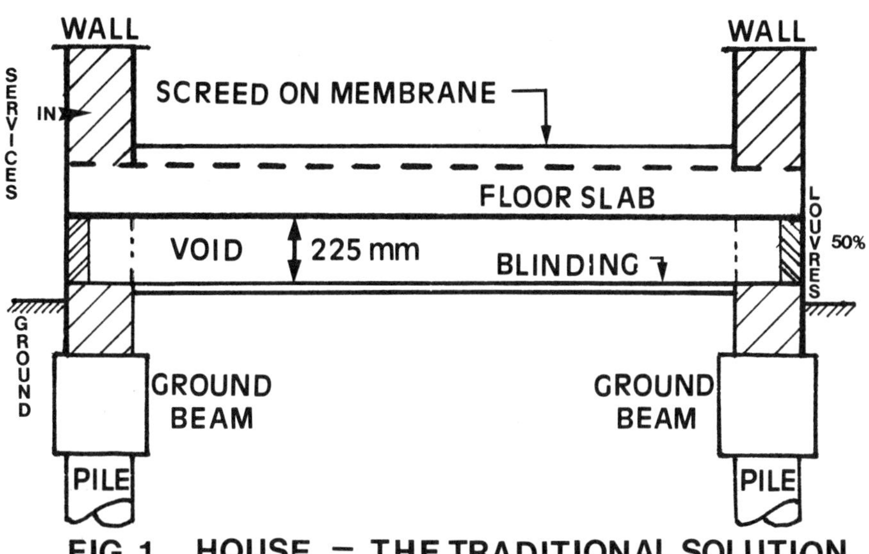

FIG. 1 HOUSE – THE TRADITIONAL SOLUTION

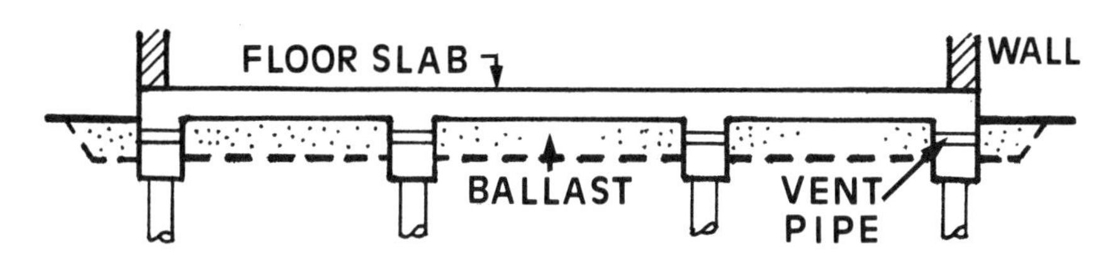

FIG. 2 RETAIL WAREHOUSE

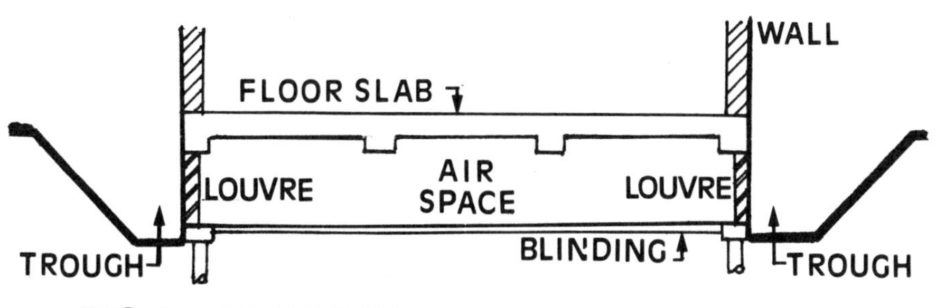

FIG. 3 FACTORY

side areas of the void to be free air space. From the data obtained from landfill gas monitoring at the site it seemed that the design shown in Figure 1 would allow the landfill gas to disperse by natural ventilation beneath the floor slab.
A full size test rig of the floor design shown in Figure 1 was constructed and the system tested by introducing methane gas through pipes into various points beneath the slab and monitoring the gas concentrations throughout the void, at the ventilator grilles and above the floor slab. It was assumed that each borehole affected an area of $10m^2$ around it and on this basis gas flow rates in excess of 100 times greater than those measured on the site were used to test the design and it was shown to be satisfactory. The methods used for measuring gas concentration and emission rates from boreholes and during the experiments on the test rig have been described in detail elsewhere[3,4]. and the success of the experiment led to the redevelopment of Londons Docklands as planned, albeit with buildings raised above ground level to accommodate the 225mm void beneath the floors. Larger buildings have also been built as part of the London Docklands redevelopment and no experimental work has been carried out to check that underfloor ventilation is as effective with larger buildings e.g. factories and warehouses. As an alternative to the ventilated void, some floor slabs have been laid directly on top of ballast which will allow any gas to migrate sideways beneath the slab to vent to atmosphere at the edges of the building. This technique, illustrated in Figure 2 must only be used where gas emission rates are low (certainly less than 0.01m/s from a 50mm diameter borehole). Experimental work carried out in conjunction with work on the test rig proved that dispersion of methane gas was much slower from a ballast layer than from the void and there is also the problem of the gas becoming entrapped between the ground beams on top of the piles. Figure 3 illustrates a solution that has been used for building a warehouse on methane affected ground[5] (not in the Docklands). In this instance the void was 350mm with venting on all four sides and the building has had no reported problems of methane ingress.

3. BUILDING REDEVELOPMENT ON DISUSED LANDFILLS

There are many reasons for building on disused landfill sites to be approached with caution[6] - ground settlement and soil contamination being two areas for concern - but it is the problem of the potential presence of landfill gas that is addressed in this paper. In my opinion, no building redevelopment must ever take place on any disused landfill site (or indeed, adjacent to any landfill site) unless a thorough site investigation has been carried out to establish the extent of the landfill gas problem that must be overcome. The gas monitoring work that London Scientific Services has carried out over the past 18 years has shown that landfill gas can still be present in sites that have been disused for 30 years and that it is invariably present in sites that have been filled with supposedly inert waste, although to a much lesser extent than sites filled with putrescible waste. The tendency to cap sites with impermeable layers, to prevent water ingress and minimise leachate production, results in the likelihood of gas being trapped within the site (unless it can migrate sideways and affect adjacent land) until such time as the cap is penetrated allowing an escape route for the gas. Such release mechanisms can be provided when piles are driven to provide the foundations for building redevelopment. This effect has been observed many times when spiking of a site with a driven rod to a depth of 0.5m to 0.75m has shown only trace concentrations of methane gas to be present, yet when boreholes are sunk to the base of the fill, very high gas concentrations and gas flow rates have been measured showing that the landfill gas is contained beneath a fairly deep capping layer of clay or some other impermeable material.

Probably the most important thing that can be established at the outset of any gas investigation is whether the landfill gas still contains any of the trace constituents that give it that characteristic landfill gas odour. If the gas is still malodourous it will not be possible to redevelop the site with buildings because gas being ventilated from beneath buildings is more than likely to give rise to an odour nuisance. In general terms, the older the site the more trace malodourous constituents have been removed from the gas and the landfill gas becomes odourless. Odourless, but no less dangerous (in fact more dangerous because it is not now so easily detectable). Once the gas is odourless, a full investigation should reveal whether the quantities of gas present within the ground will prohibit building development. A full investigation to assess the potential landfill gas problem would, ideally, consist of the following:-

(i) Gas boreholes sunk to the base of the fill on a 50 metre grid over the site

(ii) Gas standpipes set in 3m or 4m deep backfilled trial pits set also on a 50m grid but set 25m East and South of the borehole grid. Such a system would result in 16 gas boreholes and 9 gas standpipes in trial pits for a 4 hectare site.

(iii) At least 25%, preferably more, of the gas boreholes should be monitored for gas concentrations at 1 metre depths whilst the borehole is being sunk and all the trial pits should be monitored during excavation and backfilling. As well as providing useful information on the depth that landfill gas occurs, this exercise ensures that the gas monitoring points are properly installed. If this is not the case (and in my experience this happens all too frequently) the subsequent measurements that will be made will be worthless.

(iv) Measurement of gas concentration and gas emission rate from each of the monitoring locations at least once a month over a period of one year. Gas flow rates are greatly affected by atmospheric pressure, water table levels etc. and measurements need to be made over a full range of weather conditions to ensure reliable data on which to base a judgement on whether a site can be safely redeveloped.

Many landfills are now designed with gas extraction schemes and these will much more quickly reduce the quantities of gas present within a landfill. The schemes usually cease, however, once gas quality decreases and there is no viable commercial outlet. Even if gas extraction were continued just to remove landfill gas from the site as quickly as possible, there will always be residual amounts of gas retained in the site which must be taken into account if ever the site, at some distant time, is considered for redevelopment.

The fact that underfloor ventilation as outlined in Figure 1 has been proved to cope with gas flow rates well in excess of 1.2 litres/minute from a 50mm diameter borehole is not the end of the story. There are the difficulties of other buildings (sheds, added garages, extensions etc. and it is often difficult to exert any control over their construction. In this case much depends on whether any gas of significant concentration is released from the ground during spiking to 0.5m depth. From experience to date, London Scientific Services would not generally advise building redevelopment on any site where significant gas concentrations (>1% gas v.v.) had been recorded with gas emission rates in excess of 0.05m/s from a 50mm diameter borehole. Otherwise, it is felt that a satisfactory building design may be devised that will allow safe redevelopment but every site must be carefully judged on its own merits.

Where gas concentrations are generally low (<1%) but are still significant (>500ppm) and where gas emission rates are mostly less than 0.01m/s from the 50mm diameter borehole, it is possible to construct a house with the floor slab lying on top of a layer of ballast (similar to Figure 2) rather than have the floor slab raised above a void. It is still vital, however, to design a means of ventilating the area between the ground beams which could act as a collecting trap for any gas emitting beneath the slab. In Figure 2 this is accomplished by having ventilation pipes through the ground beams and another method is by employing telescopic Z shaped ventilators which will allow the area beneath the ground beams to be safely ventilated to atmosphere.

It is imperative to remember, however, that all the remedial measures are only as good as the workmanship that goes into their construction. Unless there is very careful control of workmanship during the construction of floor slabs, voids etc. these very measures, designed to safeguard, can become unintentional yet highly effective routes for landfill gas entry into buildings.

4. CASE STUDIES

CASE A A SITE IN BUCKINGHAMSHIRE

A housing estate had been built in the 1960's on a backfilled gravel pit. When an area of open space on this site was planned for new building development in 1986, geotechnical boreholes encountered decomposing refuse in the top 5 metres (the site had reportedly been filled with inert fill and completed in about 1952).
London Scientific Services were called in to carry out a site investigation to determine the extent of the presence of methane gas. Just from the appearance of garages and houses on the existing development it was obvious that a great deal of settlement had taken place. Eight gas monitoring boreholes were installed. Gas concentrations in the boreholes were of the order of 20% methane gas v.v. and 10% carbon dioxide but none of the gas emission rates from the 50mm diameter boreholes exceeded 0.01m/sec (1.2 l/min). The highest emission rate recorded was 54 l/hour and all the remaining emission measurements were less than 20 l/hours. This data was comparable with that recorded in the Docklands sites and it was possible to adopt the house design shown in Figure 1 as the solution to the problem.
Monitoring of the underfloor void was carried out and the ventilation proved satisfactory. A borehole investigation on the rest of the site that was already developed showed the presence of small quantities of methane gas. Concentrations never exceeded 3% gas v.v. and there was no emission rates greater than 0.01 m/s. Occasional monitoring is continuing but no further action on the existing houses seems likely to be required. A relief for all concerned!

CASE B A SITE IN SOUTHERN ENGLAND

This site had been a gravel pit. It was some 400m by 170m in area and filling was completed in 1977. The fill material had supposedly been mostly inert waste but it was reported that some limited quantities of putrescible waste had been deposited from time to time. Nine gas monitoring boreholes were installed. In addition, since 27 trial pits to 3 metre depth were dug so that soil samples could be taken for analysis, it was decided to install gas monitoring pipes in each of these trial pits. Only one gas emission over a period of 5 months was recorded to be greater than 0.01m/s (from the 50mm diameter pipe) - the value being 0.04m/s. Eight of the monitoring points had methane gas concentrations at various times that exceeded 5% gas v.v. and the highest concentrations were recorded in two of the boreholes that had concentrations of 21% and 23% gas v.v. respectively. Houses, proposed for the site, would need to be piled, as is usual for this type of site. It was felt that the floor slab should be of in-situ concrete with a bituthene membrane and screed and that services should be brought up outside the building into a ventilated chamber before entering horizontally through the wall. However, it was decided that gas concentrations and gas flows were such that full underfloor ventilation with the 225mm void would not be necessary. The slab could be laid on a layer of ballast and the enclosed area between the ground beams and the slab could be provided with ventilation to atmosphere through the side wall of the house using telescopic Z shaped vents. At the time of writing this particular scheme was awaiting determination of a planning application.

REFERENCES

1. CARPENTER, R J (1985), Redevelopment of land contaminated with methane gas: the problems and some remedial techniques, Proceedings of 1st International Conference (TNO) on Contaminated Soil, Utrecht, 11-15 November 1985 pp 747-757.

2. CROWHURST, D 'Measurement of gas emissions from contaminated land'. Department of the Environment, Building Research Establishment, Fire Research Station, Borehamwood, Herts, WD6 2BL.

3. PECKSEN, G N 'Methane and the development of derelict land', London Environmental Supplement No. 13 Summer 1985.

4. CARPENTER, R J, GOAMAN, H F, LOWE, G W, PECKSEN, G N 'Guidelines for the site investigation of contaminated land', London Environmental Supplement No. 12 Summer 1985.

5. RYS, L J, JOHNS, A F, 'The investigation and development of a landfill site', Proceedings of the 1st International Conference (TNO) on Contaminated Soil, Utrecht, 11-15 November 1985 pp 625-636.

6. EMBERTON, J R, PARKER, A, 'The problems associated with building on landfill sites', Waste Management and Research (1987) 5, 473-782.

CODISPOSAL OF SEWAGE SLUDGE AND DOMESTIC WASTE IN LANDFILLS

D G Craft and N C Blakey

WRc Environment, Medmenham, Marlow, UK

Introduction

In the UK disposal of sewage sludge to landfill (80 x 10^3 tonnes dry solids/yr) accounts for about 10% of all sludge disposed of to land (1). Legislative changes involving current sludge disposal practices are likely to increase pressure on waste disposal authorities and contractors to accept more sewage sludge at landfills. The work described in this paper was initiated to investigate the different aspects of this codisposal technique. Laboratory experiments were designed, using different infiltration rates, to examine the effects of sludge type and changes in mixing strategy on landfill leachate quality and gas production.

Experimental description

A series of six experiments (each duplicated) was carried out in 0.2 m^3 stainless steel drums (Figure 1). The following codisposal options were investigated along with a control containing domestic waste only;

- different sludge type - (raw dewatered, primary/mixed dewatered and liquid digested);
- changes in mixing strategy - (completely mixed, compared with layered sludge and domestic waste);
- infiltration rate - (900 mm/yr compared with 300 mm/yr).

Each drum was filled sequentially and in layers to ensure a uniform and comparative domestic waste mixture. Pulverised domestic waste was used since crude waste was inappropriate for the scale of experiment. A compaction rate of about 950 kg/m^3 (wet weight) was used in all the tests, simulating a density commonly achieved at operational landfills using steel wheeled compactors. The highest sludge to refuse ratio in the reactors (1:4.1 (wet weight) respectively) was chosen to represent the highest likely to be adopted in operational practice.

In the layered sludge reactors, a single depth of sludge was placed on top of 0.60 m of domestic waste and then covered with a thin layer (0.04 m) of additional domestic waste. Where sludge and refuse were mixed intimately this was done

ISWA PROCEEDINGS
VOLUME 1 ISBN 0-12-058451-4

before filling the drum with each layer of waste. The chemical composition of the wastes used in the experiments is shown in Table 1.

Following filling, each reactor was hermetically sealed. Infiltration rates of 300 mm/yr and 900 mm/yr (as appropriate) were simulated by adding water twice weekly to the waste in each drum via a distribution system installed inside the reactor lid (Figure 1). Temperature in all units was maintained at 30°C to reduce the effects of significant thermal variation and to simulate temperatures commonly encountered in relatively "fresh" landfilled wastes.

Samples of leachate were analysed on a weekly basis, with bulk compositional analysis of gas every three months. Gas flowrates were recorded twice weekly, at the time of watering.

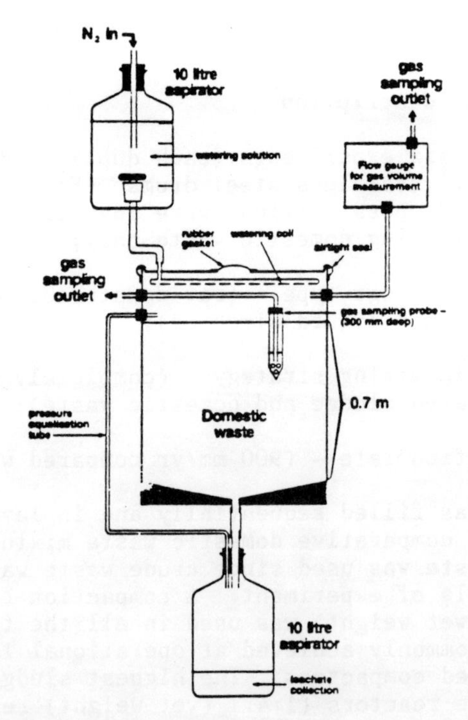

Figure 1. Experimental unit showing irrigation, gas sampling and leachate collection systems

Discussion of results

In general, the mass of individual constituents leached from the codisposed wastes (total organic carbon (TOC), biochemical oxygen demand (BOD), chemical oxygen demand

Waste	Dry solids %	Volatile solids %	<----------Total Metal Content (mg/kg Dry Solids)--------------->							
			Fe	Mn	Cu	Zn	Pb	Ni	Cd	Cr
Pulverised domestic waste	45.4	61	5 800	150	110	310	500	19	4	56
Raw dewatered sludge	30.4	65	25 000	420	680	21 000	300	50	86	180
Primary mixed sludge	30.2	52	60 000	1 700	710	2 700	860	1 700	37	4 800
Liquid digested sludge	4.3	49	15 000	700	1 250	3 500	3 200	200	100	720

Table 1. Analysis of pulverised domestic waste and sewage sludge used in the reactors.

(COD), sulphate, iron, manganese, zinc, nickel and lead) is less than that observed from the domestic waste only controls (Table 2).

To explain these observations the concentrations of COD and zinc in leachate from the codisposal reactors are used as examples to compare differences with the control tests (Figures 2 and 3). Acetogenic activity within the reactors gives rise to high levels of volatile acids (C_2-C_6) within the leachate which contribute significantly to the COD concentrations shown in Figure 2. Codisposal of sewage sludge has increased the rate of stabilisation of the waste by reducing the timescale for the onset of significant methanogenesis to about nine months (Figure 2). The corresponding increase in gas generation is shown in Figure 5. The reduction in leachate metal concentrations, exemplified by the zinc data illustrated in Figure 3, is accompanied by a corresponding reduction in sulphate concentration. It is concluded that a significant mechanism for metal immobilisation within the waste mass has been due to the precipitation of metal sulphides.

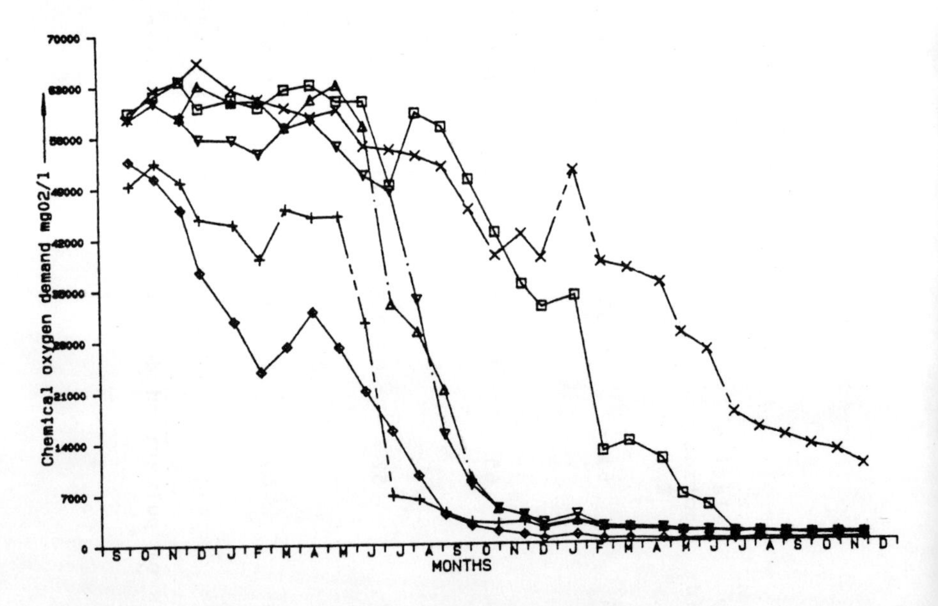

Figure 2. Leachate chemical oxygen demand

(▼ domestic waste + raw, dewatered sludge 4.1:1 wt/wt; + domestic waste + raw, dewatered sludge (homogeneous) 4.1:1 wt/wt; ◆ domestic waste + raw, dewatered sludge (high infiltration) 4.1:1 wt/wt; ▲ domestic waste + primary/mixed dewatered sludge 4.8:1 wt/wt; □ domestic waste + liquid digested sludge 9.7:1 wt/wt; × domestic waste only, control).

Test unit	Refuse/sludge ratio (wt/wt)	COD	NH_3-N*	Total Phosphorus	Fe	Zn	Ni
Refuse + raw, dewatered sludge	4.1:1	-48%	+10%	+30%	-57%	-54%	-24%
Refuse + raw, dewatered sludge (homogeneous mix)	4.1:1	-64%	+23%	+58%	-62%	-95%	-46%
Refuse + raw, dewatered sludge (high infiltration)	4.1:1	-15%	+66%	+203%	-2%	-70%	-4%
Refuse + primary/mixed dewatered sludge	4.8:1	-46%	+40%	+4%	-48%	-71%	-12%
Refuse + liquid, digested sludge	9.7:1	-7%	-8%	-2%	+5%	-68%	-4%

* Ammoniacal nitrogen

Table 2. Mass of material leached from the codisposal wastes as a percentage of that leached from pulverised refuse alone (control).

Conversely the low biological uptake of phosphorus and the absence of nitrification under the prevailing anaerobic conditions within the reactors has given rise to increased leaching of ammoniacal nitrogen (NH_3-N) and total phosphorus (Table 2). This is in direct response to the additional load present in the sludge. Figure 4 shows the temporal release pattern for NH_3-N from each of the reactors. The gradual "wash-out" of NH_3-N is in marked contrast to that of the biologically mediated pattern of release for COD.

The onset of methanogenesis in each reactor is accompanied by an increase in pH from about 5.7 to 7.2. The composition of gas collected from the headspace of each reactor ranged between 50 and 60% methane with the balance being carbon dioxide.

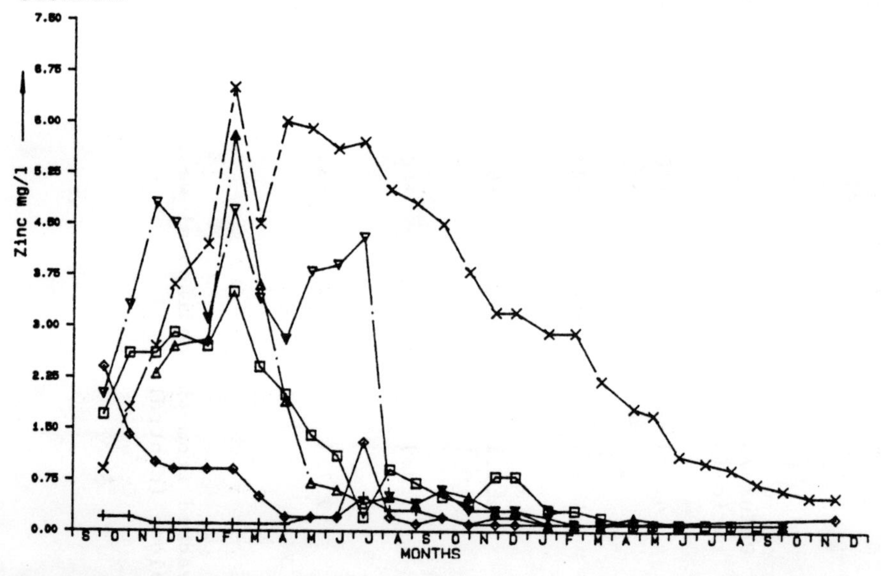

Figure 3. Leachate zinc concentrations (Key as in figure 2)

The rate of gas production is higher in the codisposal experiments with 0.45 and 0.28 m^3/tonne volatile solids/day from the raw, dewatered and liquid, digested sludges respectively. This compares with a rate of 0.05 m^3/tonne volatile solids/day for the domestic waste only controls. Total gas production in the codisposal experiments ranges from 80-150 m^3/tonne volatile solids and 36 m^3 methane/tonne volatile solids in the domestic waste only controls. In the codisposal experiments, gas generation ceases after about six to twelve months from commencement, presumably since most of the readily biodegradable material has been assimilated.

In the domestic waste only controls gas generation continues to increase twenty months after first producing gas (Figure 5).

These experimental results are comparable with Barlaz et al (2) who found domestic waste produced 80-150 m^3 methane/tonne volatile solids and Buivid et al (3) who found that domestic waste produced 90 m^3 methane/tonne volatile solids and 180-210 m^3 methane/tonne volatile solids for codisposed domestic waste and sewage sludge.

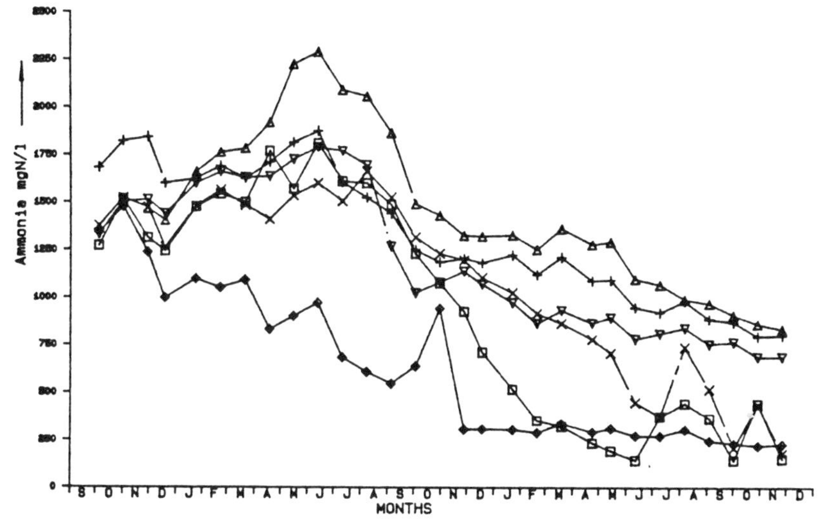

Figure 4. Leachate ammoniacal nitrogen concentrations (Key as in figure 2)

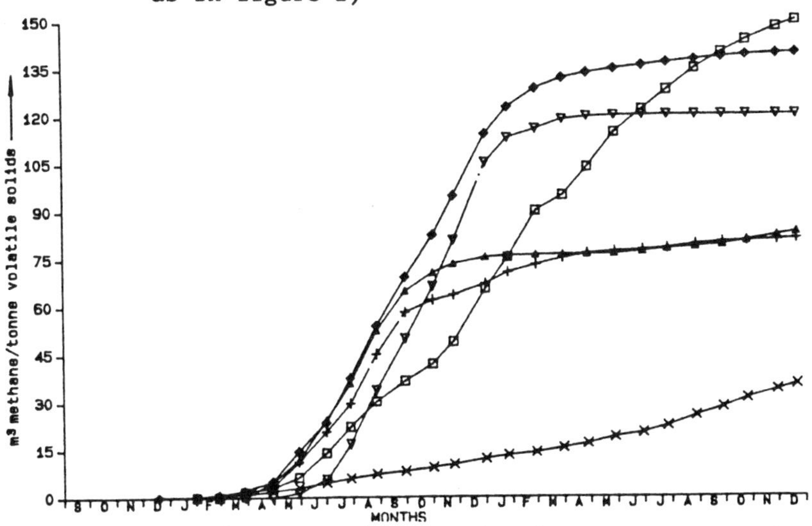

Figure 5. Cummulative methane production rate (Key as in figure 2)

Conclusions

- Codisposal of domestic waste and sewage sludge significantly increases the rate of stabilisation of the codisposed wastes, reducing the organic carbon content of the leachate; chemical oxygen demand (COD) is reduced by about 50% for codisposed dewatered sludge (domestic waste to sludge ratio 4.1:1 wet weight), compared with about 7% reduction for codisposed liquid digested sludge (domestic waste to sludge ratio 9.7:1 wet weight, under moderately low infiltration conditions (300 mm/yr). For higher infiltration rates a lower reduction can be expected.

- Codisposal significantly reduces the quantity and concentrations of metals leached (iron, manganese, zinc, nickel and lead), typically by about 50%.

- Codisposal increases ammoniacal nitrogen loads by between 10 and 40% and total phosphorus loads by between 4 and 60% under moderately low infiltration conditions (300 mm/yr). For higher infiltration conditions (900 mm/yr) loads are increased by about 70% and 200% respectively.

- Gas generation rates are greatest for codisposed wastes, 0.28-0.45 m^3 methane/tonne volatile solids/day compared to 0.05 m^3 methane/tonne volatile solids/day from domestic waste only controls. However, gas generation and hence waste stabilisation takes place over a much shorter period of time for the codisposed sewage sludge and domestic waste. Total gas production in the codisposal experiments ranged from 80-150 m^3 methane/tonne volatile solids. Infiltration in the range 300-900 mm/yr does not affect the rate of gas generation.

References

(1) Department of the Environment/National Water Council (1983) Sewage Sludge Survey - 1980 data. HMSO, London, UK.

(2) Barlaz, M. A., Milke M. W. and Ham R. K. (1987) Gas production parameters in sanitary landfill simulators. Waste Management and Research, 5, 27-39.

(3) Buivid, M. G., Wise, D. L., Blanchet, M. J. and Remedios E. C., Jenkins, B. M., Boyd, W. F. and Pacey, J. G. (1981) Fuel gas enhancement by controlled landfilling of solid waste. Resources and Conservation, 6, 3-20.

EXPERIMENTAL STUDIES ON HOUSEHOLD REFUSE AND INDUSTRIAL SLUDGES CODISPOSAL

M. Barrès*, H. Bonin*, F. Colin**, F. Lome*, M. Sauter*, *BRGM, Orléans, France, **IRH, Nancy, France

1. INTRODUCTION

A large amount of industrial waste takes a liquid form, whether this be true aqueous or non-aqueous liquid, or solid particles in suspension (sludges). Because of their nature, it is often not possible to eliminate these wastes either by incineration (since they have insufficient calorific value), or to use them for agricultural purposes (as they either have no agricultural value or contain toxic material); nor is it possible to spread them directly over the ground, bearing in mind the mobility of the liquid phase. The only practicable solution would be landfilling, providing the problem of the mobility of the liquid phase could be resolved. In the case of sludge, a possible solution consists in separating the liquid phase by mechanical dehydration, using filtration and centrifuging. However, problems are sometimes encountered. They may be technical, such as marked changes in the waste over periods of time, and the problem of how to dispose of the liquid phase if it is highly contaminated, or economic, especially in the case of small quantities of waste, where the setting up of large treatment works would not be justified.

To solve this problem, a research program was carried out with the aid of the French Ministry of the Environment. First, experiments were carried out in test containers by the "Institut de Recherches Hydrologiques"; this was followed by a full-scale experiment, by "the Bureau de Recherches Géologiques et Minières", to study the possibility of the codisposal of liquid waste and high-liquid-content waste, with solid waste which could play a useful role in fixing the aqueous phase, by absorption and imbibition, and reducing the pollutants present in the liquid phase.

2 - PART ONE : EXPERIMENTS IN TEST CONTAINERS

2.1 - Materials and methods

These tests were carried out with six water-tight reinforced concrete containers, 2 x 2m square with an effective depth of 2.5m. The base was fitted with a 20cm thick gravel filter bed, and a device for collecting and measuring the percolation fluid. The containers were mounted on a supporting frame and were exposed to the temperature and rainfall conditions of the Nancy region. The containers were filled with a mixture of household refuse and the liquid waste under study, to a depth of 2.10m. The household refuse was taken from a refuse composting plant at the pre-processing stage; it had been pulverized, cleared of iron and glass, and passed through a 30cm mesh screen. This type of waste was chosen in order to minimize any problems of heterogeneity which may have arisen during our experiment. The tests were carried out on 5 types of liquid waste, or waste with a high water-content. The source and quantity of the waste are given in the table n° 1:

ISWA PROCEEDINGS
VOLUME 1 ISBN 0-12-058451-4

Co-disposal waste	Ratio with respect to household refuse		
	kg of waste /kg of raw household refuse	kg of liquid /kg of raw household refuse	kg of liquid /kg of house hold refuse only
Sludge from grease removal of domestic water treatment plant Cell n° 2	0.193	0.095	0.38
Sediments from hydrocarbon waste prior to regeneration in a specialized processing plant Cell n° 3	0.180	0.076	0.304
Metal finishing sludge (chrome plating, galvanising) Cell n° 4	0.161	0.153	0.612
Formophenolic water (waste from the production of phenolic resin) Cell n° 5	0.128	0.126	0.504
Thick sludge from grease-extractor of a wool-combing works Cell n° 6	0.180	0.087	0.348

Table 1. Nature and quantity of sludge placed in the test containers.

Each type of waste was placed in its respective container according to the procedure which seemed a priori most suitable, taking into account the physical characteristics of each waste, with a view to developing a practical application; it was thus placed into each container either :

- by pouring the liquid waste onto the surface of the containers, previously filled with household refuse (as in the case of formophenolic water),
- by mixing the waste, outside the containers, with the pulverized waste, and then (for hydrocarbon waste and wool screening sludge) pouring the mixture onto a layer of household refuse alone previously placed in the container.
- or by placing layers of waste and household refuse over a layer of household refuse alone previously placed in the container (as in the case of grease-extraction sludge from a sewage purification plant, and metal finishing sludge).

A control container was set up, containing only household waste (Cell n° 1). All the containers filled were covered with a 20cm layer of clean washed sand. The monitoring of the containers. over a period of almost 4 years (1365 days), involved :

- the collecting of water balance data: volume of percolating water. rainfall, and potential evapotranspiration,
- analytical identification of percolating fluids. oriented according to the nature of the pollutants contained in the waste being studied.

2.2 - Results obtained

2.2.1 - Water balance and production of leachate

Figure 1 shows the curves expressing the cumulative production of leachate against time since the beginning of the experiment. A difference can be seen between each container. which is explained not only by the different quantities of easily drained water placed in each container according to the type of waste (much larger quantities were placed in containers 4 and 5). but also by a difference in the permeability of the waste in each container. The presence of grease or hydrocarbons seems to play a predominant role.

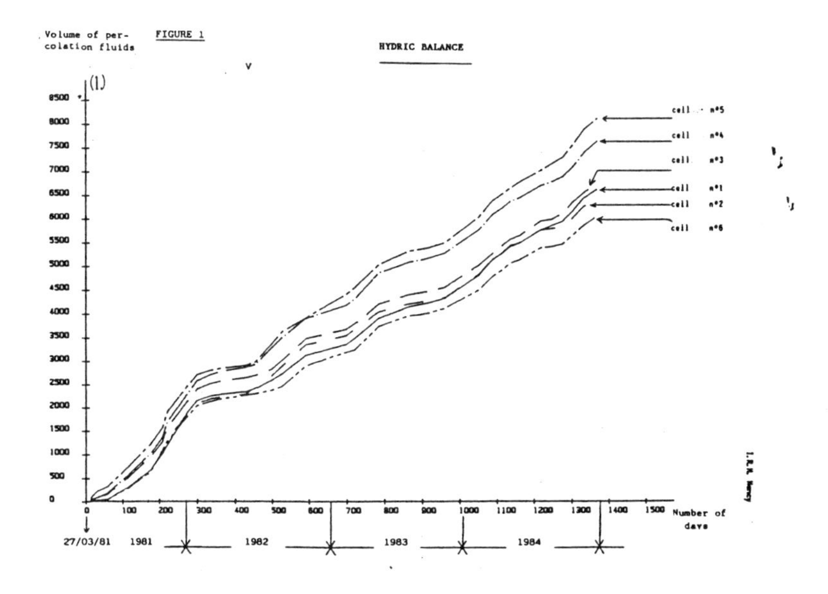

2.2.2 - Generation and transport of soluble pollution

Generally speaking. the curves representing the cumulative flow of pollutants drained out in the course of time. have a progressively decreasing slope. and tend towards an asymptotic from which. according to the type of pollutant. can be reached after 1365 days of monitoring. Overall. the pollution entrainments are best characterized by the cumulative quantities entrained (table 2.)

Substances determined / N° of containers	control	urban screener sludge	oil-treatment sludge	metal finishing sludge	formophenolic liquid	wool screener sludge
Volume of percolation liquid (1)	6621	6446	6752	7633	8108	6046
DCO(gO_2)	148112	112680	139272	213062	225558	206442
DBO (gO_2)	108703	79147	99117	156511	150274	144916
Total NH_4 (gHN_4^+)	4244	3454	-	-	-	6411
N Kjeldhal (gN)	4026	-	-	-	-	6865
Phenol g (C_6H_5OH)	66.12	-	49.75	-	13255	111
Substances extractable by chloroform (g)	3344	3042	2425	-	-	4747
Substances extractable by hexane (g)	881	594	870	-	-	852
Formol (g CH_2O)	-	-	-	-	1418	-
Cadmium (g)	0.8	0.7	0.9	1.2	-	1.1
Total chrome (g)	4.5	3.5	7.9	6.0	-	7.8
Copper (g)	1.3	1.9	1.8	1.8	-	1.4
Iron (g)	1361	949	3143	1750	-	873
Manganese (g)	600	362	1182	755	-	935
Nickel (g)	9.1	7.1	41.6	11.9	-	20.6
Lead (g)	5.4	4.7	6.5	6.9	-	6.3
Zinc (g)	129	55.9	52.5	147	-	93.2
Cobalt (g)	5.1	2.9	6.0	7.0	-	6.9

Table 2. : Dissolved pollution entrainments over 1365 days of experiments

It thus appears that the quantity of waste studied has very little effect on the quantity of common types of pollution. such as DCO. DBO_5 and forms of nitrogen extracted by household refuse. The quantities produced seem smaller in the case of containers containing liquid waste comprising liquid oil or grease. such as urban screener sludge. and hydrocarbon sediments. In contrast. for those containers containing aqueous liquid waste such as metal finishing sludge and formophenolic liquid. the quantities emitted are much greater. This excess is due not only to a greater amount of percolating fluid being produced. but also to changes within the household refuse, which are apparently speeded up by the more humid environment in the tank.

Paradoxically, the quantity of substances such as polar hydrocarbons (extractable by chloroform) and weakly polar hydrocarbons (extractable by hexane) extracted, is lower in household refuse containing urban screener sludge or oil-treatment sediments, than in household refuse alone. This is explained by the fixing of the hydrocarbons in liquid waste by the solid part of the household refuse, and also by the fact that oil or grease coats the household refuse, making it hydrophobic, and isolating it from the flowing water. Liquid waste thus has little effect on the extraction of metallic elements from household refuse. The household waste itself, on the other hand, has a fixing effect with respect to the metals carried by the liquid waste, particularly in the case of liquid metal-finishing sludge rich in chrome, iron and zinc, where the quantities extracted are minute in comparaison to the amount deposited in the container. Similar effects can be seen for other metals, such as cadmium, manganese, nickel, lead and cobalt, though, since the quantities entrained are smaller, the effects are less marked.

In conclusion, the results obtained justified a full-scale experiment, with the aim of confirming the retention phenomena observed. Such an experiment would enable practical conclusions to be drawn, concerning the exploitation of sanitary landfill, and the possible inclusion of liquid or high-water-content waste.

3 - PART TWO : FULL-SCALE IN SITU EXPERIMENT

3.1 - Materials and methods

The study was carried out on a semi-permeable landfill site, and consisted in :

- an initial phase of development, equipment, and monitoring (1982/1983) of five experimental plots (plus one control plot) each covering an area of about 300m2, filled with a 1.5m thick layer of shredded household refuse.
- a second phase (1985/1987) consisting in spreading sludge over the household refuse, and monitoring the evolution and migration of effluents through the waste and the soil.

Five piezometers were set up around the experimental site to check the quality of the groundwater (7 to 8m deep). Each plot was equipped with: porous ceramic cups for sampling the effluents at various levels, from +1m to -1.5m; a set of 5 tensiometers implanted in the waste, at a depth of between 0.2m and 1.1m; an access tube for a neutronic humidity probe, to a depth of up to 1.5m, and 3 tubes for the direct measurement of temperatures (0.4m to 1.2m deep).

The sludges and slurries spread over each plot in 1985 were of the same type and origin as those studied in containers by the IRH, i.e.:

- grease-extraction sludge from a sewage purification plant (plot A)
- wool-screener sludge: a liquid waste when warm. which solidifies at room temperature (the dry residue at 100°C is 51.6%). Plot B.
- hydrocarbon-waste sediments: a liquid waste containing over 50% heavy hydrocarbons (plot C).
- metal-finishing sludge: a liquid waste with about 50% dry residue; 32.5% of the dry weight is zinc. and 10% chrome. (plot D).
- formophenolic liquid. composed of 60g/l of phenol. and 9g/l of formol. (plot E).

Following the results obtained in the test containers (see part One). the proportions of sludge spread over the plots has been significantly reduced; the corresponding quantities of sludge are shown in table 3 :

Plot	Type of sludge	Tonnage T	Proportions spread in situ	
			kg of waste/kg solid matter	kg of waste/kg household refuse
A	Grease extraction sludge	52	0.257	131
B	Wool-screener sludge	18.4	0.082	42
C	Hydrocarbon waste	33.5	0.162	82
D	Metal finishing	16.5	0.095	48
E	Formophenolic liquid	10.7	0.936	18

Table 3. : Total proportions and quantities of sludge spread in situ (BRGM).

3.2 - Main results

3.2.1. - Behaviour of water in waste

Tensiometric profiles obtained in the initial phase showed a progressive saturation of the plots before they were covered with sludges in 1985. This was confirmed by neutronic humidity measurements made during the second phase. Several calculations of the speed of infiltration of the effluent through the ground give results in the order of 10^{-8} m/s. which seem to indicate clogging at ground level or at the base of the waste. Overall. the effect of waste spreading on the water content of the household refuse has been negligible. and can be explained by the small quantities of water carried by the waste (from 20mm to 80mm equivalent water depth). In the ground itself. the surficial water content remains relatively constant. whatever the season. Water balances were calculated on each plot using meteorological data for Orléans. and the water contents of the waste were calculated using neutronic humidity measurements. The hydrocarbon waste in plot C behaves in a distinctive way. creating a kind of barrier to evaporation (a slower process than predicted in the calculation). For the other types of sludge. in plots B. D. E and the control plot. the correlation between the calculated water contents and the average measured contents is highly satisfactory. and the effective reserve can be estimated at 45mm. Spraying sludge over the plots (which represents a maximum of a further 50mm of water). has practically no effect on the quantity of water filtered into the ground. The sludge was spread at a very favourable time of the year. with an effective reserve deficit and a very high potential evapotranspiration: 178mm in June. and 125mm in July 1985.

3.2.2 - Evolution and migration of pollutants

The effluents collected during the first phase. and which result from the natural leaching in the household refuse. are characteristic of household refuse leachates with a high organic content (DCO. forms of nitrogen. organic carbon. chlorides) which subsequently decrease. as in the case of DCO. as changes take place within the pulverized refuse. The spreading of the industrial sludges studied does not then cause any significant change in the chemical composition of the effluents. and induces only a slight and transitory further pollutant flow to that already generated by the household refuse alone. as the monitoring of the major dissolved elements shows. After the household refuse alone has remained in place for two years. the concentrations. in relation to the initial values. are multiplied by :

Ca^{++} x 4	Mg^{++} x 4	NaK^{+} x 12	Cl x 13	SO_4^- x 7	HCO_3^- x 2	NO_3^- x 4

After spreading the sludge. however. the concentrations. compared to those measured just before spreading. are only multiplied by :

Ca^{++} x 1.7	Mg^{++} x 2	$Na^{++}_2K^{++}$ x 2	Cl^- x 1.2	SO_4^- x 1.9	HCo_3^- x 2.3	NO_3^- x 1.8

Similarly. in the two plots covered with the sludge containing phenol. only very low concentrations are observed in relation to the quantities spread. Lastly. for metals. the concentrations measured in the effluents of plot D. which was covered with metal finishing sludge. as well as those measured in the other plots. are very low (about 0.001% of the quantities produced by the sludge).

4 - CONCLUSION

BRGM's in situ experimental study has enabled the results obtained by the IRH. using the same sludges on the test containers. to be confirmed on a natural scale. For the oily sludges. such as screener sludges and hydrocarbon waste sediments. the concentrations measured after spreading are lower than in the control plot; this suggests that the oily matter is absorbed or stabilized. in the form of a film. at the surface of the refuse. isolating it from the flow of liquids. For the highly liquid sludges. the quantity of metals removed is minute. which would seem to indicate that they are stabilized by the refuse. Lastly. concerning formophenolic liquids. the quantities spread (deliberately much reduced in comparison to the test containers) resulted in a marked reduction in the percolation of the phenol at depth. due to its volatility and to the degradation process.

Finally. the analysis of the behaviour of water in the waste has shown that as far as the hydric balance is concerned. the effect of the sludge is practically nil. but that this seems to be due mainly to the climatic conditions at the time the waste was spread (the period of highest evapotranspiration). and to the relatively small quantity of sludge involved.

The results of these studies certainly suggest some possibilities for the codisposal of certain types of industrial slurries or sludges with household refuse. at least in the proportions used in these experiments. and the same favourable climatic conditions (disposal of sludges during the summer).

In practice. however. it may be difficult to bring together. implement. and control these conditions in actual operations. The important question is: is the effect of retention or stabilization of certain types of pollutants. such as metals and oils. lasting? Also. can further releases take place. following changes in the environment such as the addition of more sludge. or according to the extent of changes within the household refuse?

These studies. carried out jointly over several years. were financed and directed by the French Ministry of the Environment -Department of Research and Processing of Information on the Environment (SRETIE). Soils and Wastes Division. to whom we extend our sincere thanks for the assistance they provided.

References

Barrès. M.. Bonin. H.. Lome. F. and Sauter. M. (1988). Etude des effluents issus d'une décharge mixte de déchets urbains et de boues industrielles - 2ème phase - Rapport final BRGM 88 SGN 163 3E. Ministère de l'Environnement - SRETIE - comité Sols Déchets

Colin. F. (1986). Mise en décharge mixte boues et déchets liquides. ordures ménagères - Rapport final IRH 85 143 - Ministère de l'Environnement SRETIE - Comité Sols Déchets

Lome. F. (1987). Contribution à l'étude des effluents d'une décharge mixte de déchets urbains et de boues industrielles DEA Techniques et gestion de l'environnement - Note BRGM SGN/EAU/87/13

Session 5

Sludge Disposal and Landuse

Organic substances in soils and plants after intensive applications of sewage sludge

Prof. Dr. Wolfgang Kampe

Objects*

A good level of knowledge has been obtained on the problems of heavy metals resulting from the agricultural use of sewage sludge but not yet on the potentially harmful organic substances. For example no adequate explanation has yet been put forward as to how such substances applied with sewage sludge may behave in the long term in the soil and be transferred to crops. It was the aim of a research project commissioned by the German Environmental Protection Agency in Berlin to find answers to these questions.
Soil and plant samples were taken from long-term sewage sludge trials and analyzed particularly for difficult degradable chlorohydrocarbons. It was thus an aim of the investigations to provide a contribution to environmental precautions and safeguard the nutritional quality of crops produced after the use of sewage sludge.

Details of the study

In some cases considerable sewage sludge dressings had been applied in selected field trials. A total of 84 soil and 171 plant samples from two universities (Bonn and Gießen) **and from four Agricultural Research Institutes** (Brunswick, Bremen, Munich and Speyer) were sent to the Speyer Agricultural Experimental and Research Institute for analysis. The amounts of 5 kg dry matter/ha/year permitted by the German Sewage Sludge Regulation were exceeded by as much as 200-fold.

---

* The investigations were commissioned and financed by the German Environmental Protection Agency in Berlin

ISWA PROCEEDINGS
VOLUME 1 ISBN 0-12-058451-4

The investigations in soils and crops were aimed at the following compounds:

- chlorinated hydrocarbons (CHCs), e.g. DDT
- polychlorinated biphenyls (PCBs) as 6 individual components
- polycyclic aromatic hydrocarbons (PAHs), e.g. benz-(a)-pyrene.

Investigations were also carried out into certain other organic substances more or less as spot checks, but they will not be dealt with here. The checks for the third group of substances were limited to the samples from Munich most contaminated with CHCs and PCBs.

For details of the methods of taking and handling the samples and of the chemical analysis those interested are asked to consult the original scientific work (see literature). Specially cleaned aluminium tins were used as the sampling vessels. The work on preparing the samples was done in slightly darkened rooms because some of the samples were sensitive to light. The chemical analysis was carried out using relevant or newly developed methods that had to be adjusted to the soil and plant material.

Threshold assessment data

The results must always be considered with reference to the high dressings of sewage sludge applied, which were up to 200 times the amounts presently permitted. The analytical findings should be compared with the threshold assessment data that are known from the Netherland (Leidraad Bodensamering, 1984) as guide values for organic substances, although they are not yet compulsory. Here key data are given as reference values, test requirements and clearance limits. Although these values apply to contaminated wastes, they may be used in the absence of other ways of assessment.

Chlorohydrocarbons (CHCs)

Out of this group of substances dieldrin, DDT and HCB (hexachlorobenzene) were found in the **soils to which no sewage sludge had been applied.** Other chlorohydrocarbons (aldrin, α -, β -, γ - and ξ -BHCs, chlordane, endrin, heptachlor and heptachlor epoxide) were not found. **Dieldrin and HCB occurred in approximately the same order of magnitude in** the mean and maximum values with 0.01 and 0.04 mg/kg soil dry matter, respectively. DDT (total DDT) was present in three times the amount at 0.03 and 0.09 mg/kg, respectively. In their scope these values corresponded to other results that have been obtained by the Speyer Agricultural Experimental and Research Institute in arable soils in Rhineland-Palatinate and Saarland (see literature).

After **the application of sewage sludge** the soil findings differed only slightly from the control values. This also emerges from the conversion to accumulation factors, which were 1 (= no accumulation) for dieldrin and HCB and **2 in the case of DDT. The application of sewage sludge did not result in any substantial accumulation of these** substances in the soils (Table 1).

Dieldrin and HCB were on an average about 1/10 and DDT about one half the Dutch reference value of 0.1 mg/kg dry matter. Only the maximum value for DDT (0.18 mg/kg dry matter) exceeded this limit, but differed appreciably from the test requirement (0.5 mg/kg).
A possible transfer to **crop plants** was investigated in the agricultural crops cereals, sugarbeet, potatoes, clover, corn and grass as well as in the vegetable species head lettuce, black radishes and carrots. There were

no detectable findings in 60 % of the cases; the remainder were similar to each other in the controls and after applications of sewage sludge in the trace range of mostly 0.001 mg/kg fresh matter. There was no evidence of a transfer from the soil to the crops. The use of sewage sludge did not cause any increased residue values in the harvested produce (Table 1).

Polychlorinated biphenyls (PCBs)

The mean values of all **control samples** ranged between 0.001 and 0.004 mg/kg soil dry matter, which, as in the case of CHCs, agreed with those in arable soils in Rhineland-Palatinate and Saarland. The maximum values of the individual components were lower than 0.01 mg/kg.

The **application of sewage sludge** increased the contents in the soils. Accumulations emerged in the mean values. Their degree was also determined by the chlorine fraction of the individual components (identification by rising numbers). For example PCB component K 28 was analyzed with 0.005 and K 52 with 0.017 mg/kg soil dry matter. K 101, K 138, K 150 and K 180 were present in relatively elevated orders of magnitude of 0.04 mg/kg. The maximum values of approximately 0.3 mg/kg were reached by K 101, K 138 and K 150. Calculation of the findings of controls and after the application of sewage sludge points to mean accumulation of from 5- to 17-fold.

This consideration is a relative one. It must be kept in mind that the absolute values were in low ranges in spite of the accumulation. The high application levels from sewage sludge dressings going back as many as 25 years must also be taken into account here (Table 2).

In view of the absolute data and in comparison with the **reference value** of 0.05 mg/kg the latter was not exceeded by the individual components in spite of the accumulation in the mean values. The maximum figures were **to be found in the higher chlorinated PCBs at about** 0.3 mg/kg soil dry matter above the reference value, but **they were appreciably below the margin of the test re-**quirement (1 mg/kg). The test requirement was reached with the mean values of the sum of the individual components at 1.3 mg/kg; the clearance limit of 10 mg/kg, however, is about a power of 10 higher. For this reason and on account **of the increased application rates it is justified to** assess the amounts of PCBs found as non-critical in spite of the rises.

The findings in the **crop plants** examined as for chlorohydrocarbons do not reveal any transfer of PCBs from the soil to the crops when the values from the control and after applications of sewage sludge are compared. The increases in content in the soil were not accompanied by any in the **crop plants. Nor was any difference caused by the species** of crop plant. At the same time the absolute values indicated very low contents that scattered around the figure of 0.001 mg/kg fresh matter. In many cases there was no positive finding at all where there was absence of evidence (Table 2).

Polycyclic compounds

Some polycyclic compounds are substances that constitute a health hazard. Benz-(a)-pyrene, for example, can be carcinogenic. The **control findings** obtained scattered as mean values between 0.005 and 0.04 mg/kg, the maximum values between 0.015 and 0.1 mg/kg soil dry matter. As a rule all the values were below the reference limits or just reached this level.

The **application of sewage sludge** increased the contents in the soils in the mean and extreme values 5- to 10-fold. The average values for benz-(a)-pyrene were found to be about 0.2 and for the other polycyclic compounds 0.03 to 0.3 mg/kg. In comparison with the Dutch reference values they were closer to the reference values (0.05 or 1.0) than to the test requirement (1 or 20). No critical situation is apparent, with the reservation that the threshold assessment variables are provisional.

The transfer to crop plants followed various patterns. Crop produce growing in the soil such as sugarbeet, potatoes, carrots and black radishes had slightly higher contents in the trace range as a result of the application of sewage sludge. The findings were less than 1 μg (microgram) and mostly 0.1 μg/kg fresh matter. Parts of crops harvested above the ground such as wheat straw, beet tops, clover and head lettuce did not exhibit any increased transfer of these substances; rather, there was substantial agreement between the values from the control and those after the application of sewage sludge. It may be concluded from this that air-borne sources are the main cause of the content; at least they masked any transfer from the soil to the plant. The analytical data and the relationships emerging from these show that the applications of sewage sludge did not constitute a substantial source of contamination.

Summary

Soil and plant samples were analyzed from a joint project with samples from trials carried out with dressings of sewage sludge between 1959 and 1980. Chlorohydrocarbons did not accumulate in the soils; there was no transfer to crop plants. Polychlorinated biphenyls were found to have increased on an average 5- to 17-fold and polycyclic compounds 5- to 10-fold after applications of sewage sludge. No regular transfer to crop plants was detectable, where the order of magnitude was in the µg/kg range. The results prove that, with a high degree of certainty, the organic substances that were investigated do not need to be limiting factors for the use of sewage sludge in agriculture. It must be pointed out with reference to the investigations presented here that the sample material was removed from soils to which high, in some cases extremely high, levels of sewage sludge had been applied. The conclusion drawn is justified not least for this reason.

Literature

Kampe, W. (1987). Schwer abbaubare organische Stoffe in Ackerböden von Rheinland-Pfalz und Saarland 1986, Wissenschaft und Umwelt

Kampe, W. (1987). Organische Stoffe in Böden und Pflanzen nach langjährigen, intensiven Klärschlammanwendungen, Korrespondenz Abwasser, 8, 820 to 827

LUFA Speyer. Schadstoffe im Boden insbesondere Schwermetalle und organische Schadstoffe aus langjähriger Anwendung von Siedlungsabfällen, Forschungsvorhaben des Umweltbundesamtes, Berlin, FV-Nr. 10701003/1087

Niederländisches Ministerium für Wohnungswesen, Raumordnung und Umwelt (1984). Leidraad Bodemsaanering

Table 1 **Chlorohydrocarbons in arable soils after the application of sewage sludge**

- Summary of 106 findings from 6 locations in 1985 -

Variants	mg/kg **soil dry matter** or mg/kg plant moist matter					
	Dieldrin		Total DDT		HCB	
Control						
Maximum	**0.033**	0.008	**0.092**	0.012	**0.041**	0.008
Minimum	**0.001**	n.d.	**0.001**	n.d.	**0.002**	n.d.
Mean	**0.009**	0.001	**0.032**	0.002	**0.010**	0.001
Application of sewage sludge						
Maximum	**0.030**	0.005	**0.182**	0.007	**0.052**	0.005
Minimum	**0.022**	n.d.	**0.001**	n.d.	**0.002**	n.d.
Mean	**0.011**	0.001	**0.058**	0.001	**0.013**	0.001
Accumulation factor						
Maximum	**1**	1	**2**	1	**1**	1
Minimum	**2**	1	**1**	1	**1**	1
Mean	**1**	1	**2**	1	**1**	1

Remarks: accumulation factor = $\frac{\text{application of sewage sludge}}{\text{Control}}$

n.d. = nothing detectable

Locations: Bonn, Brunswick, Bremen, Gießen, Munich and Speyer

Table 2 **Polychlorinated biphenyls in arable soils after the application of sewage sludge**

- Summary of 133 findings from 6 locations in 1985 -

- PCB peak sample analysis -

Variants	mg/kg **soil dry matter** or mg/kg plant moist matter											
	K 28		K 52		K 101		K 138		K 150		K 180	
Control												
Maximum	**n.d.**	<0.001	**0.002**	<0.001	**0.008**	0,002	**0.009**	0.002	**0.008**	0.002	**0.006**	<0.001
Minimum	**n.d.**	n.d.	**0.001**	n.d.	**0.001**	n.d.	**0.001**	n.d.	**0.001**	n.d.	**0.001**	n.d.
Mean	**n.d.**	<0.001	**0.001**	<0.001	**0.003**	<0.001	**0.004**	<0.001	**0.003**	<0.001	**0.003**	<0.001
Application of sewage sludge												
Maximum	**0.011**	<0.001	**0.123**	<0.001	**0.299**	0.002	**0.339**	0.002	**0.303**	0.003	**0.197**	<0.001
Minimum	**0.001**	n.d.	**0.001**	n.d.	**0.001**	<0.001	**0.001**	n.d.	**0.001**	n.d.	**0.001**	n.d.
Mean	**0.005**	<0.001	**0.017**	<0.001	**0.036**	<0.001	**0.049**	<0.001	**0.044**	0.001	**0.031**	<0.001
Accumulation factor												
Maximum	**11**	1	**61**	1	**37**	1	**38**	1	**38**	1	**33**	1
Minimum	**1**	1	**1**	1	**1**	1	**1**	1	**1**	1	**1**	1
Mean	**5**	1	**17**	1	**12**	1	**12**	1	**14**	1	**10**	1

Remarks: accumulation factor = $\frac{\text{application of sewage sludge}}{\text{control}}$

n.d. = nothing detectable

Locations: Bonn, Brunswick, Bremen, Gießen, Munich and Speyer

SEWAGE SLUDGE RECYCLING IN ENERGY FORESTRY

Kenth Hasselgren

Department of Civil Works
Eslöv Community
Box 1100
S - 241 26 ESLÖV
S W E D E N

INTRODUCTION

Energy cultivation systems based on fast growing deciduous trees with short rotation, so called energy forestry, have been identified as promising energy sources in the future in Sweden. Species of alder (*Alnus* spp.), birch (*Betula* spp.), poplar (*Populus* spp.) as well as willow and sallow (*Salix* spp.) have been thoroughly investigated during the last ten years. To date, willow species of *Salix* seem to be most competitive in a national perspective. Full-scale *Salix* plantations are established for evaluations of growth potentials, planting and harvesting techniques, irrigation and fertilization requirements, economic and environmental consequences, etc. A realistic scenario shows that there will be about 200 000 hectares of energy forestry by the year 2000 with an average yield of 15 tons of dry matter (6-7 toe) per hectare and year.

The nutrient requirement in energy forestry is estimated to be equivalent to or slightly exceed the need of fertilizers in normal, intensive agriculture. Rough calculations indicate that addition of commercial fertilizers to energy forestry will correspond to about 20 % of the total energy production cost considering the whole production chain, *i.e.* cultivation, harvest, storage, transports, chipping and conversion. The possibilities to utilize different municipal waste products rich in nutrients as alternatives or complement to artificial fertilizers in energy forestry have been discussed. For example investigations are carried out concerning municipal wastewater (Hasselgren, 1984), sanitary landfill leachate (Hasselgren *et al*, 1986 and 1987), compost from

ISWA PROCEEDINGS
VOLUME 1 ISBN 0-12-058451-4

household waste (Gajdos, 1988 and Hasselgren, 1988), and ashes from peat and wood incineration plants (Bramryd, 1985).

At present about 60 % of the produced sewage sludge in Sweden is utilized in the traditional agriculture and about 10 % is used as a soil conditioning agent in parks and for landscaping purposes. The rest is landfilled. In recent years, sludge utilization in agriculture has been frequently discussed due to the risk of accumulation of toxic substances in the soil. Evaluations of sludge use on farming land will probably result in further restrictions mainly due to the content of trace organics in sludge. For the time being there is not any evaluated and tested alternative to agriculture application where simultaneously the valuable resources in the sludge can be recovered. In this respect sludge recycling in energy forestry could come to play an important role.

In this paper are presented some preliminary results from a six-year field study on sewage sludge fertilization of Salix stands. The main project objectives are to evaluate the possible Salix yield and the distribution of nutrients and heavy metals in soil, Salix and groundwater. The project started in 1983 and will be completed in spring 1989. A final report will be published during 1989.

EXPERIMENTAL PROCEDURES

The test field is situated in the south of Sweden and was formerly used as pasture land. The soil consists of a 0,3 m top-soil overlying a 1-1,5 m deep sandy loam. Deeper soil layers consist of silty clayey loam. The area is drained naturally and the groundwater table is fluctuating within the sandy loam layer.

The test field was divided into plots, each with a size of 750 m^2 (15 m x 50 m). Cuttings, 20 cm in length, were taken from a high-productive osier willow species, Salix viminalis (Q 77082), and planted in spring 1983 after soil preparation and sludge application. The planting pattern, 0,75 m distance between the rows and 0,50 m distance between the plants in the rows, resulted in about 2,7 plants/m^2 (27 000 plants/ha).

Digested and dewatered excess sludge from a conventional activated sludge process was applied in spring 1983 before planting and after harvest in spring 1984 and 1986 respectively. Two sludge application rates were used, i.e. 5 tons and 25 tons of dry matter per ha, resulting in an average application of 2,5 tons and 12,5 tons of dry matter per ha and year respectively during the six-year study period. The chemical composition of the sludge is presented in Table 1.

A perforated 50 mm plastic tube for registration of the groundwater level and sampling of superficial groundwater was installed in the center of each parcell. Groundwater analyses are carried out every two months concerning nutrients, metals, pH and chloride.

Soil samples are taken out annually before and after the growth period (April and October) from the layers 0-0,1 m, 0,1-0,3 m and 0,3-0,6 m and analyzed concerning pH, nutrients and metals. The content of nutrients and metals in plant tissues is measured each fall.

The plant growth is measured when the plots are cut or harvested. After the first growth period the plants were cut for stimulation of shoot sprouting the following year. A first harvest was taken after three growth periods with two-year old shoots. A second harvest is planned during the winter 1988/1989 after six years with three-year old shoots.

PRELIMINARY RESULTS

Plant growth

As is shown in Table 2, sludge application resulted in increased *Salix* growth already during the first year. Plants in sludge treated plots were twice as high and yielded ten times more compared with control plots. The cutting after the first year resulted in that the sludge applied plants produced twice as many shoots as non-treated plants indicating the clear stimulating effect of sludge fertilization.

The stem production of two-year old shoots on three-year old roots amounted to 24-27 tons of dry matter per ha, which was 50-60 % higher than for control stands. These figures are well in accordance with production rates reported from other energy forestry projects in the south of Sweden, where Salix is fertilized conventionally (Perman, 1987).

Plants receiving the higher sludge yield seem slightly favoured but not in correspondence to the increased amount of nutrients. This may indicate that the lower sludge application rate so far is close to the optimum sludge yield. However, the higher sludge yield seems to result in a more evenly distributed plant growth considering the lower coefficient of variance for shoot length in established stands.

It was assessed that the root system was fully developed after three growth periods when the ongoing three year period will reflect a fully established energy forest with possibilities to evaluate the optimum production of Salix stands fertilized with sewage sludge.

Effects on groundwater

The results indicate clear differences in groundwater quality but these seem to depend more on natural variations within the test field area rather than being a result of sludge application. Two or three months after sludge spreading, however, a slight increase of the nitrate and chloride concentrations was detected in superficial groundwater beneath the plot receiving the highest sludge yield. No distinct change of groundwater quality was registered for the other plots.

The effect of metals on groundwater is limited. For example, the cadmium concentration in superficial groundwater has not at any time so far exceeded the value 5 ppb (detection limit), which is a tenth of the drinking water quideline. The concentration of lead has on a few occasions exceeded the quideline for potable water by a factor of 3 or 4.

Thus, a preliminary conclusion at this stage is that sludge application at actual rates does not significantly affect groundwater quality in terms of nutrients and heavy metals.

According to the energy forestry concept, the trees are supposed to be harvested in winter after defoliation. This means that sludge fertilization for practical reasons will be carried out in spring before resprouting. Compared with sludge spreading in the traditional agriculture which preferably is done in fall before plowing, this would result in a better use of sludge nutrients and a less risk for groundwater pollution considering an active and extensive root system.

Effects on soil

The concentrations of nitrogen and phosphorus in the measured 0,6 m soil profile have clearly increased during years when sludge have been applied. By and large, the increase could be related to the sludge amount. During years with no sludge applications, the concentrations have declined correspondingly.

The plant available fraction of potassium and to some extent also the more fixed potassium have steadily decreased in all measured soil layers indicating that the potassium content in sludge is low compared with the content of the other macro nutrients. Hitherto, the potassium supply in the soil has been sufficient but in the long run addition of K-fertilizers will be required for optimum growth.

In general, the concentration of heavy metals have increased in the upper soil layer (0-0,1 m) after sludge application. The transport of metals to deeper soil layers seems limited. This may be explained by the neutral soil pH and that metals are fixed to the organic material in applied sludge and in the native top-soil.

Effects on Salix

The uptake of nutrients and some of the heavy metals in Salix stems is summarized in Table 3. The nutrient requirements seem to be well supplied according to the plant chemical analyses. During the first years the nutrient content in plants fertilized with sludge was higher compared with control plants. This has most likely contributed to the higher biomass production on sludge applied plots. The nutrient concentrations are decreasing as the plants are aging which may be due to a gradually decreasing share of bark in relation to more liqnified parts of the stem. The main part of the nutrients is transported upwards through the bark in which also nutrients are stored during the winter. More or less, this is also the case concerning heavy metals.

The results obtained so far indicate that the uptake of heavy metals in Salix seems unaffected by sludge application. However, the metal concentration in Salix tissues is high by nature and for example about ten times higher than in coniferous trees (Bramryd, 1985). This offer a further application of energy forestry, namely the function as a biological purification filter to treat waste products or contaminated land. By balancing the metal input from sewage sludge or other metal contaminated waste products with the metal uptake rate by Salix it would be possible to obtain an environmentally acceptable solution where at the same time valuable nutrient resources can be reused.

REFERENCES

Bramryd, T. (1985)
Peat and wood incineration ashes as fertilizer. Effects on production, nutrient balances and uptake of heavy metals. The National Environment Protection Board, PM 1997. Stockholm, Sweden. (In Swedish)

Gajdos, R. (1988)
Plant cultivation with compost. Reforsk Report FoU No 22. (English summary)

Hasselgren, K. (1984)
Municipal wastewater and reuse in energy cultivation. In: Water Reuse Symposium III, pp. 414-429. American Waterworks Association Research Foundation. Denver, Colorado, USA.

Hasselgren, K. (1988)
Research programme on utilization of compost in energy forestry. The Swedish Association of Public Cleansing and Solid Waste Management. Malmö, Sweden. Unpublished

Hasselgren, K., Bramryd, T. and Anshelm, L. (1986)
Sanitary landfill leachate application in soil-plant systems. Progress Report No 2. The Swedish Association of Local Authorities. Stockholm, Sweden. (In Swedish)

Hasselgren, K., Bramryd, T. and Anshelm, L.(1987)
Resource conserving treatment of sanitary landfill leachate. To be published by Reforsk. (English summary).

Perman, G. (1987)
Energy forestry in South Scania. Progress Report EO-88/6. The National Energy Administration. Stockholm, Sweden. (English summary)

Table 1 Chemical composition of applied sludge to energy forestry parcells.

Parameter		Applied digested and dewatered sewage sludge Year 1983	Year 1984	Year 1986
pH, units		7,5	7,2	7,5
Dry solids(DS), %		21,5	21,6	24,0
Organic part, % of DS		61,7	62,2	58,3
Total N,	" - "	3,40	3,58	3,55
NH_4 -N,	" - "	0,84	0,46	0,58
Total P,	" - "	1,82	1,63	1,73
K,	" - "	0,165	0,148	0,112
Ca,	" - "	5,85	6,02	7,59
Mg,	" - "	0,741	0,552	0,412
Cu,	" - "	0,167	0,163	0,152
Zn,	" - "	0,186	0,189	0,107
Mn,	ppm of DS	120	164	165
Cr,	" - "	55	57	35
Ni,	" - "	102	75	39
B,	" - "	19	21	-
Co,	" - "	5	5	3
Cd,	" - "	4,4	6,5	5,1
Hg,	" - "	6,8	5,4	4,6
Pb,	" - "	165	185	148

Table 2 Growth data for Salix with one-year old shoots on one-year old roots and two-year old shoots on three-year old roots respectively. (DM=dry matter, CV=coefficient of variance)

	Con- trol	5tDM sludge/ha	25tDM sludge/ha
Cutting after year 1			
Length of shoots (x), m	1,01	1,98	2,06
(CV), %	34	17	19
Number of shoots per plant	2,38	2,68	2,53
Production, t DM as stems/ha	0,26	2,45	2,73
Harvest after year 3			
Length of shoots (x), m	3,91	4,28	4,50
(CV), %	11	11	3
Number of shoots per plant	4,52	8,54	9,10
Production, t DM as stems/ha	16,9	24,3	26,7

Table 3 Uptake of macro nutrients (mg/g DM) and some heavy metals (μg/g DM) in *Salix* stems. (0=Control, 5=5 t DM sludge/ha, 25=25 t DM sludge/ha).

		Age of plants, shoot/root (years)				
		1/1	1/2	2/3	1/4	2/5
Nitrogen,	0	7,6	6,4	7,2	9,3	4,7
	5	8,8	10	5,8	8,3	4,3
	25	12	8,8	6,2	8,6	3,9
Phosphorus,	0	1,1	1,1	0,9	0,9	0,7
	5	1,1	1,3	0,7	0,9	0,7
	25	0,8	1,2	0,7	0,9	0,7
Potassium,	0	2,6	1,7	1,7	2,8	1,4
	5	4,8	3,2	1,6	2,8	1,7
	25	2,2	2,9	1,5	3,2	1,5
Zinc,	0	98	170	119	84	71
	5	50	108	53	65	50
	25	74	105	56	71	64
Cadmium,	0	2,0	2,1	2,2	2,5	3,2
	5	2,0	1,3	0,8	0,9	1,4
	25	3,1	2,0	1,0	1,1	1,0
Lead,	0	2,0	5,5	4,0	-	3,0
	5	1,0	6,0	2,0	4,6	2,5
	25	2,0	8,5	2,5	4,4	1,8
Copper,	0	6,0	5,2	9,3	5,8	3,4
	5	2,0	5,1	10	5,8	3,4
	25	4,0	4,4	8,0	4,8	3,1
Chromium,	0	0,2	1,1	0,4	1,7	0,5
	5	0,4	1,8	0,4	0,4	0,4
	25	0,5	1,9	0,4	0,6	0,3
Nickel,	0	-	0,8	0,8	1,0	0,8
	5	1,0	0,5	0,7	1,2	0,8
	25	4,0	0,4	0,8	1,1	1,0

TRANSFORMATION OF SEWAGE SLUDGE IN LANDFILLS

Th. Lichtensteiger and P.H. Brunner
Swiss Federal Institute for Water Resources and Water Pollution Control (EAWAG/ETH), CH-8600 Dübendorf, Switzerland

INTRODUCTION

Modern concepts of sanitary landfilling are based on the main objective to minimize longterm risks of such deposits. There are two approaches to reach this goal: Waste materials can be disposed of in water proof containments such as underground salt mines where they may be excluded from the hydrological cycles for geological time scales. Or wastes can be transformed to materials which have no adverse effect on the environment, even for long periods of time. The transformation of a waste material to a material with such "ultimate disposal quality" may be achieved either by a technical process (e.g. incineration with chemical treatment of the residues) or possibly in a biochemical reactor such as a sanitary landfill. Our investigations focused on the transformation of sewage sludge in reactor type landfills. The investigation of sludge deposits has the main advantage that a landfill with sludge only is a less complex and more homogeneous reactor which can be sampled, analyzed and interpreted with less difficulties than a municipal solid waste (MSW) landfill.

In this study, the longterm effects of mechanical dewatering and lime addition were taken into account, too.

METHODS

While laboratory experiments such as leaching tests yield information which is limited to short time periods only, field experiments on recent and altered sediments of natural or anthropogenic origin can be a key to understand the transformation from sediments to stones (diagenesis).

Existing sludge landfills were used to study the transformation during the first 30 years: Samples were taken from sludge-only landfills with liquid sludge, dewatered sludge and limed sludge of various ages, constant landfill practice and well known sources (Table 1).

We assessed the longterm behavior (10^3 to 10^6 years) by comparing the potential transformation of the landfilled sludge to the evolution of natural organic sediments, such as peat, sapropels or organic soils, which can be looked at as natural analoga to anthropogenic organic sediments.

The samples from sludge-only landfills were analyzed for:

1. chemical/biochemical parameters: volatile matter, C, H, S, N, O, Cl, P, Zn, Cu, Fe, Cd, Hg and pH-value as "indicator parameters"; xenobiotic organic substances,

ISWA PROCEEDINGS
VOLUME 1 ISBN 0-12-058451-4

such as polychlorinated biphenyls (PCB) and detergent derived aromatic compounds (alkylbenzenesulphonates [ABS], nonylphenol ethoxylates [NPEO] and nonylphenol [NP]; see GIGER et al., 1984)

2. physical/geotechnical parameters, such as shear strength, bearing capacity, water content, density and texture. The texture was analyzed by scanning electron microscope to determine the relationship between physical/geotechnical and chemical/biochemical parameters.

A literature study on the development of peat or sapropel during diagenesis focused on the same parameters. Here we emphasized the early diagenesis (various transformation stages of peat). Because erosion and weathering compete with ongoing sedimentation the effects of these three processes were assessed.

LANDFILL SITE	SEWAGE SLUDGE CHARACTERISTICS	Tm(1)
Menden-Bösperde	liquid raw sludge	15 30
Edewechterdamm (Bremen)	liquid digested sludge	0.5 2 5 10 13
Essen-Werden	liquid digested sludge	0.5 2
Essen-Eicken-scheidterbach 1	liquid digested sludge	20
Essen-Eicken-scheidterbach 2	digested sludge, dewatered by centrifuge	2 3
Hagen	digested sludge, treated with ferric chloride and lime, dewatered by chamber filter pressure	0.0 1 4
Erkrath	digested sludge, treated with ferric chloride and lime, dewatered by chamber filter pressure	0.0 0.5
	digested sludge, dewatered by centrifuge, mixed with fine-grained lime	0.0 1
Langenfeld	digested sludge, cond. by polyelectrolytes, dewatered by chamber filter pressure	2 4
Erlimoos	digested sludge, dewatered by centrifuge, mixed with fine-grained lime	0.5

Table 1: Sampled sewage sludge from the Federal Republic of Germany and Switzerland (Erlimoos)
(1) Tm = Mean Residence Time [years];
(for definition of Tm see Baccini et al., 1987)

RESULTS

1. Chemical/biochemical data:

In sewage sludge and peat from low lying moors (NAUCKE, 1980) the major components defined as volatile matter, C, O, H, N and Si are similar in concentration. In both materials, the matrix is mainly composed of carbohydrates (cellulose and hemicellulose), lignin, proteins and inorganics such as silicates. 30 years after disposal in a sludge-only landfill up to 1/3 of the initial organic matter (defined as volatile matter) is mineralized, a further fraction is transformed into refractory organic substances. The transformation rate in sludge deposits without lime addition is highest between 2 and 10 years and then slows down significantly. During this stage, electron micrographs confirm that the cellulose fibres are only partly decomposed. The carbon to nitrogen ratio increases from C/N=11 (1-2 years) to C/N=17 (>15 years), which is appropriate for optimum bacterial growth in both periods. The pH-value is nearly constant at about 7. Below a surface layer of up to 40 cm, anaerobic conditions are predominant. Phosphorus, sulfur and mercury concentrations do not alter within the time interval sampled; in old samples, they are found to be slightly higher. The chlorine concentration is constant or slightly decreasing. Cadmium, copper and zinc are significantly higher concentrated in old samples. Since this cannot be explained by an enrichment due to the decomposition of organic matter, it is assumed that the metal input in recent sludges has decreased. The xenobiotic organic chemicals (PCB, ABS, NPEO, NP) are only partly or not at all decomposed. Thus, sludge-only landfills may serve as historical records for the flux of metals and synthetic refractory organic substances from the anthroposphere to waste water. It should be noted that in old sludges refractory branched ABS are present.

2. Physical/geotechnical data:

The relationship between dry matter and shear strength is not linear. In sludge pond landfills (LICHTENSTEIGER et al., 1988), depending on the thickness of disposal, the shear strength increases 2 to 10 times during the first 15 years. According to electron micrographs, the increase of shear strength coincides with a textural homogenization. The shear strength of mechanical dewatered sludge is 5 to 10 times higher than without dewatering and is slightly increasing with increasing residence time.

3. Effects of lime addition:

Assuming that 20-30% of total lime added is recycled to the sewage treatment plant during chamber filter pressuring (PESCHEN, 1988), 25-30% of dry matter found in landfills are composed of lime or lime and iron which were added. The carbon content increases with increasing residence time while the pH-value decreases from 13 to 8.3, both indicating a carbonatization. The volatile matter does not decrease within the sampled time intervals. The stage of high transformation rate is thought to be postponed due to the reduced biological activity at the high pH-value in the beginning. The initial shear strength is 30 to 50 times higher than in sludge without any lime addition. Shear strength was found to be higher in sludge treated with ferric chloride and lime before dewatering than in sludge mixed with lime after dewatering. With increasing residence time the shear strength of limed sludge is slightly decreasing. Large inhomogenities in terms of water content variations were observed in sludge landfills.

4. Leachates and supernatant water:
Sludge-only landfills have lower permeabilities and interparticular porosities than MSW landfills. Thus, landfill gas and leachate control are more difficult and less effective than in MSW landfills. In sludge ponds, sedimentation of the particulate fraction of the liquid sewage sludge results in a separation of sludge water as supernatant water which for many purposes is comparable to leachates. Dissolved substances such as nitrogen and phosphorus reach the supernatant water which consequently has to be treated. Heavy metal concentrations (Cr, Cu, Ni, Zn, Pb, Cd and Hg; others were not determined) in the surface water of the sampled Edewecht sludge ponds comply with the FRG quality requirements for surface waters to be upgraded to drinking water (SEYFRIED et al., 1984). COD-values in leachates from landfills of sludge mixed with lime after dewatering were found to be 5 to 10 times higher than from other sewage sludge landfills.

5. Longterm behavior of organic sediments:
The "van Krevelen-diagram" (van Krevelen, 1961) may be used to classify organic materials, sediments and stones. Digested sewage sludge in such a graph of atomic H/C versus O/C is positioned in the field of peat (Figure 1). During peat diagenesis, the initial organic substances and their metabolites are either mineralized or transformed further into insoluble, refractory organic compounds. However, even in the highest transformation stage of peat, there exists still cellulose, although cellulose degrades relatively easy (TISSOT and WELTE, 1978). With increasing transformation, the carbon content increases relatively to the oxygen content which decreases. In digested sewage sludge the organic carbon content of volatile matter is of the same order of magnitude as in poorly decomposed peat. During the first 30 years in a landfill the organic carbon of the volatile

Figure 1:
Van Krevelen-diagram:
Evolution of organic substances and sediments during diagenesis.
The diagenetic potential of digested sewage sludge corresponds to the potential of peat.

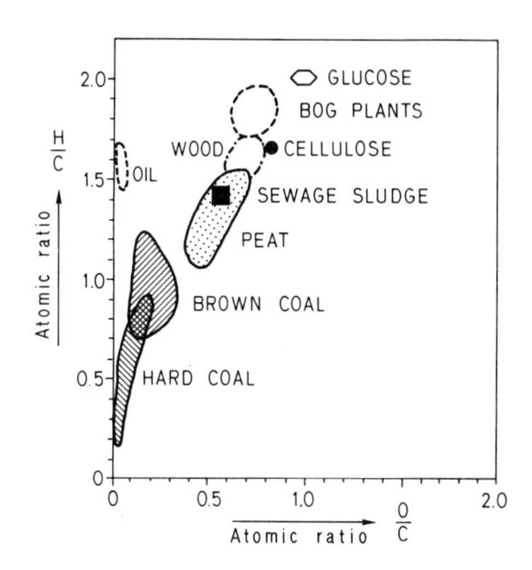

matter does not change significantly. Thus, the landfilled sewage sludge is still in an early peat stage.
The evolution path of natural organic sediments cannot be correlated to a certain time scale. As a function of climatic and tectonic conditions, peat turns into brown coal in 1-20 x 10^6 years. The geotechnical stability of peat is low and settling due to the ongoing degradation is common. The longterm stability of landfilled sewage sludge will be in the same order of magnitude.
Most of the organic sediments in a non marine environment are eroded within the first 10^6 years before they reach an early coal stage. Assuming an erosion rate of 5-100 cm / 10^3 years (JUDSON, 1968), a sludge landfill might be eroded within 10^3 to 10^5 years. Under anaerobic conditions, geological time scales (10^6 years) will be necessary to decompose the organic material to water, gas and insoluble refractory organics. Therefore, due to erosion and weathering it may well be for some landfills that aerobic mineralization is the more important longterm (10^3 to 10^5 years) process than anaerobic transformation.

CONCLUSIONS

The following hypotheses can be made: 1) The transformation of organic material in a sludge landfill lasts for geological time scales (10^3 to 10^7 years). 2) Due to biochemical transformations of the organic material, it is expected that the body of the landfill will continue to settle for 10^3 to 10^7 years. 3) The geotechnical stability will be in the range of peat or other organic soils for 10^1 to 10^7 years. 4) Lime addition postpones the biochemical transformation for a few years only. 5) After 10^1 to 10^2 years, there are still xenobiotic organic substances which are only partly or not at all degraded in a sludge landfill.

ACKNOWLEDGMENTS

This project was supported by COST 681 (Bundesamt für Bildung und Wissenschaft, N. Roulet). We thank the group of W. Giger (EAWAG) for the organic analysis. The elements C, H, N, S and Cl were analyzed by R. Müller (EMPA). We appreciated the help provided by Freie Hansestadt Bremen, Ruhrverband, Bergisch Rheinischer Wasserverband, Rippstein AG, M. Langmeier and G. Henseler (EAWAG) and M. Müller (Labor für Elektronenmikroskopie, ETH). We are **grateful** to M. Schärer (Inst. für Grundbau und Bodenmechanik, ETH) and U.Loll for placing geotechnical instruments at our disposal.

REFERENCES

BACCINI, P., HENSELER, G., FIGI, R. and BELEVI, H. (1987). Water and Element Balances of Municipal Solid Waste Landfills. Waste Management and Research, 5, 483-499.

GIGER, W., BRUNNER, P.H. and SCHAFFNER, CHR. (1984). 4-Nonylphenol in sewage sludge: Accumulation of toxic metabolites from nonionic surfactants. Science, 225, 623-625.

JUDSON, S. (1968). Erosion of the land, or what's happening to our continents? American Scientist, 56/4, 356-374.

KREVELEN, D.W. VAN (1961). Coal. Typology, chemistry, physics, constitution. Elsevier, Amsterdam, 400-414.

LICHTENSTEIGER, TH., BRUNNER, P.H. and LANGMEIER, M. (1988, in press). Transformation of sewage sludge in landfills, in: Treatment of Sewage Sludge: Thermophilic Aerobic Digestion and Processing Requirements for Landfilling. Eds. A.M.Bruce, F.Colin and P.G.Newman, Elsevier Applied Science, London.

PESCHEN, N.(1988), Bundesverband der deutschen Kalkindustrie E.V., personal communication

SEYFRIED, C.F., ROSENWINKEL, K.H., WEBER, B. and STRüNCK, U. (1984). Qualitative Begutachtung von abzuführendem Wasser aus der Schlammdeponie "Edewecht".

TISSOT, B.P. and WELTE, D.H. (1978). Petroleum formation and occurrence. Springer Verlag, Berlin, New York, p. 538.

NAUCKE, W. (1980). Chemie von Moor und Torf. In: Moor- und Torfkunde (Ed. Göttlich, K.). E.Schweizerbart'sche Verlagsbuchhandlung, Stuttgart, 173-195.

Session 6

Hazardous Waste

Innovative Technologies for the Treatment of Hazardous Waste

John H. Skinner, Director
Office of Environmental Engineering and Technology Demonstration
U.S. Environmental Protection Agency, Washington, DC

Introduction

In many countries there is considerable interest in treatment technologies that can permanently destroy or detoxify hazardous waste. In the United States, the Resource Conservation and Recovery Act (RCRA) requires hazardous wastes to be treated with the best demonstrated available technologies. Also, the Comprehensive Environmental Response, Compensation, and Liability Act (more commonly referred to as Superfund) requires that to the maximum extent possible, hazardous waste sites should be cleaned up using treatment technologies that permanently and significantly reduce the volume, toxicity, or mobility of hazardous constituents.

As a result of these policies many innovative technologies to treat hazardous waste are being developed. A national research, development, demonstration, and training program entitled the Superfund Innovative Technology Evaluation (SITE) Program has been established by the United States Environmental Protection Agency (EPA) to promote the application of these technologies. A primary objective of the SITE program is to demonstrate at full-scale, the more promising innovative technologies in order to establish reliable performance and cost information. The demonstration should establish treatment or destruction efficiency,

ISWA PROCEEDINGS
VOLUME 1 ISBN 0-12-058451-4

air and water emissions, operating parameters, and approximate capital and operating costs.

Project funding is cost shared with the developer paying the cost to construct and operate the equipment during the demonstration. EPA independently evaluates the technology and pays for the costs of sampling, analysis, and quality control of data. At the completion of the demonstration the successful technologies are promoted through a technology transfer and technical assistance program. The SITE program established in 1987 includes over 20 projects that are discussed in the remainder of this paper.

Thermal Treatment Technologies

The most common form of thermal treatment of hazardous waste is incineration. This technology is commercially available and liquid injection units and rotary kilns are the most widely used incinerators. However, fixed hearth, fume and fluidized bed units are used as well. Hazardous wastes are also burned in industrial furnaces, cement and asphalt kilns, and blast furnaces.

All hazardous waste incinerators in the United States must achieve 99.99% destruction and removal efficiency (DRE) for the principal organic hazardous constituents and meet requirements for hydrogen chloride and particulate matter control. Incinerators that burn polychlorinated biphenyls (PCBs), chlorinated dibenzo-p-dioxins and chlorinated dibenzofurans must achieve a 99.9999% DRE. Many performance tests have shown that properly designed and operating incinerators are capable of achieving or exceeding these requirements. Innovative technologies include:

Mobile Incineration. EPA has built and operated a mobile incinerator mounted on four trailers. The system consists of a rotary kiln which vaporizes waste at 1000°C, a secondary combustion chamber operating at 1200°C, a water spray for cooling the flue gas, a filter for particulate removal, and an alkaline scrubber for acid gas neutralization. The incinerator can process 4,000 kg of solids or 280 liters of liquid per hour. It was used at a site in Missouri to burn over 1.5 million kg of dioxin-bearing solids, including soil, drums, and trash, and 150,000 kg of dioxin-contaminated liquid wastes. The system demonstrated 99.9999% DRE on the dioxin constituents.

Electric Infrared Furnace. This process, developed by Shirco Infrared Systems, Inc., uses a primary chamber where infrared energy provided by electrically powered silicon carbide rods achieves a temperature of 1000°C at residence times from 10 to 90 minutes. A secondary combustion chamber operates at temperatures up to 1300°C and 2 seconds gas residence time. Exhaust gases pass through a venturi scrubber spray tower.

At Times Beach, Missouri, a portable pilot unit with a processing rate of 15 to 50 kg per hour achieved a 99.999996% DRE on soil containing 227 ppb of 2,3,7,8-TCCD. At Joplin, Missouri, this unit decontaminated creosote sludges and achieved a DRE of 99.99999% on pentachlorophenol and 99.99% on naphthalene. Results from SITE demonstrations on the pilot unit and on a larger 100 to 250 metric tons per day unit will be available later this year.

Circulating Bed Combustor. This advanced fluidized bed incinerator is owned by Ogden Environmental Services. High velocity air suspends a bed of solid particles and creates a highly turbulent combustion zone with a temperature of 800°C. Solids have a residence time of 30 minutes while gases have a residence time of 2 seconds. There is no afterburner. The suspended bed materials (and waste solids) recirculate through the furnace and are recovered in a cyclone.

A transportable pilot unit with solids feed rates of 180 kg per hour achieved a DRE of 99.9999% on soils contaminated with PCBs at 10,000 ppm. Test burns were also conducted on a mixture of organic liquids and several halogenated compounds. DREs of at least 99.99% were achieved for Freon 113, carbon tetrachloride, ethylbenzene, and xylene. A SITE demonstration will be conducted once a suitable location is found.

Electric Pyrolyzer. This system, developed by Westinghouse Waste Technology Services Division, pyrolyzes organic hazardous wastes. The pyrolyzer operates at temperatures up to 1800°C. The system reduces metals to their elemental state and all inorganic oxides, sulfides, and halides are fixed as liquid silicates. Solids fall into a molten bath which when cooled forms a vitrified material. A prototype system exists which can process from 5 to 20 metric tons per day of solid material with up to 10% organics and 25% water.

Pyroplasma System. The pyroplasma system developed by Westinghouse Plasma Systems uses a plasma arc torch which produces temperatures of more than 5000°C. Waste liquids injected into the plasma are reduced to their atomic states and the atoms then recombine to produce hydrogen, carbon monoxide, nitrogen, and other gases. The system has been operated on pumpable liquid organic wastes including methyl ethyl ketone, methanol, ethanol, carbon tetrachloride, and Askarel. A mobile unit mounted on a 15 meter trailer is available and can process 8 to 12 liters per minute.

Oxygen Enriched Burner. A proprietary burner developed by American Combustion Technologies, Inc. uses pure oxygen in combination with air and natural gas. The use of oxygen produces temperatures up to 2500°C which improves the kinetics of waste destruction, allows higher waste throughput and reduces stack gas volume. Liquid waste particles must be less than 1 mm and blending may be required to reduce viscosity. A SITE demonstration of this burner was carried out in late 1987 at the EPA Combustion Research Facility in Jefferson, Arkansas. Contaminated soil from the Stringfellow Acid Pit Superfund site in California was burned during the 5 week demonstration. Results will be published later this year.

Centrifugal Reactor. A spinning centrifugal reactor heated with a plasma torch has been developed by Retech, Inc. The system which reaches 1500°C vaporizes organics and melts solids into a glassy or metallic liquid. The molten liquid is removed by tilting the reactor to pour the slag into a cooling mold. This reactor similar in design to a metal melting furnace will have a power rating of 75 kW and a waste feed rate of 70 kg per hour.

Stabilization and Solidification Technologies

Stabilization and solidification technologies reduce the release of hazardous constituents from a waste. Stabilized waste hardens into a monolithic block and its physical and handling characteristics are improved. Most processes mix wastes with cement, lime, and other pozzolanic materials. Thermoplastic materials and vitrification techniques are also used. Stabilization techniques are most effective on wastes containing heavy metals and inorganic salts. However, some of the processes also encapsulate organic constituents.

Cementation and Pozzolanic Processes. Two cement based processes are being evaluated under the SITE program. In one process developed by Hazcon, Inc., liquid wastes or sludges are mixed with a cementation material (such as fly ash or cement kiln dust) and a proprietary reagent. The mixture becomes hardened within 10 to 15 minutes and the hazardous constituents are micro-encapsulated with significantly reduced leaching potential. Under the SITE program, an evaluation has been carried out to determine the applicability of this process to organic sludges and soils contaminated with organic materials.

A second cement based stabilization process developed by Soliditech, Inc. uses microblending to mix a reagent into the waste stream and microencapsulation to coat the waste particles with pozzolanic materials and seal the resulting matrix. The sealing process significantly reduces the leaching potential of the waste. The Soliditech process has been tested at bench scale and will be scaled up to a 20 cubic meter batch reactor for the SITE program.

Stabilization Using Silicates. A stabilization process developed by Chemfix Technologies, Inc. uses soluble silicates and silicate setting agents. It has been successful in stabilizing heavy metals in electroplating wastes, refinery wastes, contaminated soil, electric arc furnace dust and municipal sewage sludge. It has also been used on high molecular weight organic wastes such as refinery wastes, creosote, and wood treating waste. A SITE demonstration is being planned.

Thermoplastic Stabilization. One stabilization process developed by Waste Chem Corporation mixes wastes with liquid asphalt or other plastic binders. The mixture is discharged to a container or form for cooling and solidification. The binder coats the waste particles and reduces leaching potential. Organics that volatilize are collected through carbon adsorption and treated using ozone. A SITE demonstration is being planned.

In Situ Stabilization. A SITE demonstration is being planned to evaluate the *in situ* stabilization and solidification of soils contaminated with PCBs using a process developed by International Waste Technologies. Special drilling equipment will be used to inject and blend the stabilization media into contaminated soils. Permeability, density, and leaching tests will be conducted.

In Situ Vitrification. The *in situ* vitrification process developed by Battelle Memorial Institute converts contaminated soil or sludge into a chemically inert, stable glass and crystalline product. Electric potential is applied to an array of electrodes inserted into the ground. The power heats the soil to over 2000°C, far above its melting and fusion temperature. Organic components are pyrolyzed. A hood placed over the soil area draws released gases into a scrubber and filter. Normal processing rates are 3 to 5 metric tons per hour at depths of 15 meters.

Separation and Extraction Technologies

There are various physical or chemical processes that can be used to separate wastes so that the components can be treated or disposed separately. Distillation, steam or air stripping, precipitation, and adsorption are used routinely. Innovative new approaches include:

Vacuum Extraction. Terra Vac, Inc. has developed a process for *in situ* vacuum extraction of volatile contaminants from soils and groundwater. A vacuum source is applied to extraction wells placed in the unsaturated soil zone. Volatile organic compounds are drawn off and passed through an activated carbon adsorption system.

Vacuum extraction was used to clean up a spill of carbon tetrachloride from a ruptured underground storage tank. The extraction rate of carbon tetrachloride reached 100 kg per day from an unsaturated soil zone 100 meters deep. After 30 months of operation, more than 70% of the spilled volume was removed. In other applications, methylene chloride and gasoline were successfully extracted from contaminated soils. A SITE demonstration is underway to evaluate this process on soils contaminated with volatile organic compounds, hydrocarbons, and solvents.

Sludge Extraction Using Aliphatic Amines. The Basic Extraction Sludge Technology (B.E.S.T.) process, developed by Resources Conservation Co., uses the universal miscibility properties of aliphatic amines to break down oily wastes into the distinct fractions of water, oil, and organics, and dry solids. The technology is applicable to pumpable wastes with solid particles less than 6.4 mm. Heavy metals are isolated by the process into the solids as nonleachable metallic hydroxides. EPA used this technology to conduct a PCB cleanup at a Superfund site in Savannah, Georgia. The results of this cleanup are currently being evaluated. A SITE demonstration is being planned.

Soil Washing. EPA has developed a mobile soil washing system that separates contaminants form soils by high energy mixing with solvents, additives, and surfactants. The system consists of three components: a rotating drum screen (18 cubic meters per hour capacity), a countercurrent extraction chamber (two cubic meters per hour capacity) and a dewatering unit. A bench scale unit removed 98.6% of phenol from inorganic soil and 88.4% from organic soil. A full-scale unit removed 90.7 percent lead from contaminated soil at an Alabama site.

Ion-Exchange. Sanitech, Inc. has developed synthetic siderophore materials that selectively remove heavy metals from contaminated water. The materials operate over a wide-pH range, have high adsorption capacities and appear unaffected by most organic contaminants and changes in temperature and pressure. A SITE project is planned.

Liquefied Gas Extraction. This extraction technology developed by CF Systems Corporation uses liquefied gases near their critical conditions to remove organic constituents from sludges, solids or liquid wastes. Liquefied gases have lower viscosities and higher diffusivities allowing high rates of extraction compared to other solvents. Two mobile trailer-mounted units are available: a 100 liter per minute full-scale unit and a 4 liter per minute pilot-scale unit. Applications for the SITE program will use liquid carbon dioxide or liquefied propane.

Biological and Chemical Treatment Technologies

Biological and chemical processes are commonly used to treat hazardous waste. Chemical oxidation using chlorine and caustics is used to destroy cyanide wastes. Chemical reduction of hexavalent chromium wastes by sodium sulfate is an established practice. Neutralization of acidic or basic wastes is quite common. The activated sludge process is used widely to treat domestic and industrial wastewaters. Innovative biological and chemical treatment processes include:

Naturally Occurring and Special Strains of Microorganisms. The common white rot fungus, *Phanerochaete chrysosporium*, is capable of degrading halogenated hydrocarbons such as PCP, DDT, 2,3,7,8-TCDD and lindane. A special strain of the bacterium *Pseudomonis cepacia* isolated by A. M. Chakrabarty can utilize a persistent contaminant 2,4,5-T as its sole source of carbon and energy and dechlorinate a variety of chlorophenols. One notable example of an organism showing remarkable competence toward all PCB congers is a *Pseudomonas putida* species isolated by researchers at the General Electric Co. and University of Texas. Within two years, sufficient scientific and engineering progress should be made to demonstrate these microorganisms in the field.

Biodegradation in sludges and soils using proprietary, naturally-occurring microorganisms developed by Detox Industries, Inc. will be evaluated in a SITE demonstration. Detox has adapted microorganisms to biodegrade PCBs, PCPs, creosote, oils, phenolics, and PAHs. During a test project 200 kg of PCB-contaminated sludges were biodegraded to PCB levels below 4 ppm.

Fluidized Bed Biological Reactor. A mobile fixed-film, fluidized bed biological reactor developed by Air Products and Chemicals, Inc. can be used to treat organic aqueous waste. Biomass buildup occurs on a fluidized sand or carbon support media high in surface area. Long solids retention time provides maximum opportunity to destroy slowly degradable compounds. A SITE demonstration has been proposed in a pilot unit with a capacity of 400 to 1,200 liters per day for the sand system and 4,000 to 20,000 liters per day for the activated carbon system.

KPEG Chemical Detoxification. Potassium polyethylene glycolate (KPEG) reagents are effective dehalogenators of aromatic and aliphatic organic materials. The reaction produces an innocuous ether and potassium chloride salt. A mobile unit has been constructed consisting of a 10,200 liter batch reactor mounted in a 14-meter trailer. It has been field tested at a wood preserving site on 30,000 liters of oily liquid pesticide waste containing 84 ppm of dioxins and furans. The mixture of waste oil and KPEG reagent was heated to 150°C for 90 minutes. Dioxins and furans were destroyed below the detection limit of one part per billion (ppb). Processing costs were less than 10% of the estimated cost of incineration. At a second site 21,000 liters of spent solvents contaminated with 120 ppb of 2,3,7,8-TCDD were decontaminated to below a 0.3 ppb detection limit. Field tests on soils contaminated with PCBs will be conducted at a site in Guam in 1988.

Liquids Solids Contact Digestion (LSCD). The LSCD system developed by MoTech, Inc. biodegrades organics in sludges or soils. High energy aerators mix water and emulsifiers with contaminated wastes. Organics are solubilized and acclimated seed bacteria are added and aerobic biological oxidation occurs. Final biological treatment takes place in a polishing cell where organic waste concentrations are reduced to less than 1 ppm. The system uses 100,000 to 200,000 liter portable tanks or 300,000 to 400,000 liter lined *in situ* earthen digesters. A SITE demonstration project is being planned.

Powdered Activated Carbon Treatment (PACT). In this process developed by Zimpro Environmental Control Systems powdered activated carbon is added to the aeration basin of a biological wastewater treatment system. Treatment effectiveness depends on carbon dosage and other process parameters, such as hydraulic retention time of the wastewater and the solids residence time of the carbon and biomass. Wet air oxidation at elevated temperature and pressure further treats effluents. The system has been used for many years to treat industrial wastewater. Recently it has been used to treat leachate from the Stringfellow site in California and to treat contaminated groundwater from the Bofors Nobel site in Michigan. A SITE demonstration is being planned pending receipt of additional information from Zimpro.

Summary

The SITE program is in the early stages of implementation but in the next few years comprehensive data on the performance of innovative treatment technologies will be available. Demonstrations of thermal treatment systems to destroy hazardous organic constituents will be completed. Solidification and stabilization of both inorganic and organic wastes will be evaluated. Advanced extraction and separation techniques and innovative biological and chemical treatment units will be tested. There are a number of very promising technologies and significant advances in the state-of-the-art are anticipated.

EXPERIENCES IN THE TREATMENT OF HAZARDOUS WASTES BY WET OXIDATION

C. Collivignarelli (*), G. Bissolotti (**)

(*) Civil Engineering Department - University of Brescia (Italy)
(**) S.I.A.D. S.p.A. - Bergamo (Italy)

1. THE WET OXIDATION PROCESS

By working at high temperatures and pressure (up to 350°C and 250 bar), wet oxidation treatment allows the oxidation of many otherwise difficult to degrade substances. To date this technology has found applications in the treatment of industrial effluents with high COD, in the sludge conditioning, in the reclaiming of activated carbon (even pulverized) [1] [2].
In its commonest embodiment, the process employs air as oxidizer; the use of catalysts to accelerate the kinetics and allow the practical removal of particular compounds is frequent. Pure oxygen has been used recently with interesting results.
Among the problems which have restricted the diffusion of this method, those related to corrosion are particularly relevant. These phenomena compromise the operating lifespan of the equipment or force the choice of very expensive materials with consequent increases in the extent of the investments.
The range of convenient application of the process relates to wastes with high dissolved organic content (COD from 20 to 200 g/l) which cannot be treated biologically.
The literature mentions several experiences on the possibility of oxidating many even difficult-to-degrade substances which have shown that for contact times of approximately 20'+30' and for temperatures of 280-300°C it is

ISWA PROCEEDINGS
VOLUME 1 ISBN 0-12-058451-4

possible to achieve very high removal rates (even > 99%) [3] [4] [5] [6] [7].

One of the most interesting prospects is, according to the authors, the use of wet oxidation as pretreatment before a final biological treatment [1]. In this case, since the main aim is not the COD high-rate removal, it may be hypothesized that the increase in biodegradability of the waste may be obtained by working in less severe conditions. The advantage would be, therefore, to significantly reduce the dangers of corrosion of the reactor and the investment and running costs of the process.

The aim of this work is indeed to verify, for some wastes treated by wet oxidation, the increase in biodegradability as the operating conditions vary.

2. THE EXPERIMENTAL PLANT

Testing was performed in a continuously-operating wet oxidation installation. Pure oxygen was used as oxidizer.

The main components of the apparatus are: oxygen supply unit; waste compression volumetric pump; tubular heat exchanger for the heating of the waste fed to the reactor (volume: 20 l); tubular reactor (volume: 40 l); tubular exchanger for the cooling of the treated waste (volume: 20 l); group of waste depressurization valves; control and monitoring panel.

The waste compressed by the volumetric pump is mixed with the gaseous pure oxygen and continuously fed to the plant. The mixture is first heated to the required temperature and then sent to the reactor. At the output from the reactor the treated waste is cooled, depressurized and then discharged.

3. EXPERIMENTAL RESULTS

3.1. Treatment of metal painting waste

The waste used in this series of tests derives from the painting (by cataphoresis) to which the body of cars is subject in a large Italian car industry. This process consists of two phases: in the first, the bodies are painted by cataphoretic immersion; in the second they are washed to remove the excess paint. To delay the aging of the cataphoretic bath, due to the metals which accumulate therein, it undergoes ultrafiltration, from which two phases are obtained: a "concentrate" containing the cataphoretic paint, which is recirculated in the cataphoresis bath; a "permeate" constituted by an aqueous solution of metals and solvents. The permeate of the ultrafiltration, used for the washing of the bodies, is then recycled to the ultrafiltration phase. A part, however, is purged and sent to the discharge: the wet oxidation treatment is performed on this waste. Tests were carried out on samples from two different cataphoretic baths having the qualitative characteristics listed in Table 1.

Waste no. 1 was treated in two different operating conditions: in the first test the temperature was 285°C, the pressure was 100 bar and the waste flow-rate was 200 l/h; the second test was performed at a temperature of 300°C, a pressure of 100 bar and a flow-rate of 150 l/h; for both tests the flow-rate of O_2 introduced in the reactor was 20 kg/h.

Waste no. 2 was also treated in two different operating conditions: the first test was carried out at 300°C, the second at 310 °C, while the pressure (100 bar) and the flow-rate (150 l/h) remained constant.

Samplings of the liquid effluents were performed for each test after one hour of permanence of the system in operating conditions (constant T and P): the results are listed in

TAB. 1 - Characteristics of the cataphoretic bath wastes.

PARAMETERS		WASTE N°1	WASTE N° 2
COD	(mg/l O_2)	27,819	41,235
TOC	(mg/l C)	10,400	14,730
BOD_5	(mg/l O_2)	1,740	2,725
BOD_{20}	(mg/l O_2)	4,170	6,703
Suspended Solids	(mg/l)	95	130
Dry matter at 600°C		0.5%	1%
epossidic matter		0.4%	0.9%
Organic matter (prevailingly solvents)		1.7%	1.7%

Table 2. The tests yielded COD removals between 64% and 80% with the highest value at the temperature of 310°C; it should be noted that the increase of the temperature from 285°C to 300°C causes the increase of the COD removal by 14%, while there are no significant increases by passing from 300°C to 310°C.

Similar considerations can be proposed about TOC removal.

During the oxidation reaction the pH of the waste tends to lower, passing from 5.1. to values below 4.

The most interesting result, however, relates to the increase of the BOD_{20}/COD ratio of the effluent, which passes from 15-16% to 82%. The BOD_{20}/COD ratio is furthermore almost constant for the various temperatures. This may mean that, beyond a certain temperature, the increase of the COD removal occurs at the expense of the oxidation of biodegradable matter and that, therefore, it is not convenient to increase the temperature any further if the main aim is the increase of the biodegradability.

At last, an important reduction of the ($COD-BOD_{20}$) was obtained.

3.2. Treatment of landfill leachate

The waste employed is the leachate extracted from a large urban solid waste landfill active in Lombardy (northern Italy). Its main qualitative characteristics are listed in Table 3.

The treatment of this waste currently occurs in an activated-sludge biological plant after mixing (in small doses) with domestic sewage. The biodegradability of landfill leachate is, as is known, a very severe obstacle for this type of treatment, especially for leachates from "old" landfills (in which the biodegradable rate of organic matter has been proved very small). It becomes therefore interesting, in view of the basic aim of our research, to verify if (and how much) a wet oxidation process is capable of improving the biological treatability of this waste, constituting, in this case, an interesting possibility for leachate pretreatment.

Experiments were carried out in three separate tests performed respectively at T = 275°C, 295°C and 305°C, while pressure (= 100 bar), flow-rate (= 150 l/h) and oxygen dose

TAB. 3 - Main characteristics of landfill leachate.

pH	7.06	
Electric conductivity	22,100	μS/cm
Dry matter at 105°C	1.5%	
Dry matter at 600°C	0.9%	
COD	19,400	mg/l
TOC	7,934	mg/l
BOD_5	7,236	mg/l
BOD_{20}	13,497	mg/l
NH_4	2,045	mg/l
P	28	mg/l
Oil	140	mg/l
Fe	210	mg/l
Pb	5	mg/l
Zn	10	mg/l
Ni	4.5	mg/l
Phenols	4	mg/l
SO_4^{--}	450	mg/l
Cl^-	1,830	mg/l

(= 20 kg/h) remained constant. A further test was also carried out to verify the result of a simple heat treatment (at 275°C with Q = 150 l/h) without the introduction of oxygen.

The results of this series of three oxidation tests, given in table 4, suggest the following considerations: - the COD removal varies from 50 to 66% with the maximum value at 295°C; - similar yields occur for the TOC (53-68%); - the biodegradable fraction (BOD_{20}/COD) is subject, due to the wet oxidation treatment, to a marked increase: from 69% in raw waste to values comprised between 88% and 95% for the effluent after wet oxidation; - the portion of bioresistant organic matter (COD - BOD_{20}) is also subject to very significant reductions (decreases from 81% to 93% of the initial bioresistant COD); the effect of T on the biodegradability increase seems to be significant, if one observes the behaviour of the BOD_{20}/COD ratio and of the difference (COD-BOD_{20}).

The heat-only treatment test indicated negligible decreases of COD and TOC (12 and 14% respectively) which had evidently occurred, moreover, at the expense of the already

TAB. 2 - Metal painting wastes: results of wet oxidation treatment

		T °C	pH	COD (mg/l O_2)	TOC (mg/l C)	BOD_5 (mg/l O_2)	BOD_{20} (mg/l O_2)	$\frac{BOD_{20}}{COD} \cdot 100$	$COD-BOD_{20}$ (mg/l O_2)	$\frac{BOD_5}{BOD_{20}} \cdot 100$
WASTE n° 1	RAW	–	5.1	27,819	10,400	1,740	4,170	15%	23,649	41.7%
	1^ EFFLUENT	285	4.3	10,061	3,700	3,020	8,154	81%	1,907	37.1%
	2^ EFFLUENT	300	4.4	6,017	2,350	3,167	4,918	81.7%	1,099	64.4%
WASTE n° 2	RAW	–	–	41,235	14,730	2,725	6,703	16.3%	34,532	40.6%
	1^ EFFLUENT	300	3.9	10,062	3,650	3,050	8,235	81.8%	1,827	37%
	2^ EFFLUENT	310	–	8,384	3,060	3,260	6,870	81.9%	1,514	47.4%

TAB. 4 - Landfill leachate: results of wet oxidation treatment

	T °C	COD (mg/l O_2)	TOC (mg/l C)	BOD_5 (mg/l O_2)	BOD_{20} (mg/l O_2)	$\frac{BOD_{20}}{COD} \cdot 100$	$COD - BOD_{20}$ (mg/l O_2)	$\frac{BOD_5}{BOD_{20}} \cdot 100$
RAW	–	19,400	7,934	7,236	13,497	69%	5,903	54%
1^ EFFLUENT	275	9,702	3,716	4,536	8,580	88%	1,122	53%
2^ EFFLUENT	295	6,664	2,532	3,240	6,070	91%	594	53%
3^ EFFLUENT	305	7,310	2,778	3,660	6,910	95%	400	53%
HEAT TREATMENT (without O_2)	275	17,052	6,820	6,372	10,750	63%	6,302	59%

biodegradable organic part, since the BOD_{20}/COD ratio decreases with respect to the raw waste.

3.3. Treatment of oil mill waste

The waste employed derives from a large oil-mill in southern Italy; as is known, this waste is very rich in organic matter. Two samples underwent wet oxidation in the same operating conditions (P = 100 bar; T = 280°C; Q = 150 l/h; QO_2 = 20 kg/h). The results (see Table 5) show that the COD removal varied from 82% to 86% and the BOD_{20} removal was equal to 98-99%. As can be seen, in this case practically all the biodegradable fraction was removed, which makes the comparison of the BOD_{20}/COD ratios meaningless. In any case the net reduction of bioresistant organic matter (COD - BOD_{20}) in the effluents is significant. The results of these tests give support to the hypothesis that the operating conditions were in this case even too severe for the essential aim of our testing: in particular, tests at lower T and P have been scheduled to verify the possibility of the removal of the bioresistant organic matter without affecting the BOD at all.

4. CONCLUSIONS

The results of our tests have shown that from the wet oxidation treatment applied to some wastes it is possible to obtain an important biodegradability increase in the effluent. Fig. 1 shows the comparison between the bioresistant organic matter in the effluents and the same quantity (COD - BOD_{20}) in the raw waste: reductions between 74% and 96% were estimated.

TAB. 5 - Oil mill waste: results of wet oxidation treatment.

		T °C	COD (mg/l)	BOD_5 (mg/l)	BOD_{20} (mg/l)	COD-BOD_{20} (mg/l)
WASTE N° 1	RAW	-	9,243	2,000	2,900	6,343
	EFFLUENT	280	1,328	40	40	1,288
WASTE N° 2	RAW	-	22,390	5,200	7,200	15,190
	EFFLUENT	280	4,103	40	40	4,063

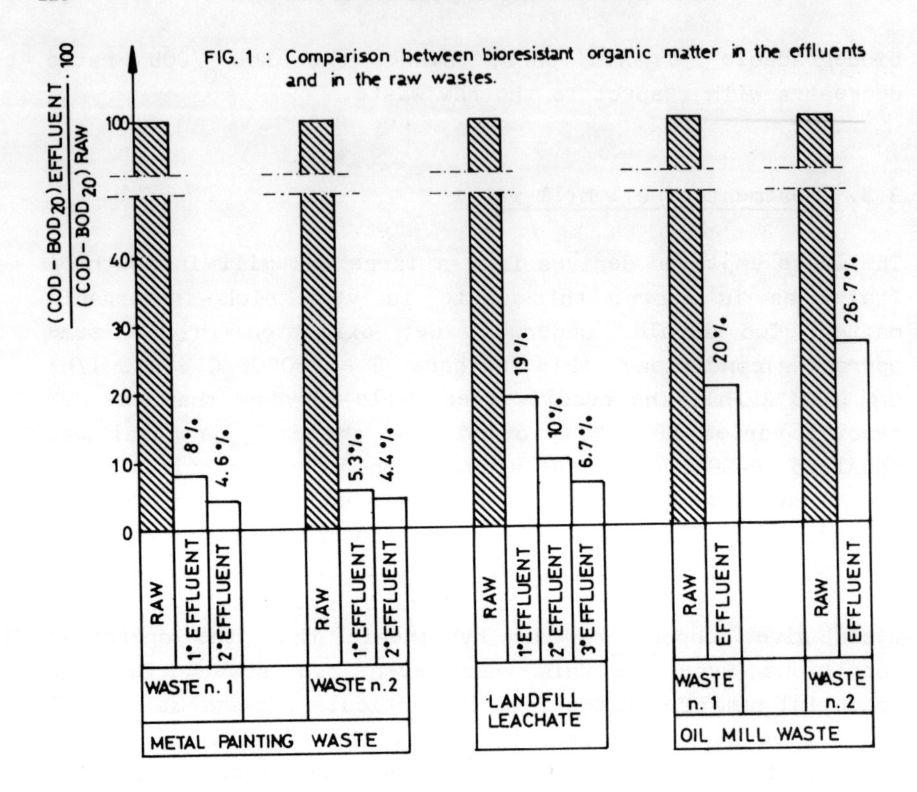

FIG. 1 : Comparison between bioresistant organic matter in the effluents and in the raw wastes.

REFERENCES

[1] Baldi M., Berbenni P., Bissolotti G., Collivignarelli C. and Fortina L. (1985). L'ossidazione ad umido come trattamento dei reflui idrici. Applicazioni e prospettive di sviluppo. Inquinamento 7/8 Luglio-Agosto, 33-46.

[2] Berbenni P., Bissolotti G., Collivignarelli C. and Fortina L. (1987). L'ossidazione ad umido come trattamento dei reflui. Rassegna di alcune soluzioni brevettate. Inquinamento, Gennaio-Febbraio, 48-56.

[3] Cannel P.J. and Schaeffer P.T. (1983). Detoxification of hazardous industrial wastewater by Wet Air Oxidation. Spring National AlChE Meeting, Houston.

[4] Prandt L.A. (Zimpro Inc.) (1972). Developments in Wet Air Oxidation. Chemical Engineering Progress, dicembre.

[5] Perkiw H., Steiner R. and Volmuller H. (1981). Wet Air Oxidation: a review. Chem. Eng. 193-201.

[6] Canney P.J. and Schaefer P.T. (1983). Wet oxidation of toxics: a new application of existing technology 1983. Toxic and hazardous wastes 277-284. Proceedings of the fifteenth Mid Atlantic Industrial Waste Conference.

[7] Baillod R.C., Lamparter R.A. and Leddy D.G. (1980). Wet oxidation of toxic organic substances. 34° I.W.C. 206-213.

THERMAL INCINERATION OF PCBs

J. Kálmán, Gy. Pálmai, I. Szebényi
Department of Chemical Technology
Technical University of Budapest, Hungary

The polychlorinated biphenyls are very toxic compounds and harmful to the environment. Although their synthesis was known already in 1881 their commercial use started in 1930. No preventive measures were made in the course of their manufacture and utilization for a long time. When already attention was paid to lipo-soluble compounds accumulating in the human body and food chain, more than one million tons of PCB have been used and come into the environment.

In Hungary, there are 30-50 tons of polychlorinated biphenyls in out-of-order and still operating synchronous condensers. This hazardous waste problem can be solved by disposal of this amount for ever because production and importation of PCBs is recently prohibited.

Several processes were developed for the disposal of PCBs, e.g. thermal, salt melt and wet oxidation methods. Among them the thermal methods are of the greatest importance in practice. Incineration in clinker rotary kiln seems to be the most suitable disposal method because of its low investment costs, economy and reliability. This paper deals with some theoretical and practical aspects of this method.

The first question to be answered is the expectable efficiency of the decomposition. The main factors determining the reaction rate of PCB decomposition are:

- the reaction temperature
- the residence time (in the high temperature zone)
- the efficiency of mixing (spraying, dilution)
- the ratio of PCBs to air.

From these factors the first two and the second two are correlated, that is, in the case of higher temperature a shorter residence time is sufficient to the same degree of decomposition, and the efficiency of mixing is correlated with the air quantity.

ISWA PROCEEDINGS
VOLUME 1 ISBN 0-12-058451-4

Data relating reaction temperature and time, **necessary** for PCB decomposition can be found in the literature within wide limits: temperatures between 600 and 1300 °C, times between 0.1 and 2.5 seconds.

The most reliable data can be found in the paper of D.S. Duvall et al (1). They investigated the thermal decomposition of PCBs very thoroughly on behalf of the US Environmental Protection Agency. **Figure 1** shows the temperature dependancy of PCB decomposition when the reaction time is one second.

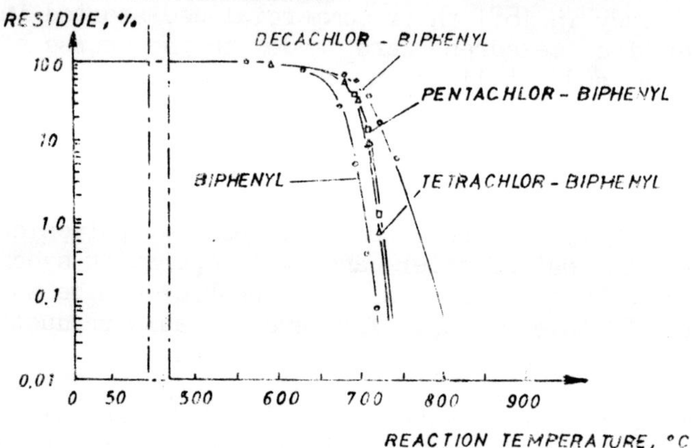

Figure 1 Temperature dependance of thermal decomposition of biphenyls

Their experiments were carried out in gas phase and the PCB was evaporated and mixed with air before the reaction. Therefore, we have to take into consideration the time requirement of evaporation and mixing if we want to use these results in case of industrial burners or furnaces. This time can be evaluated by determining the temperature and rate of evaporation.

From our derivatographic m**easurements** it was stated that the least volatile commercial product having the highest chlorine content (Aroclor 1262) evaporates completely until 185°C.

The empiric equation of T. Kurabayashi (2) for the evaporation rate is:

$$x_t^2 = x_o^2 + c \cdot t$$

where:

x_o - the initial drop size, mm
x_t - drop size in t time, mm
t - residence time, s
c - rate constant of evaporation, its value is 0.7-1.0 mm^2/s

At usual **spraying** the size of sprayed drops is less than 0.2 mm and the time necessary for evaporation calculated from this equation is 0.04-0.06 second. This means that the data presented in Figure 1 are suitable for describing the processes occurring in furnaces equipped with spray-burners. Consequently, the PCBs can be decomposed practically completely at 800-850°C in a second. This is proved by disposal experiences in clinker rotary kilns as well (**Table** 1)

Table 1: Efficiency of PCB decomposition in clinker rotary kiln

Place	Efficiency(%)
St. Lawrence, Canada	PCB is not detectable in flue gas
Stora Vika, Sweden	99.99998^x
San Juan, Puerto Rico	PCB is not detectable in flue gas
Slemmersted, Norway	99.9999^x
Detroit, USA	PCB is not detectable in flue gas

^{x}Calculated value on basis of detection limit of the used analytical method.

At qualification of PCB disposal technologies an important aspect is whether the extremely harmful compounds, the polychlorinated dibenzodioxins (PCDDs) and polychlorinated dibenzofurans (PCDFs) are formed during disposal.

This question can be answered by means of **thermochemical equilibrium data of burning and pyrolitic** decomposition of PCBs. PCDDs and PCDFs are formed only under **oxygen-deficient, pyrolitic circumstances. That is the reason** why the minimum oxygen content of the flue gas is specified in the PCB disposal facilities by the environmental agencies of several countries (e.g. in U.S.A.,Japan, G.F.R.)

The question can be answered also from reaction kinetics point of view. The formation of PCDDs and PCDFs can be limited to the evaporation time after spraying.This time is 0.04-0.06 second according to the above evaporation equation. From the kinetic equation of decomposition rate of

dibenzodioxins (2) derives that at a temperature of 1400°C 2.6×10^{-4} and 3.9×10^{-4} s are necessary for conversion of 99.99 % and 99.9999 %, respectively.
It follows from this that the very small amount of PCDD incidentally formed in the surroundings of spraying decomposes during the residence time in the clinker rotary kiln. The above kinetic equation is suitable for describing of decomposition of PCDFs as well.

The results of calculations were supported by the experimental results of Duvall **et al who have** got similar temperature dependance curves for thermal decomposition of dibenzofuran and dibenzo-p-dioxin **as those in Figure 1.**

The equilibrium and reaction kinetic calculations were proved by industrial experiences as well: PCBs, PCDDs and PCDFs were never detected in the flue gases of clinker rotary kilns.

Table 2 summarizes specifications required by the authorities of some countries in case of PCB thermal disposal installations.

Table 2: Required parameters for thermal disposal of PCBs

Country	Temperature °C	Residence time s	**Oxygen** excess v/v %
USA	1200±100	2	3
	1600±100	1.5	2
Japan	1200±100	2	3
G.F.R.	1200	0.2	6
Circumstances in clinker rotary kilns	1450-1800	4-8	4-10

The theoretical considerations and the experimental results prove convincingly the suitability of clinker rotary kilns for disposal of PCBs. Of course, due foresight and observance of severe conditions during disposal are needed but these circumstances can be economically created and guaranteed in clinker rotary kilns.

Finally, two case studies are presented. Utilization of PCB-containing used oil was carried out in a powdered coal-fired power plant and in a cement works. After the event calculations were made on the basis of supply of data for investigation of environmental impact caused by incineration of the used oil. The main parameters of incineration, the calculated maximum emission values (E_{max}) and its ratio to the permissible emission values (E_n) are presented in **Table 3.**

Table 3: Circumstances during incineration of PCB-containing used oil

Characteristics		Powdered coal-fired power plant	Cement works
Max. mass flow of used oil	(kg/h)	750	200
PCB content of used oil	(mg/kg)	6870	6870
Temp. of burning zone	(^oC)	min. 1,100	min. 1,400
Residence time	(s)	min. 2.2	min. 5.7
Air excess	(%)	min. 20	min. 11
Conversion	(%)	99.99998	99.99998
E_{max}	(kg/h)	$1x10^{-6}$	$2.8x10^{-7}$
E_n	(kg/h)	$2x10^{-3}$	$1.2x10^{-3}$
E_{max}/E_n		$5x10^{-4}$	$1.2x10^{-4}$

The value of E_n relates to benzpyrene because no limit value for PCB is in force in Hungary. This value was taken as a basis for safety's sake because it is certainly more severe than the limit to be expected. This is justified by **Table 4** in which air quality standards are compared.

Table 4: Air quality standards for PCBs and benzpyrenes.

Country	Air pollutant	Air quality standard
Japan	PCB	0.5 $\mu g/m^3$
USA	PCB	10 $\mu g/m^3$
Hungary	benz(a)pyrene	0.001 $\mu g/m^3$

According to our calculation no harmful air pollution attributable to PCB or PCDD was caused by the test incinerations in the power plant and cement works, in fact, the actual emission values are less by several orders of magnitude than the allowable ones. The PCB concentrations at ground level due to test incineration were calculated taking into account the meteorological parameters and the results of our calculations for the most unfavourable cases in power plant are presented in **Figure 2.**

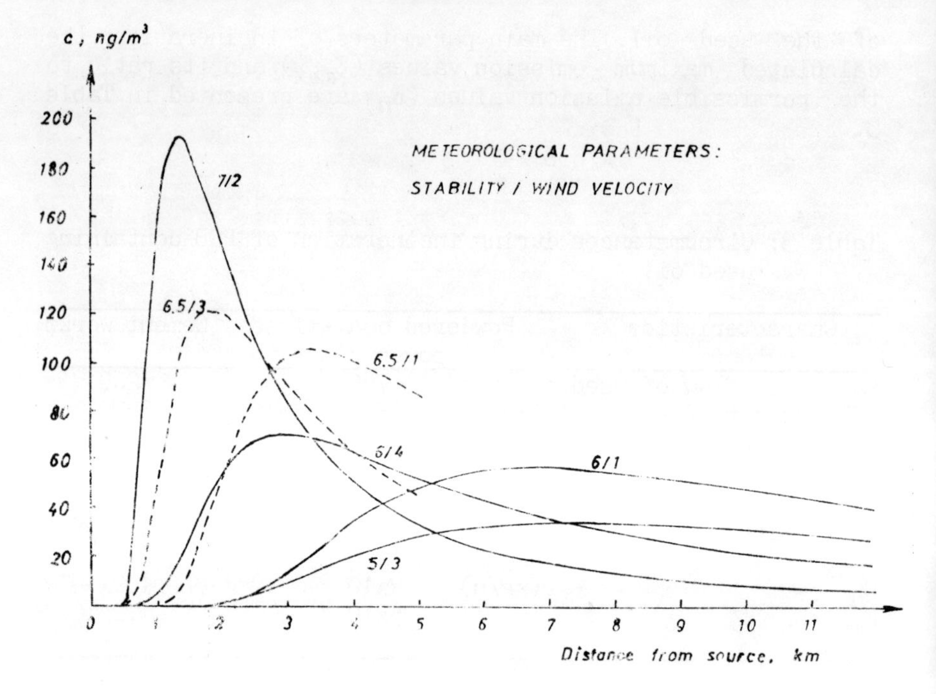

Figure 2 PCB concentration in air at ground level

Another important question is what harmful effect on the environment would have the disposal of max. 50 tons of PCB accumulated in Hungary in clinker rotary kilns.

The **emission** would be 0.1 g/h if the decomposition efficiency and the mass flow of PCB disposed are 99.9999 % and 100 kg/h, respectively. The allowable emission values of PCB for a given Hungarian cement works would be:

0.6 kg/h according to Japanese and

12 kg/h according to American

air quality standards. It means that the actual emission of 1 g/h is only a small part of the allowable values in USA and Japan. In case of disposal rate of 100 kg/h the total PCB amount to be found in Hungary can be disposed in 500 operating hours. The total emitted PCB would be less than 50 g. It may be supposed with reason that more than fifty **grams of PCB has got and will get to the environment during** waiting and storage.

REFERENCES:

1./ **Duvall, D. S. et al, (1977), Laboratory evaluation of** high temperature destruction of polychlorinated biphenyls, Dayton Univ. Ohio, Research Inst. PB-279.139, EPA/600/2-77/288, Dec.1977.

2./ Air Pollution Division, Air Quality Bureau, Environment Agency (1986), Outline of test results on high temperature thermal destruction of liquid waste PCB, Japan.

Performance tests carried out at new incineration plant of AVR Chemie CV

Ir. D. den Ouden

Summary

Performance tests carried out at a new incineration plant of AVR Chemie CV.
At the end of 1986 a totally computerised incineration plant was put into operation. The plant consists of a rotary kiln, after burner, energy recovery system and flue gas cleaning system. Total investmentcosts are 45 million US dollars. The performance tests are summarised in this article. The destruction efficiency is more than 99.999% and the dioxins and benzofurans emissions independent of the composition of the hazardous wastes and in absolute term very low. The emission of metals is within the permit conditions, only the mercury emission can exceed permit conditions. The efficiency of the scrubber for mercury amounts to 92%.

Introduction

The AVR Chemie CV was founded as a central facility in the Netherlands to give environmentally acceptable solutions for the final treatment of hazardous wastes. The main activities are secure landfill and

ISWA PROCEEDINGS
VOLUME 1 ISBN 0-12-058451-4

high temperature incineration. Shareholders of AVR Chemie are 45% eight multinationals, 45% city of Rotterdam and 10% national government.

At the moment a marketing study is being carried out to catalogue the market of hazardous wastes in the Netherlands. Special attention is being given to future effects on quantity and quality. It is expected that the quantities of hazardous wastes will reach a maximum of 600.000 ton per year in the period from 1990-1995. After this period a decrease of quantity and quality will take place.

Although in the last three years the AVR Chemie has invested f 120 million (65 million US dollars) in new projects (rotary kiln incineration and secure landfill), a new investment program is being prepared in order to fulfil the future market needs for the treatment of hazardous wastes.

In this article, the performance tests of the newest incineration plant of AVR Chemie CV are summarised. The rotary kiln incineration started operation end of 1986.

Description of the incineration plant

An outline of the incineration plant is given in figure 1.
End of 1986 the incineration plant was put into operation. The incineration plant is entirely computerised and consists of a rotary kiln, after burner, energy recovery system and flue gas cleaning system (electrostatic precipitator and wet scrubber).

The rotary kiln has a capacity of 40.000 tons of hazardous waste per annum. Wastes can be brought into the incinerator by means of lances (liquids and sludges), grabs (solids) or a barrel supply system. The combustion temperature in the kiln is about 1200°C. In the after burner the minimum temperature is 1000°C and is raised to a level of 1200°C during incineration of wastes containing chlorinated aromatic compounds (e.g. PCB's). Per year 10.000 different types of wastes are incinerated.

The total investment of the plant amounts to f 86 x 10^6 (ca. 45 million US dollars). The project management was executed by Akzo Engineering and the incineration plant was a turnkey delivery by a consortium W+E Umwelttechnik AG/HCG.

Programme for the performance tests

The object of the programme is the implementation of both guarantee measurements and additional measurements. The guarantee measurements have to show the efficiency of the incineration, the flue gas cleaning and the waste water purification. The results are used for testing guarantee terms of the contractor and environmental permit conditions. Additional measurements have to enlighten the formation of dioxins (PCDD) and benzofurans (PCDF) and are also meant for showing the distribution of metals over the various process outlets.

Incineration conditions during the various periods

Period	Incineration conditions
1	charge 60% 1000°C in after burner incinerated wastes: "standard" *)
2	charge 100% 1000°C in after burner incinerated wastes: "standard" *)
3A	charge 130% 1200°C in after burner incinerated wastes: "standard"
3B	charge 130% 1200°C in after burner incinerated wastes: "standard" + wastes with high contents of PCB's and chlorin[illegible] benzenes
3C	charge 130% 1200°C in after burner incinerated wastes: "standard" + wastes with high contents of chlorophenols
4A	charge 130% 1200°C in after burner incinerated wastes: "standard" + high contents of Pb and Zn
4B	charge 130% 1200°C in after burner incinerated wastes: "standard" + high contents of Pb, Zn and Cl
5	charge 100% 1000°C in after burner incinerated wastes: "standard" + wastes with high contents of various metals

*) "standard": wastes of which the chemical composition does not involve any specific dosage limitations

In period 3 the destruction efficiency is determined. Furthermore a comparison was made of the formation of PCDD/F with and without the combustion of chlorinated aromatic compounds. During period 5 wastes with high metal contents were incinerated and the distribution of 13 metals over the plant was investigated.

The entire programme was executed by TAUW Infra Consult in co-operation with the Rheinisch-Westfahlischer TÜV (Essen, FRG). The University of Tübingen provided the PCDD/F analysis.

Main results

Testing conditions

The performance of the incineration plant was tested under the most unfavourable incineration conditions. In this way, it was possible to examine the flexibility of the plant. For instance the testing of the destruction efficiency was carried out under maximum plant charge combined with a batch dosage of the wastes containing chlorinated aromatic compounds. Because of a large quantity of these (oil-like) wastes in each barrel, the oxygen percentage in the combustion gases was greatly reduced for a short period each time after dosing a barrel.

Comparison with permit conditions

Concentrations (at standard conditions and 11% O_2) in flue gases and comparison with permit conditions

Pollutant	Permit conditions	Average over testing peri[od]
Dust in mg/m3 (wet) at 11% O_2	50	13
HF idem	2	0,6
HCl idem	100	<10
Pb in mg/m3 (dry) at 11% O_2	1,5	< 0,3
Zn "	50	< 3
Cd "	0,05	< 0,01
Hg "	0,05	0,10
Cn "	2	< 0,1
Cr "	5	< 0,2
Ni "	1	0,33
Sn "	20	< 0,2
Co "	1	< 0,02
Ag "	0,1	< 0,02
As "	0,5	< 0,002

Concentrations in effluent and comparison with permit conditions

llutant	Permit	Average over testing periods
D in mgO_2/l	200	96
rticulate matter in mg/l	50	31
CL "	0,1	0,002
"	0,1	0,003
"	2	0,04
"	1	0,03
"	3	0,05
"	0,01	0,03
"	3	0,4
"	3	0,1
"	3	0,01
"	3	0,7
"	1	<0,01
"	1	0,05

The emission of mercury is a problem. The average removal efficiency of the scrubber over the testing periods was 92%. It must be noted that during the testing periods wastes with a very high Hg content were incinerated. Additional experiments have shown that the mercury content in effluent can comply with the permit conditions by means of an additional and very well controlled dosage of chemicals specifically for the deposition of Hg.

Destruction efficiency

The contractor guarantees a destruction efficiency for chlorinated aromatic compounds of at least

99.999%. Destruction guarantees were calculated by comparing the total amount of a component in all the process outlets (slag, fly ash, waste water and flue gases) with the amount in the incinerated wastes. The destruction efficiency for chlorophenols, chlorobenzenes and for the sum of 7 PCB's meets the requirement (η>99.999%).

Dioxines and benzofuranes

Total outgoing PCDD/F quantities in the testing periods 3B (PCB's and chlorobenzenes) and 3C (chlorophenols) are larger than in 3A ("standard"). The larger total outgoing PCDD/F quantities in 3B and 3C compared with 3A can be traced entirely to larger PCDD/F quantities in the fly ash of the electrostatic precipitator.
The emission of PCDD/F towards atmosphere appears to be independant from the contents of PCB's, chlorobenzenes, chlorophenols or PCDD/F in the incinerated wastes. The emission is very low for each of the three considered periods. The emission to atmosphere amounts:

Sum of PCDD (4-8)	1 mg/h
Sum of PCDF (4-8)	4 mg/h
Toxicity equivalents	0,01 mg/h

Distribution of metals over the process outlets

For mercury, only 1% is recovered in slag and fly ashes. From the remaining 99% about 92% is removed by the scrubber. Approx. 7% leaves the plant by means of the stack.

For Sn at least 94% is recovered in slag and fly ashes. Subsequently a certain part of the remainder is removed in the scrubber.

For each of the remaining 11 metals at least 95% is recovered in slag and fly ashes. The average recovery in slag and fly ashes amounts to an even 99%.

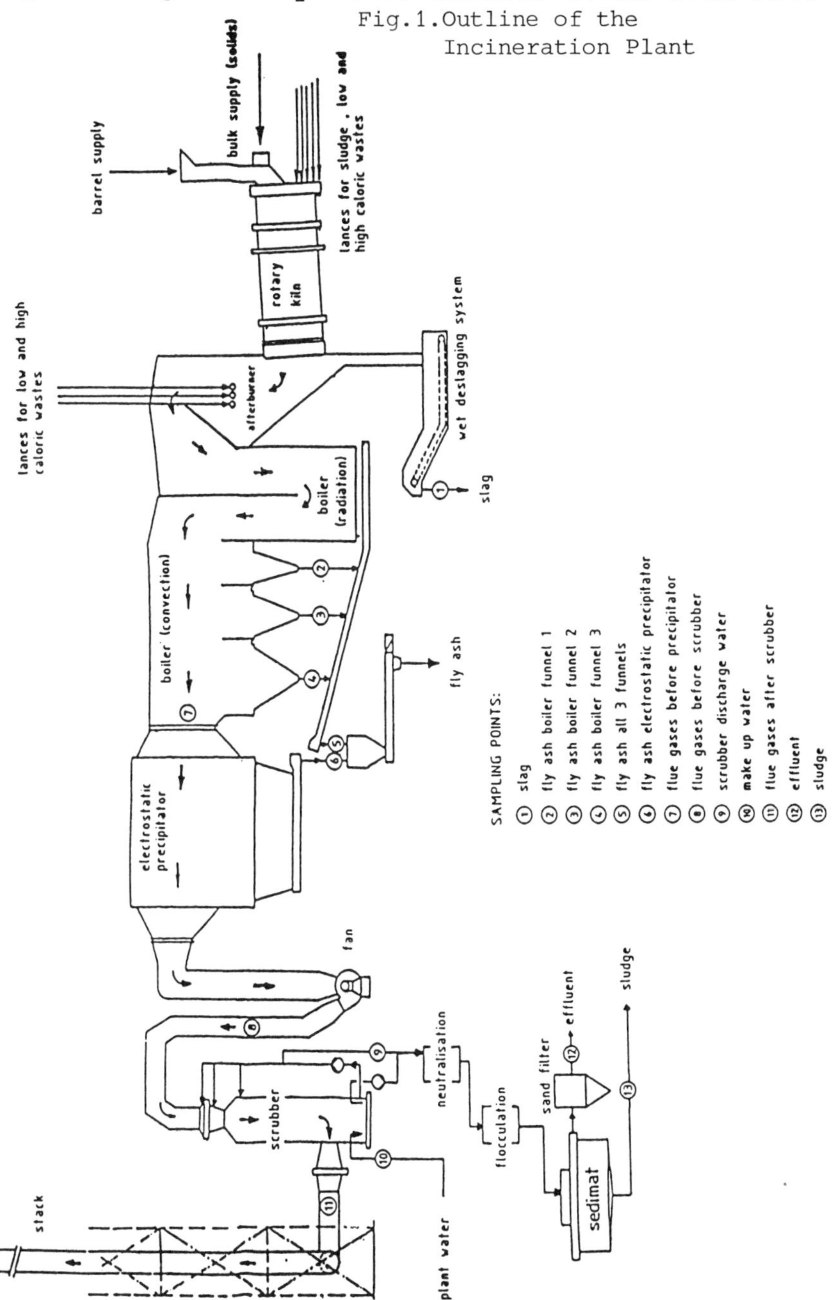

Fig.1.Outline of the Incineration Plant

COMPARISON BETWEEN LEACHING TESTS PERFORMANCES AND TOXIC WASTE BEHAVIOUR IN LANDFILL DISPOSAL.

Misiti A., Rolle E., Gavasci R., Majone M., Sirini P..
University of Rome "La Sapienza".

1. INTRODUCTION

The identification of the best methods for the disposal of hazardous wastes represents one of the subjects for research that in Italy today receives great attention. This is largely determined by the fact that Italian territory, with its elevated production of toxic waste, unlike other countries is lacking in adequate sites for correct disposal. This in the long run could cause serious damage to the environment. A law recently passed in Italy for the disposal of solid wastes allows refuse containing toxic metals in elevated concentration to be deposited directly on the soil as long as these wastes have small leaching characteristics and the soil has the right requisites.

Technical regulations up to now set tests which evaluate the behaviour of wastes in leaching; they also define certain principles both for the choice of sites and the evaluation of suitable soils. However, there is a great deal of perplexity as to the real ability of the tests to predict the behaviour of refuse during leaching action in conjunction with leachate rainfall. The technical regulations already in act are not considered sufficient to guarantee the protection of the natural environment. This problem is the subject of the present research, which has been divided into the following stages:

a. a laboratory study of the phenomena of leaching; particular attention has been given to the influence of environment variables such as pH, contact time, presence of complexing substances, etc.;
b. a further laboratory study of interaction between soil and leachate for different physical and chemico-physical characteristics during the contact phases;
c. the development of an overall model of behaviour in disposed waste according to hydraulic and chemico-physical sub-models. The latter are defined according to the results of the laboratory tests;
d. verification and validation of the models by means of continual tests on pilot plant;
e. analysis, on pilot scale, of the modification to the physical, chemico-physical and geotechnical characteristics of the leaching on leached soils.

The tests conducted at the moment are interested in the first two aspects. Both for the waste and the soil, reference is made to real samples so as to obtain, during the primary stages of the research, the best possible information. This facilitates research on simulated samples, vital for the

ISWA PROCEEDINGS
VOLUME 1 ISBN 0-12-058451-4

correct modelling of the phenomena.

2. EXPERIMENTAL STAGE

The following refuse were studied:

a. sludge resulting from treatment of waste water from tanneries;
b. ashes from the incineration of sludge from municipal waste water treatment.

The chemico-physical characteristics of this refuse are reported in Tables 1 and 2.

Tests for the study of leaching phenomena were carried out following the experimental procedure established by the Italian law (acetic acid EPA test).

The tests made provision for these alternatives:

a. periodical renewal of the liquid phase in contact with the same solid phase (refuse).
b. periodical renewal of the solid phase (refuse) in contact with the same extracting solution;

TAB. 1 - TANNERY SLUDGE - CHARACTERISTICS.

PARAMETER	VALUE
MOISTURE	79%
DRY RESIDUE	21%
VOLATILE SOLIDS	45%
COD	600 mg/g dry sludge
CALCIUM	152.4 " "
IRON	47.4 " "
MAGNESIUM	4.2 " "
SODIUM	14.4 " "
POTASSIUM	0.9 " "
CHROMIUM (III)	19.5 " "
ALUMINUM	3.0 " "

TAB. 2 - INCINERATOR ASHES - CHARACTERISTICS.

PARAMETER	VALUE
MOISTURE	30%
ALUMINUM	46.80 mg/g dry weight
CHROMIUM	0.11 " "
IRON	19.00 " "
SODIUM	20.00 " "
CALCIUM	70.00 " "
LEAD	16.20 " "
CADMIUM	0.03 " "
ZINC	4.59 " "
NICKEL	0.12 " "
COPPER	0.77 " "

For the study of interaction between leachate and soil, the leachate produced according to the experimental procedure established by the leaching test was brought into contact with two different types of soil: a clay loam taken from the bottom of an operating landfill and montmorillonite.

3. RESULTS

Leaching tests. Leaching tests were carried out to verify legal procedure tests in environmental situations different from those to which the standard procedure refers. In particular, the following influences were analysed:

a. differences in pH above and below the value of 5 imposed by the standard procedure of maintaining this value during contact with solid phases and liquid phases;
b. periodical renewal of solid phase (refuse) in contact with the liquid phase. This simulates what happens in the refuse disposal when leachate, formed by coming into contact with the superficial layers of refuse, comes into contact with the lower layers;
c. periodical renewal of the liquid phase in contact with the solid phase (refuse). This bears witness to the ability of the leaching test to identify the leaching fraction of refuse by means of a single contact operation.

The results obtained, as regards tannery sludge, are shown in Tables 3 and 4. The state of leaching in the different cases was evaluated by means of chromium, which constitutes the metal present in the highest concentration in tannery sludge. Environmental danger in this sludge depends largely on this. The results obtained as regards waste matter are shown in Tables 5-8.
In the case of incinerator ashes, the process of leaching in the various stages was followed by analysing the principal toxic metals present i.e. chromium, lead, cadmium, nickel, copper, zinc and aluminum.

Tests for Soil-Leachate Interaction. As regards the soil leachate interaction test, the data was limited to the system composed of leachate from tannery sludge and the two soil samples: clay loam taken from the bottom of an operating landfill and montmorillonite.
The complete picture of the tests carried out on constant pH=4.5 (tests are being carried out at present on a higher pH) are shown in Table 9.
The initial pH of the solutions in samples 1, 2, 4 and 5 was altered before contact to pH=4.5.
The various tests proceeded along different lines:

a. 1 and 4: verify the amount of soil interaction with a solution of chromium in nitric acid;
b. 2 and 5: verify role played by other cations (calcium, sodium, iron) in the solution in contact with the soil compared with chromium-soil interaction (in this case too, acidity is ensured by means of nitric

acid). This as regards previous tests.

c. 3 and 6: verify the part played by organic substances contained in the real leachate of tannery sludge on chromium-soil interaction and comparing it to previous tests.

A complete picture of the results obtained is shown in Table 10.

4. ANALYSIS OF RESULTS

Leaching tests. Tests carried out on tannery sludge bring us to the following conclusions:

1. consistent extractions of chromium are obtained only by tests conducted on constant pH=4.5. The amount of acid added to ensure such conditions is even now in such proportion as to reasonably expect that this is an extreme condition, rarely found in landfill;
2. In the tests renewing the solid phase, we can see how the leachate renews chromium at each extraction and the quantity extracted the second time is regularly higher than that of the first extraction. This testifies that a dissolution of some compounds initially found in the sludge tends to increase the leaching capacity of the liquid phase;
3. Referring to the tests carried out renewing the liquid phase, we can confirm that, with the exception of the extreme conditions produced by pH at 4.5, the extraction of leaching fractions of refuse is more or less complete at the first test.

As far as regards incinerator ashes, what has been stated above for the test on pH at 4.5 is valid in this case as well.

TAB. 3 - LEACHING TESTS - SOLID PHASE RENEWAL: CHROMIUM DETERMINATION.

pH	Extr. N°	$pH_{fin.}$	wet weight sludge (g)	CH_3COOH m.eq./100g	Extracted Cr Conc. (mg/L)	Extraction %
	1	4.51	20.00	420	13.2	7.8
	2	4.57	17.65	340	31.5	10.6
4.5	3	4.63	15.51	337	43.5	7.0
	4	4.55	11.95	368	62.5	11.1
	5	4.55	9.12	373	90.0	16.1
	1	5.16	20.00	230	0.5	0.3
5.0	2	5.19	16.70	204	2.2	1.0
	3	5.05	12.90	264	3.0	0.5
	1	5.85	20.00	200	0.1	---
5.0	2	5.74	18.50	200	0.3	0.1
IRSA	3	5.66	16.30	200	1.2	0.6
	1	5.58	20.00	148	0.1	---
5.5	2	5.59	15.30	150	0.1	---
	3	5.81	11.60	120	0.1	---

TAB. 4 - LEACHING TESTS - LIQUID PHASE RENEWAL: CHROMIUM DETERMINATION.

pH	Extr. N°	$pH_{fin.}$	wet weight sludge (g)	CH_3COOH m.eq./100g	Extracted Cr Conc. (mg/L)	Extraction %
4.5	1	4.51	20.00	420	13.2	5.9
	2	4.55	20.00	10	4.0	1.9
	3	4.64	20.00	20	2.0	1.0
	4	4.61	20.00	---	0.8	0.4
	5	4.52	20.00	3	0.6	0.3
5.0	1	5.16	20.00	230	0.5	0.2
	2	5.16	20.00	10	0.1	---
	3	5.03	20.00	---	0.1	---
5.0 IRSA	1	5.85	20.00	200	0.1	---
	2	5.04	20.00	22.5	0.2	0.1
	3	4.99	20,00	2.5	0.3	0.1
5.5	1	5.58	20.00	148	0.1	---
	2	5.57	20.00	20	0.1	---
	3	5.78	20.00	---	0.1	---

For the other aspects relating to the chemical behaviour of different metals, quite varied indications emerge from the experimental tests.

In lead, for example, we notice that the renewal of solid phase determine stronger and stronger extraction in relation to the complexing power exerted on this metal by the ion acetate, on which the liquid phase is progressively enriched.

A contrary process is seen with copper, while nickel and chromium do not appear to have significant reactions in the renewal of the solid phase.

As for the renewal of the liquid phase, we notice a leaching of nearly all the metals even in later tests. This would indicate, in this case, a poor capacity in the standard procedure to determine the real leaching fraction. An explanation of this behaviour could lie in the chemical inertia of the metals, present in the waste matter exclusively under form of oxides. This might determine a lower kinetic of dissolution.

Interaction tests. As we mentioned above, interaction tests are at the moment limited to those carried out at pH=4.5 on tannery sludge. Even within these limits, the results obtained indicate several conclusions of noteworthy interest.

For instance chromium is almost completely adsorbed in the test with simulated leaching; on the other hand, hardly anything results as regards the adsorbing capacity, either in clay loam soil or in montmorillonite when in the solution organic substances and acetic acid are present. Evidently they compete with solid matter in order to bind the chromium.

TAB. 5 - LEACHING TESTS - INCINERATOR ASHES: pH = 4.5.

LIQUID PHASE RENEWAL															
Extr. Nº	Cr	%	Pb	%	Cd	%	Ni	%	Cu	%	Zn	%	Al	%	$pH_{fin.}$	CH_3COOH m.eq./100g
1	0.4	12.5	36.0	7.5	0.8	95	0.7	20.0	3.5	15.5	165.0	95.0	132.0	10.0	4.47	576
2	und.	--	3.0	1.0	0.1	--	0.4	13.0	1.0	5.0	0.7	--	10.0	1.0	4.50	22
3	und.	--	2.0	0.5	0.1	--	0.3	11.0	1.0	5.0	0.7	--	4.5	0.5	4.50	8.5

SOLID PHASE RENEWAL																
Extr. Nº	Cr	%	Pb	%	Cd	%	Ni	%	Cu	%	Zn	%	Al	%	$pH_{fin.}$	CH_3COOH m.eq./100g
1	0.4	12.5	36.0	7.5	0.8	95	0.7	20.0	3.5	15.5	165.0	95.0	132.0	10.0	4.47	576
2	0.8	12.0	102.0	21	1.5	70	1.0	11.0	5.5	8.0	215.5	28.4	218.0	6.0	4.52	480
3	1.2	12.5	133.0	28	2.0	86	1.7	17.0	8.3	13.0	353.0	77.9	365.0	11.0	4.50	560

TAB. 6 - LEACHING TESTS - INCINERATOR ASHES: pH = 5.0.

LIQUID PHASE RENEWAL																
Extr. Nº	Cr	%	Pb	%	Cd	%	Ni	%	Cu	%	Zn	%	Al	%	$pH_{fin.}$	CH_3COOH m.eq./100g
1	0.2	4.7	19.5	3.7	0.8	81	0.4	10.7	2.5	9.9	103.8	69.0	32.3	2.1	4.99	272
2	und.	--	2.6	0.5	und.	--	0.2	4.3	0.8	3.1	12.8	26.0	2.9	0.2	4.97	19
3	und.	--	1.4	0.3	und.	--	und.	--	0.6	2.6	3.8	10.0	1.4	--	4.99	7

SOLID PHASE RENEWAL																
Extr. Nº	Cr	%	Pb	%	Cd	%	Ni	%	Cu	%	Zn	%	Al	%	$pH_{fin.}$	CH_3COOH m.eq./100g
1	0.2	4.7	19.5	3.7	0.8	81	0.4	10.7	2.5	9.9	103.8	69.0	32.3	2.1	4.99	272
2	0.4	7.5	51.3	6.0	1.5	67	0.9	11.5	3.5	4.1	193.3	60.0	125.7	6.1	4.97	284
3	0.7	7.2	84.2	6.2	1.9	48	1.3	10.7	4.9	5.5	278.3	57.0	206.6	5.3	4.99	284

. concentrations are expressed in mg/L; und. = undeterminable.

TAB. 7 - LEACHING TESTS - INCINERATOR ASHES: pH = 5.0 IRSA.

Extr. Nº	Cr	%	Pb	%	Cd	%	Ni	%	Cu	%	Zn	%	Al	%	$pH_{fin.}$	CH_3COOH m.eq./100g
LIQUID PHASE RENEWAL																
1	und.	--	24.2	5.1	0.6	69	0.2	6.6	2.1	9.4	61.3	46.0	4.7	0.3	5.80	200 (max)
2	und.	--	20.0	3.8	0.1	31	0.1	2.6	1.4	5.9	13.2	16.0	4.4	0.2	5.03	34
3	und.	--	0.9	0.2	und.	--	und.	--	0.2	0.7	2.5	3.0	und.	--	6.02	8
SOLID PHASE RENEWAL																
1	und.	--	24.2	5.1	0.6	69	0.2	6.6	2.1	9.4	61.3	46.0	4.7	0.3	5.80	200 (max)
2	0.2	5.0	28.9	1.0	0.9	40	0.4	5.4	2.9	3.6	138.1	57.0	18.9	1.0	5.56	200 (max)
3	0.3	3.7	38.2	2.0	1.2	32	0.6	5.4	4.0	5.0	207.1	51.0	23.2	0.3	5.55	200 (max)

TAB. 8 - LEACHING TESTS - INCINERATOR ASHES: pH = 5.5.

Extr. Nº	Cr	%	Pb	%	Cd	%	Ni	%	Cu	%	Zn	%	Al	%	$pH_{fin.}$	CH_3COOH m.eq./100g
LIQUID PHASE RENEWAL																
1	und.	--	3.8	0.8	0.5	58	0.2	5.3	1.3	5.6	50.5	37.0	3.6	0.2	5.51	162
2	und.	--	0.6	0.1	und.	--	und.	--	0.4	1.6	9.8	10.0	2.9	0.2	5.50	22
3	und.	--	0.5	--	und.	--	und.	--	0.4	1.6	7.2	8.0	2.5	0.1	5.51	22
SOLID PHASE RENEWAL																
1	und.	--	3.8	0.8	0.5	58	0.2	5.3	1.3	5.6	50.5	37.0	3.6	0.2	5.51	162
2	und.	--	5.5	0.3	0.9	41	0.3	4.4	2.2	3.9	86.1	25.0	4.7	0.1	5.50	143
3	0.1	3.0	21.1	3.2	1.2	32	0.5	3.6	2.8	2.4	116.8	22.0	7.6	0.2	5.50	170

. concentrations are expressed in mg/L; und. = undeterminable.

The most difficult interpretation is the effect of other cations in relation to the special experimental conditions under which the tests were conducted. A more complete picture could result when the programme of planned tests has been terminated.Because of the many metals present and for the unusual thermodynamic and kinetic behaviour, incineration ashes is capable of giving us precious indications about the influences of the tests and the simulations.Proposals for modifications can then be directed along the best possible lines.

TAB. 9 - SOIL-LEACHATE INTERACTION. SAMPLE CHARACTERISTICS.pH = 4.5 CONSTANT.

TEST Nº	CHARACTERISTICS
1	60 g of clay loam-600 cc of solution with 10 mg/L Cr
2	60 g of clay loam-600 cc of solution with 10 mg/L Cr+other cations
3	60 g of clay loam-600 cc of leachate from leaching test at pH=4.5
4	12 g of montmorillonite-600 cc of solution with 10 mg/L Cr
5	12 g of montmorillonite-600cc of sol. with 10 mg/L Cr+other cations
6	12g of montmorill.-600cc of leachate from leaching test at pH=4.5

Samples 4,5 and 6 were placed in contact 2 days later than samples 1,2 and 3.

TAB. 10 - SOIL-LEACHATE INTERACTION. VARIOUS pH CONSTANT SOLUTIONS: pH = 4.5.

TEST	CHROMIUM					IRON				
Nº	t=0	4d	8d	12d	16d	t=0	4d	8d	12d	16d
1	11.0	0.1	0.1	0.1	0.1	und.	2.7	3.0	4.2	3.6
2	8.0	und.	0.2	0.2	0.2	270	5.6	2.7	2.0	1.8
3	9.5	10.8	11.4	12.1	12.2	270	350	360	480	590
4	10.2	und.	0.2	0.2	0.2	und.	und.	und.	und.	und.
5	8.0	und.	und.	und.	0.3	270	120	100	93	82
6	9.5	6.3	6.1	6.9	7.0	270	210	205	170	160

TEST	CALCIUM					SODIUM				
Nº	t=0	4d	8d	12d	16d	t=0	4d	8d	12d	16d
1	und.	11,000	10,700	10,700	10,000	und	32	28	27	23
2	1,200	16,000	13,900	10,200	10,000	300	340	250	220	195
3	1,400	12,000	10,200	10,000	12,400	120	130	140	140	140
4	und.	200	240	260	260	und	160	195	200	200
5	1,200	2,090	1,500	1,500	1,500	300	630	410	410	405
6	1,400	1,600	1,870	1,620	1,620	120	280	360	350	340

concentrations are expressed in mg/L; und. = undeterminable.

REFERENCES

- E.P.A. (1984). Solid waste leaching procedure manual, Mun. Env. Res. Lab., Cincinnati, Ohio, March..
- Rolle, E. and Gavasci, R., (1987). Leaching tests of pollutants from leachate to soil, in Proc. of the Int. Sym. on Sanitary Landfill, Cagliari.
- Berbenni, P. (1984). Leaching tests for toxic metals, Inquinamento, Oct..
- Griffin, R.A., Shimp, N.F. (1978). Attenuation of pollutant in municipal landfill leachate by clay minerals, Illinois St. Geological Survey, Urbana.

INVESTIGATION OF LEACHING FROM SOLIDIFIED WASTES

J.A. Stegemann, P.L. Côté
Environment Canada, Burlington, Ontario, Canada

Introduction

A study was conducted by Environment Canada, in cooperation with the United States Environmental Protection Agency, Alberta Environment, and fifteen companies, to investigate the suitability of fifteen methods for characterizing the intrinsic properties of solidified wastes (1), and develop a database of solidified waste properties achievable with state-of-the-art technology.

One of the tests applied to the solidified wastes was a dynamic leach test which was adapted from an American Nuclear Society Test (2). A cylindrical specimen was immersed in distilled water, which was renewed at intervals over nine days. The amount of contaminant leached in each interval was used to calculate an apparent diffusion coefficient which provides a measure of the contaminant's mobility. In this paper the results from testing are summarized, and their validity is examined.

Theory

The dynamic leach test is based on solution of a diffusion model (3) to yield the following equation:

$$\left(\frac{\sum a_n}{A_o}\right)\left(\frac{V}{S}\right) = 2\left(\frac{D_e}{\Pi}\right)^{1/2} t_n^{\ 1/2} \qquad \text{Equation 1}$$

where a_n = contaminant loss during leaching period n, mg
A_o = initial amount of contaminant in the specimen, mg
V/S = volume to surface area ratio of the specimen, cm
t_n = time at the end of leaching period n, s
D_e = apparent diffusion coefficient, cm^2/s.

This solution is based on the following assumptions:

1) the mobility of a contaminant is limited by diffusion,

2) the specimen behaves as a semi-infinite medium,

3) the surface concentration of a contaminant is zero, and

4) the equilibrium between the immobile form of a contaminant (C_{im}) and its mobile form (C_{mo}) is controlled by a linear adsorption isotherm; $C_{im}/C_{mo} = k$, a constant.

This latter assumption is often true for adsorbed organic compounds. Metals are usually immobilized as ionic compounds, with a constant, very low, mobile concentration controlled by a solubility product. Since the concentration of the immobile form of a contaminant changes very little over the course of a test, the ratio C_{im}/C_{mo} is constant for a particular solidified product.

The apparent diffusion coefficient in Equation 1 is a function of the molecular diffusion coefficient of the contaminant, the tortuosity of the sample matrix, and the coefficient of adsorption, k. Thus the test measures both physical and chemical retardation of leaching.

Approach

Five wastes were sent to each of the fifteen participating companies for solidification by their proprietary processes. The test methods were applied to the resulting solidified products. The dynamic leach test was carried out in two laboratories, for solidified products from three wastes. The waste types, the number of solidified products tested for each, and their contaminant concentrations are listed in columns 1, 2, 3, and 4 of Table 1.

The dynamic leach test was conducted by immersing a 117 cm^3 specimen of solidified waste in distilled water at a leachant volume to specimen surface area ratio of 10 cm. The leachant was renewed at intervals which were calculated based on a diffusion model such that the expected concentration in the

TABLE 1 - Wastes, Contaminant Concentrations and Apparent Diffusion Coefficients

Waste	Number of Products Tested	Contaminant	Range of Contaminant Concentrations in the Different Solidified Products (mg/kg, w/ww)	Number of pD_e's Calculated (see Note)	Range of pD_e's in the Different Solidified Products	Median pD_e
WTC Synthetic Solution	11	Arsenic	500 to 1700	11	9.3 to 14.7	11.5
		Cadmium	1000 to 3300	3	11.9 to 16.1	14.0
		Chromium	300 to 1100	8	10.5 to 13.3	12.4
		Lead	1800 to 5800	9	9.3 to 14.6	12.8
		Phenol	800 to 2100	11	6.3 to 11.9	7.7
Aluminum Coil Plating Sludge	11	Arsenic	10	0	-	-
		Chromium	300 to 900	11	9.8 to 13.2	11.8
		Lead	10 to 40	9	6.9 to 10.4	9.1
		Thallium	5 to 15	0	-	-
		Total Cyanide (TCN)	500 to 1500	11	6.3 to 11.8	8.6
Wood Preservation Waste Soil	12	Pentachlorophenol (PCP)	1900 to 9300	11	7.6 to 10.9	8.6

Note: The pD_e was not calculated if the contaminant concentration in the leachate from more than one interval was below detection. A concentration below the detection limit can be the result of successful immobilization, or a low initial concentration in the specimen, as is the case for arsenic and thallium in the aluminum coil plating sludge.

leachant for each interval was above the analytical detection limit, but not near equilibrium (to satisfy assumption 3). An incremental form of Equation 1 (3) was used to calculate an apparent diffusion coefficient from the fraction leached in each interval.

Results and Discussion

The results of the test for different contaminants are presented as the negative logarithm of the apparent diffusion coefficient (pD_e) in Columns 6 and 7 of Table 1. The mobility of a contaminant increases as the pD_e decreases. It appears that immobilization of cadmium was most successful, followed by chromium, lead, arsenic, and cyanide and pentachlorophenol.

The validity of the assumption of diffusion control was evaluated by plotting the cumulative fraction leached vs the square root of time for each solidified product (where data from each interval was available), since Equation 1 shows a linear relationship between the two. The plots for the different solidified products showed four different trends:

1) diffusion control,
2) initial resistance to leaching, followed by diffusion,
3) initial rapid washoff followed by diffusion control, and
4) linear relationship between the cumulative fraction leached and time.

Examples of these plots are shown in Figures 1 to 3; the observed release mechanisms are summarized in Table 2. It is postulated that the fourth leaching mechanism was observed for arsenic because it was mobilized by reaction of basic calcium arsenite with the carbonates in CO_2 saturated water. Thus the arsenic release rate was controlled by the rate of renewal of CO_2 saturated leachant (4).

Examination of the cumulative fraction leached after the

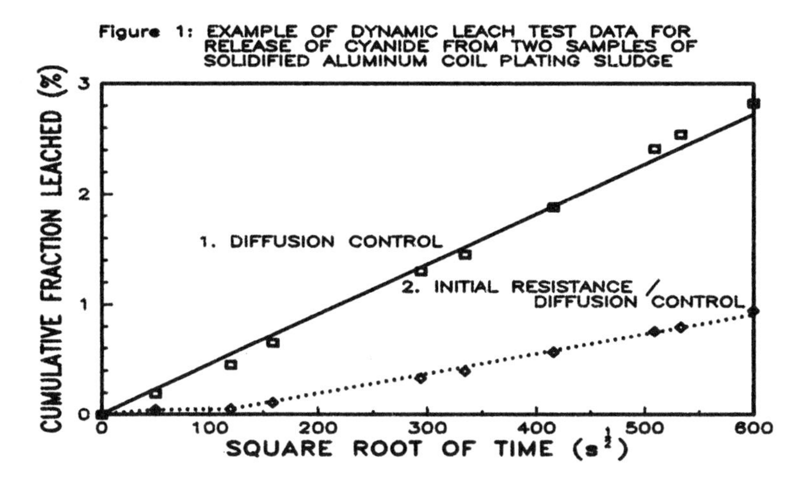

Figure 1: EXAMPLE OF DYNAMIC LEACH TEST DATA FOR RELEASE OF CYANIDE FROM TWO SAMPLES OF SOLIDIFIED ALUMINUM COIL PLATING SLUDGE

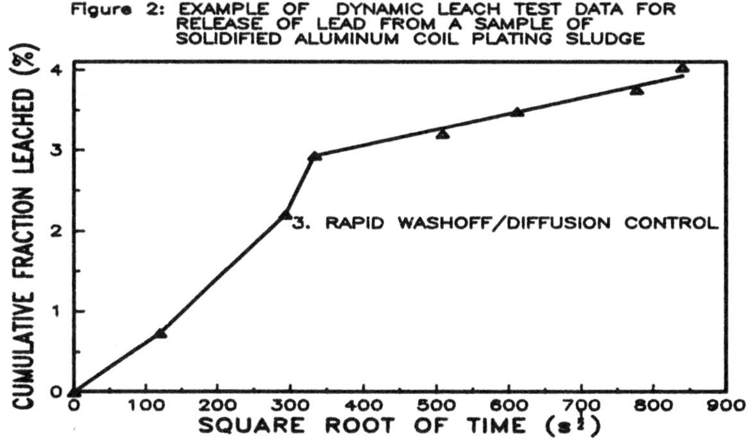

Figure 2: EXAMPLE OF DYNAMIC LEACH TEST DATA FOR RELEASE OF LEAD FROM A SAMPLE OF SOLIDIFIED ALUMINUM COIL PLATING SLUDGE

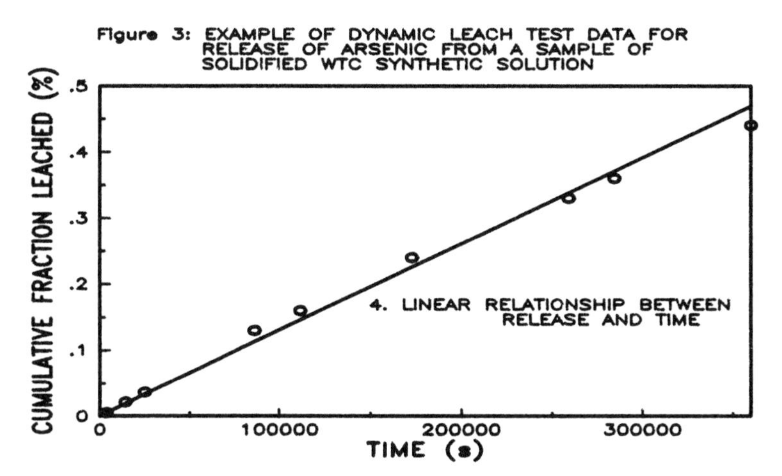

Figure 3: EXAMPLE OF DYNAMIC LEACH TEST DATA FOR RELEASE OF ARSENIC FROM A SAMPLE OF SOLIDIFIED WTC SYNTHETIC SOLUTION

TABLE 2 - Number of Solidified Products Exhibiting Release Mechanism for the Various Contaminants

Contaminant	Diffusion	Resistance/ Diffusion	Washoff/ Diffusion	Linear with Time
Arsenic	0	0	0	11
Cadmium	2	0	0	0
Chromium	18	0	1	0
Lead	3	2	8	0
Phenol	5	4	0	0
Cyanide	6	4	0	0
Pentachlorophenol	8	2	1	0

last interval showed that the majority of specimens do behave as a semi-infinite medium, since the cumulative fraction leached exceeded 20% in only six instances (2). Depending on the contaminant, apparent diffusion coefficients for the different solidified products ranged over three to five orders of magnitude. Comparison of the apparent diffusion coefficients measured in the two laboratories for replicate samples of each of the solidified products showed a difference greater than one order of magnitude (ΔpD_e=1) in only two instances, thus, in general, the dynamic leach test is able to distinguish between solidified products of different qualities.

References

1. Wastewater Technology Centre, Investigation of Test Methods for Solidified Waste Characterization, Environment Canada Report in Preparation, 1988.
2. American Nuclear Society (ANS), "Measurement of the Leachability of Solidified Low-Level Radioactive Wastes", ANSI/ANS 16.1, 1986.
3. Godbee, H.W. and Joy, D.S., "Assessment of the Loss of Radioactive Isotopes from Waste Solids to the Environment, Part 1: Background and Theory," TM-4333, Oak Ridge National Laboratory, Tennessee, 1974.
4. Côté, P.L., Constable, T.W., and Moreira, A., "An Evaluation of Cement-Based Waste Forms Using the Results of Approximately Two Years of Dynamic Leaching", Nuclear and Chemical Waste Management, Vol.7, pp 129-139, 1987.

THE CO-DISPOSAL OF CHEMICAL WASTES IN TEST CELLS IN HONG KONG: DESIGN DETAILS AND INTERIM EXPERIMENTAL RESULTS

D J V Campbell, Harwell Laboratory,
M P Pugh, Binnie and Partners

Introduction

In 1985 the Environmental Protection Department (EPD) of the Hong Kong Government commissioned Binnie and Partners (HK), in association with Harwell Laboratory, to undertake a study, including laboratory and field experiments, on the co-disposal of chemical wastes with municipal and commercial wastes. Although similar experiments have been undertaken in the UK and elsewhere, the impact of factors such as differences in waste composition and climate, justified the obtaining of locally derived maximum loading rate data. These would then be incorporated in wider study objectives to develop guidelines on acceptable co-disposal practice as part of future waste management and disposal options in Hong Kong. The paper briefly discusses basic principles of co-disposal, summarises laboratory experiments and reviews the construction and interim results of field experiments.

Basic Principles of Co-disposal

It is well known that a variety of physical, chemical and microbial processes occur within landfilled wastes. These assist in the breakdown, retention or elution of inorganic and organic components of wastes, resulting in the 'typical' concentrations of liquid and gaseous products normally observed. It is argued that these same processes may be used beneficially to treat, degrade, precipitate, or otherwise attenuate contaminants contained in and released from selected chemical wastes co-deposited in refuse. Many such contaminants are already present in municipal and commercial wastes and the acceptability of co-disposal should be judged by whether (or not) minimal impacts on leachates and gases are produced. To achieve this, chemical wastes suitable for co-disposal must be carefully defined by type, concentration and quantity. It is also preferable that normal landfill processes are already well established before co-disposal is permitted, to take full advantage of the maximum 'treatment' capabilities provided by other wastes.

Design of Laboratory Experiments

19 columns were constructed to allow study of a wider selection of chemical wastes than was feasible in field experiments. Each column (Figure 1A) contained up to 28 kg of pulverised refuse with chemical wastes added as a discrete layer or irrigated on the surface. Simulated rainfall irrigation (100mm/day) was then continued on the column surfaces for periods of up to 18 months. Sampling facilities allowed leachates to be collected and analysed from the centre funnel or the column base, and gases to be sampled at the column surfaces.

There are limitations inherent in all such small experiments, in particular the limited waste depths available may be insufficient to establish fully methanogenic conditions, resulting in an acid pH environment where solubilisation of metals will be more likely. Despite conditions not always being favourable, as would occur in a landfill, substantial or complete attenuation was achieved in all columns, as summarised in Table 2.

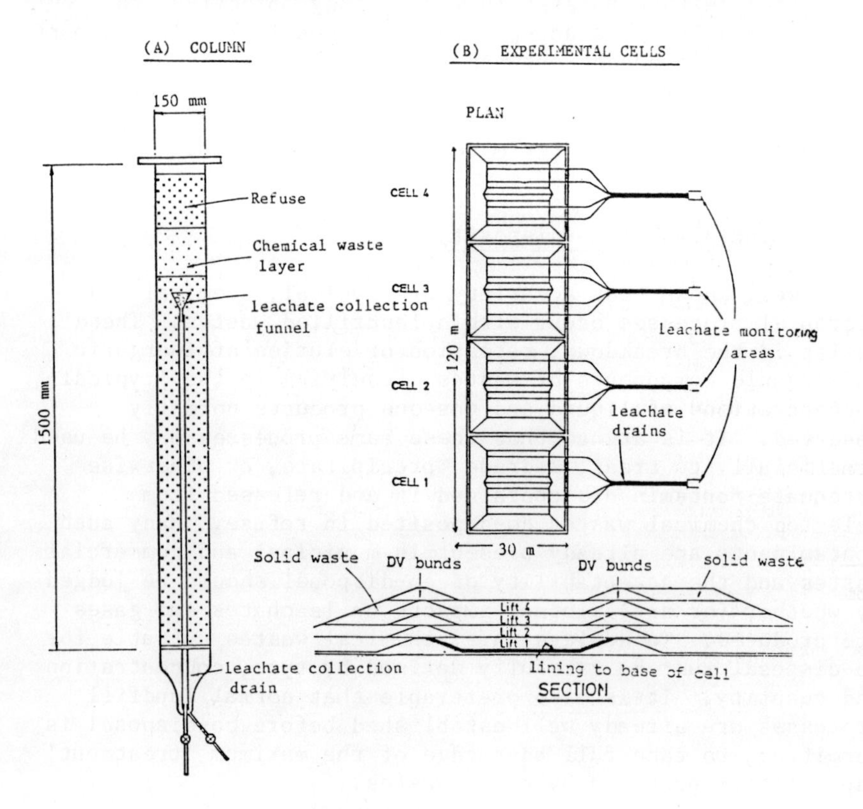

FIG 1 COLUMN AND TEST CELL DESIGN

In the absence of locally established data chemical waste loading rates used in experiments were based on existing maximum recommended rates at UK co-disposal sites. Thus zinc, copper, lead, iron and chromium bearing wastes were limited to 100g metal per tonne of refuse, and 10g t^{-1} for nickel and cadmium wastes. To determine the mass of particular chemical wastes to be used, leaching or solubility tests were undertaken. The maximum concentrations and total mass of readily leachable metal or other contaminant (in acetic acid solution) were measured and used in calculating loading rates, rather than using the obvious, but less relevant, total mass of contaminants in the waste.

Loading rates are also limited by such factors as the absorptive capacity of refuse (e.g. for liquids), or bio-degradability or volatility (e.g. for some organic wastes). Wastes containing significant solvent concentrations should not normally be co-deposited with refuse but there are currently, in Hong Kong, limited alternative disposal options for such wastes. In the summary of chemical wastes used in both laboratory and field experiments (Table 1), halogenated and non-halogenated solvent to refuse ratios (by weight) were arbitrarily selected as 1:100 or 1:200 respectively.

Column*	Chemical Waste	Amount added	Column	Chemical Waste	Amount Added
1	(Control)	-	11	(Control)	-
2	Synthetic metal soln	2 litres*	12	Gasoline sludge	6.9 kg
3	Non-halogenated solvent	300 g	13	Lube oil sludge	250 g
4	Tannery off-cuts	6.2 kg	14	Metal sludge B	7.6 kg
5	Crushed batteries	5.9 kg	15	Varnish and paint	154 g
6	Oil	0.8 kg	16	Printing waste	154 g
7	Paint	1 kg	17	Halogenated solvent	108 g
8	Metal sludge A	3.36 kg	18	Non-halogenated solv	172 g
9	Sewage sludge	6 kg	19	Cyanide and mercury	86 mg(Hg) 216 mg(Cn)
10	Ash (metal oxide)	406 g			
Cell 1A	Sewage sludge (lift 4)	77.13t	Cell 3A	Fuel and lubricating oils	14.05t
1B	Sewage sludge (lifts 2,3,4)	95.70t	3B	Tannery off-cuts	51.34t
2A	Boiler Grit	24.82t	4A	Control	-
2B	Solvent/paint	11.5 t	4B	Control	-

Notes: 1. Each column contained 20-28 kg of refuse.
2. Each half cell contained 1100-1200 tonnes of mixed wastes (excludes inerts/ soils etc) beneath added chemical wastes.
* contained 51.8 g of mixed metal salts.

TABLE 1: Chemical Wastes Used in Column and Cell Experiments

Parameter	Controls(2)	Co-disposal Columns(17)
pH	6.31-6.88	6.31-7.07
Ammoniacal nitrogen	100-2300	190-2300
Sulphate	67-820	34-15000*
Chloride	25-2500	27-4500*
Total Organic Carbon	2100-31000	500-34000
Zinc	0.13-26	1.3-200
Copper	0.01-0.8	0.01-3.5
Lead	0.01-0.18	0.01-1.8*
Chromium	0.01-0.3	0.01-140*
Nickel	0.05-1.2	0.08-120
Cadmium	0.01-0.07	0.01-0.03
Mercury	0.05-0.05	0.05-0.22*
Cyanide	0.01-0.6	0.01-0.28
Toluene	0.01-3.7	0.01-40*
Hexane	0.01-8.8	0.01-29*
Methyl ethyl ketone	0.53-31	0.01-32

(All data except pH in mg/1)
* Either isolated figures or deviation from control in one or two columns only.
- Major deviation from control in several columns.

TABLE 2 Abstract of Selected Parameters Measured in Column Leachate

Construction of Field Experiments and Interim Data

The potential limitations of small scale experiments, including difficulties in extrapolation of results to field conditions made it essential to construct a limited number of small 'co-disposal landfills'. Within these isolated cells natural landfill processes could be expected to occur and allow resultant liquid and gaseous products to be monitored and analysed. In setting criteria for cell design two basic principles were to be met. Firstly typical refuse deposits should be used as encountered at a potential co-disposal site, with no restrictions on normal site practice including operation of waste compaction plant. Secondly the total waste depths should be at least 10m to ensure that normal waste degradation processes would be established, and to provide a reasonable zone within which chemical waste contaminants may be attenuated without short-circuiting to the 'landfill' base.

Four cells were constructed as shown in Fig 1(B), isolated from each other but sub-divided at the base to allow 8 separate experiments. Each whole cell contained upwards of 4500 tonnes of refuse placed in five layers. The base and sides of the initial layer were protected by a liner to permit collection of leachate via 2 drains in each half cell. The plate below shows the initial layer of waste being deposited in one cell. Chemical wastes, as described in Table 1, were added to selected waste layers and the whole block of four cells capped with soils. Leachate quality and flow monitoring was supported by gas and temperature data obtained via probes buried in the wastes. Local climatic and waste settlement data were also recorded. The loading rates used were either controlled by limitations on chemical waste arisings, or by factors previously described for the columns.

Figures 2 and 3 provide interim data on selected parameters over the 18 months monitoring period to date. There have been no adverse impacts detected on landfill processes or leachate and gas composition for any of the six co-disposal half cells when compared with the control cell. If a proposed extension to the study is agreed further chemical wastes will be added to the cells in a further attempt to establish maximum loading rates.

Plate: Initial Placement of Waste in Test Cell

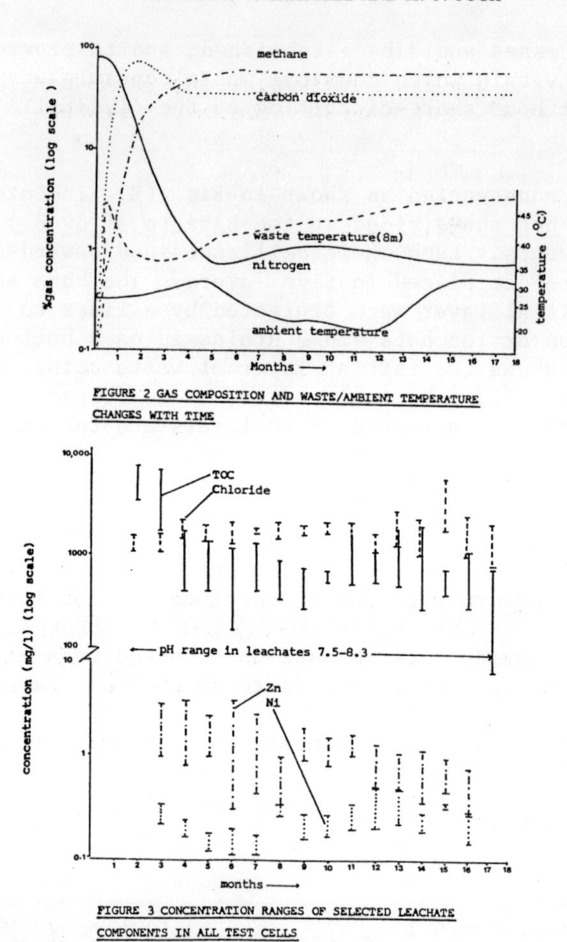

FIGURE 2 GAS COMPOSITION AND WASTE/AMBIENT TEMPERATURE CHANGES WITH TIME

FIGURE 3 CONCENTRATION RANGES OF SELECTED LEACHATE COMPONENTS IN ALL TEST CELLS

Conclusions

The interim results available have supported the proposition that municipal and commercial wastes can assimilate a wide diversity of chemical wastes without giving rise to adverse effects as evidenced by leachate and gas quality.

Acknowledgement

This paper is presented by the kind permission of the Director of Environmental Protection in Hong Kong, and the authors wish to thank staff from EPD, Binnie and Partners and Harwell for their assistance in experimental work.

Hazardous waste - Study of a new sampling procedure for control at the entrance of treatment centres

Jean-Bernard LEROY
Groupe Générale des Eaux
52, rue d'Anjou
75008 PARIS
(France)

INTRODUCTION

Presentation

The admission of a dangerous waste to a treatment centre includes two different stages :

- a preliminary study that may take several weeks, enabling the choice of the most suitable treatment for the waste concerned, both from a technical and an economic point of view. This study is generally materialized in the form of a contract or an exchange of letters mentioning the price of the treatment and the waste delivery rate,
- a second stage consists in checking the characteristics of the waste as it enters the site, at the time of delivery.

It is this second operation that will be addressed in this paper. Some simple arithmetic immediately shows where one of the difficulties lies. In the case of a plant receiving 600 tons of waste daily in truckloads weighing 10 tons (i.e. an annual capacity of 120,000 T) , there would be a total of 60 trucks during the ten opening hours, meaning 6 trucks each hour, or one every ten minutes ! The operator therefore has only a quarter of a hour to decide whether he will accept the load or not, and this decision is irrevocable.
The purpose of this study, performed within the scope of a general plan decided by the EEC, is to collect a small representative sample of one or two decilitres, giving as faithful a picture as possible of the several cubic metres of liquid transported in a tanker truck.

ISWA PROCEEDINGS
VOLUME 1 ISBN 0-12-058451-4

Purpose of study

The most usual method used is a manual process. A small amount is sampled at two or three spots on the tanker, either at the time of unloading or beforehand via a manhole if there is one. Another more scientific technique has also been contemplated i.e., batch sampling at an adjustable rhythm, on the flowing liquid stream. This kind of sampling is done with the help of a diaphragm type, pneumatic recirculating pump, fed by a branch tee on the unloading pipe.

Wastes studied

The experiment was conducted at the waste treatment Centre in **Limay (France)**, operated by **SARP INDUSTRIES** that processes yearly 150,000 tons of hazardous waste from a wide variety of industries, mainly from the Paris region, but not exclusively (fig. 1).

fig. 1 - the treatment centre in Limay (France), operated by SARP INDUSTRIES, processes yearly 150,000 tons of hazardous waste (doc. Generale des Eaux)

The 4 following types were analysed :

a) waste to be incinerated, chlorinated or other wise, but of which the "average" chlorine content constitutes an important parameter.

b) mixed water-hydrocarbon waste of which the ratio of the two elements has a direct influence on the technique upon which the treatment sequence will be based.

c) acids and alkalis used for neutralisation

d) mineral sludges to be solidified.

TEST DEVELOPMENT

Choice of a pump

Some of the conditions under which an industrial waste sampling pump is called upon to operate are :

- the waste may have a high and variable rate of viscosity ;
- there are a great deal of suspended solids, sometimes abrasive or adhesive.

Finally a **diaphragm type pneumatic recirculating** pump was chosen (engineered by IRH (**Institute of Hydraulic Research**, Nancy, France) (fig. 2). This pump performs the three following functions :

- suction by negative pressure
- constitution of the elemental sample at an adjustable frequency ;
- pneumatic discharge of sample.

Operating Method

In current **practice, analyses must be done quickly** and follow either the AFNOR methods or faster techniques derived from the latter.

However, as far as this study is concerned, the analyses were performed the same way on samples collected at plant entrance and on those performed by the automatic samples on the unloading pipe. This was done in order to compare exclusively the sampling methods. These analyses comply with AFNOR standards except for the measurement of hexavalent chromium and cyanides (quick method using capsules of reagent). The results obtained therefore supply information on the difference of the sampling methods. Manual sampling consists of two or three samples of about 250 ml each, collected from the top of the truck, when it is equipped with a manhole, and at different heights when the waste has a tendency to separate into several phases. If there is no manhole, the operator proceeds by tapping from the existing valve.

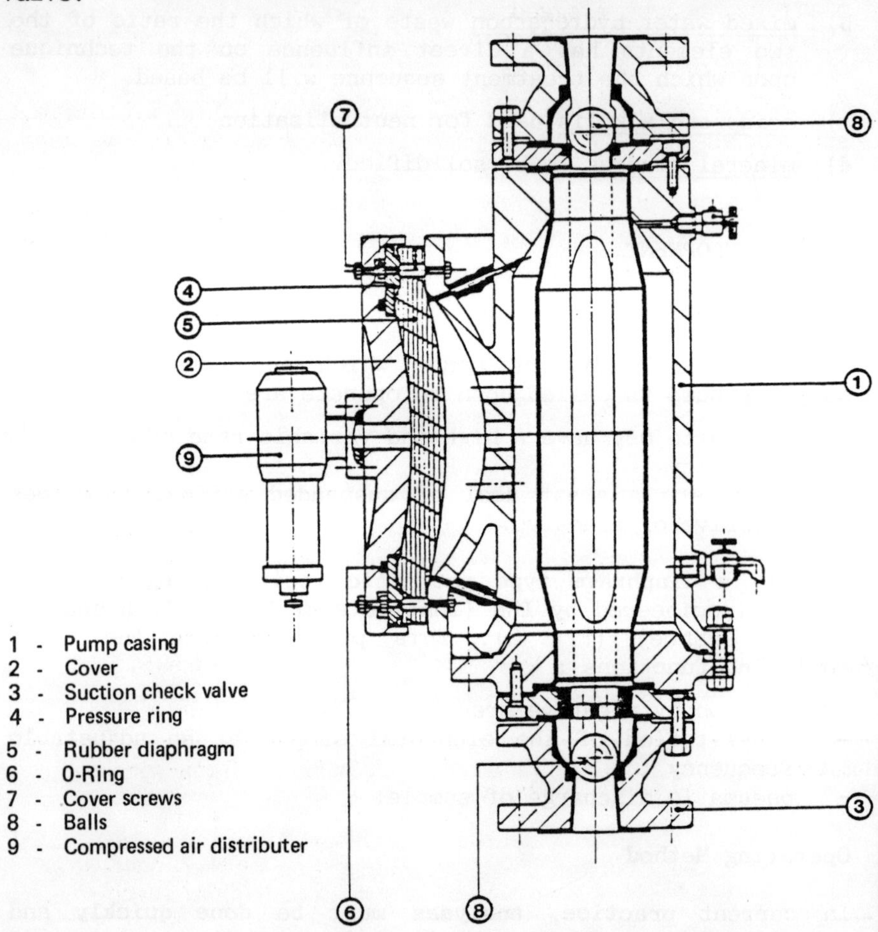

Fig. 2: Scheme of the IRH pump

Sampling with the IRH pump was easy with regard to the liquids to be incinerated, except for a few problems in connection with priming, when priming was necessary. The frequent presence of stones and gravels after unloading showed that our fears concerning the performance of the equipment were justified and it was always necessary to have an operator in attendance.

RESULTS

On practical test details :

1) The IRH pump worked very satisfactorily subject to three cautions with 3 qualifications :
   - . it may become clogged up after the unloading
   - . priming was difficult at times, when working on multiple-phase waste
   - regulation of effluent flow rate was not provided

These are only minor defects inherent in any prototype, and would be easy to adjust if the pump is to be mass produced.

2) We, at last, dispose of 200 litre samples out of which about 1 litre must be sent to the laboratory. This is extra handling that is not always a convenient process !

3) The really significant observations remain limited (Table 1).

Type of waste	Number of sample	Analyses giving distinctly unlike results		Samples for which analysis gives higher values
		Type	Number	
Acids and Alkalis	24	Hexavalent Chromium	4	admission
		TC	4	IRH
Incineration	23	Chlorine	3	3 admission
		NCV	2	1 IRH admission
		Sedimentation :		
		. water	4	1 IRH - 3 admission
		. oil	3	1 IRH - 2 admission
		. sediment	5	4 IRH - 1 admission
Oil - water	23	oil	5	1 IRH
Sludges	26	TC +COD	3	1 admission - 2 IHR
		phenols	2	1 admission

On the advantages of the IRH Pump

1) Despite its rustic appearance, the customary method can be maintained for everyday operations, providing of course that it is used by experienced staff. The sampling must be done by a chemical engineer able to be quickly aware of the possible existence of sudden difficulties and to resolve them.

2) The IRH is an interesting piece of equipment for waste composed of several phases, especially if the latter separate during transportation. It is important not to overlook that these very heterogenous waste products are always hard to analyse at the level of a treatment plant where, we recall, available time is a restrictive factor.

3) The IHR pump does however present a serious drawback : it can operate only during the whole time the tanker is being emptied and this requires a large number of intermediate storage facilities with retaking equipment should this technique be systematically used. And how could the waste be reloaded in the truck if it is impossible to treat ? Despite its indubitable qualities, it is a method that can only be used as an exceptional measure.

4) It appears as a result that automatic sampling during unloading will never be able to replace point sampling in the normal state of the technique, although it can render great services for exceptional controls or general analysis on the initiative of the treatment centre or on administration request - in which case it is only fair that the latter should lover the costs involved - to test a particularly difficult waste substance and readjust the analyses made on admission.

5) A feasible solution might be to place a fast stirrer inside the tanker from which the IHR pump would collect a sample on a by-pass connection without emptying it. This is technically possible but would require a compulsory standardisation of all collection vehicles and this seems a difficult measure to apply in the present economic situation.

The above tests have in any case made it possible to check the value of the IRH equipment and of our chemical engineers' expertise, that they have been able to acquire with time in the form of a real understanding of waste, even enclosed in a tanker.

INFLUENCE OF INDUSTRIAL WASTE LANDFILLS UPON THE SOIL AND GROUNDWATER

A.Jedrczak, E.S.Kempa
College of Engineering, 65-246 Zielona Gora, Poland

INTRODUCTION

The storage of solid waste on ground, although most often in use (~95% of all waste generated in global scale), inspires more and more anxiety. The impact of the landfills reveals in pollution of the surrounding environment, i.e. in pollution of ground- and surface waters, of soil and air. The symptoms of impact are: a change of pH values in water and soil, the increase in soil salinity, higher concentration of micro- and macroelements; sometimes the pollution is manifested by substances normally not present in the nature.

The paper discusses the impact on the environment of two selected storage yards supplied mainly with industrial wastes and studied by the Authors for many years. The first of the stockyards in question has been located under favourable hydrogeological conditions (K=0.1-1 m/d), but it is not properly operated; the second is directly adjacent to the groundwater layer.

DESCRIPTION OF THE STOCKYARDS & SCOPE OF STUDY

First Case Study

The storage yard with an area of 0.05 square km is located in a local cavity. The volume of industrial wastes and municipal refuse which is tipped off there every year, amounts to 124,000 cbm and 21,000 cbm, respectively. Sludges and deposits from electroplating shops, paint and varnish wastes, scales from mechanical cleaning of ship hulls, wastes polluted by oil-bearing compounds and fishing industry waste of high salinity should be considered as hazardous.

ISWA PROCEEDINGS
VOLUME 1 ISBN 0-12-058451-4

The surrounding of the tip is cropland. These are lixiviate brown soils belonging to the good rye-complexes. The terrain slopes to NE. Rain waters flow through drainage ditches into a river. Around the yard up to a depth of 1 meter occur sands of various grain-size distribution and clayey sands; below this thin layer occur clays and clayey sands with interbeddings of dusty clays and silts. Groundwater occur as interbeddings inside the clay layers, as tense water level or as very shallow free level waters inside the sandy-gravel deposits of depressions of the sandy clays roof. The real aquiferous layer one can find at a depth of some 40 meters below the surface.

The operation procedures at the yard do not follow those prescribed for sanitary landfills. The different kinds of wastes delivered are not storaged acc. to their physical and chemical properties but mixed together. No covering layer with inert soil is practiced, though the tip is a source of dust. The heap is now 3m high over the surface. There is no outflow of leachates which fill the bottom of the tip and infiltrate in part to the ground. The composition of these leachates is shown in Table 1; data of leachates from refuse landfills are attached for comparison.

Table 1: Composition of leachates

Parameter	Unit	Case study 1	Case study 2	Leachates from refuse
pH		7.26-7.28	3.05-5.9	7.30-7.85
Alkalinity	meq/L	133-160		84-122
Hardness	meq/L	90-140		78-122
Conductiv.	S/m	8.8-10.3		0.93-1.25
TDS	g/m3	116-124		693-866
COD	gO2/m3	3.5-11.8		0.98-2.49
Chlorides	g/m3	50-61.7	0.24-1.38	1.34-2.63
Sulphates	ppm	155-265	658-1660	32-105
Phenol	ppm	0.34-2.0	-	0.04-1.20
Na	ppm	36.0-51.0		0.85-2.65
Zn	ppm	2.2-4.3	0.02-3.20	0.90-5.2
Hg	ppm	0.5-1.3	0.03-0.19	0.03-2.0
Pb	ppm	0.2-1.3	0.10-0.34	0.10-2.0
Ni	ppm	0.2-2.9	0.21-0.84	0.06-1.6
Cd	ppm	0.07-0.22	0.01-0.09	0.0 -0.1
Cr	ppm	0.01-0.16	1.52-4.84	0.0 -0.11

Table 2: Surface and groundwater characteristics

Parameter	Unit	Pond	Drainage ditch	Piezo-meters
pH		7.7-8.3	6.9-7.6	7.1-7.2
Alkalinity	meq/L	2.7-4.0	4.4-4.8	1.0-1.6
Hardness	meq/L	9.2-10.6	16.5-17.8	14.7-20.8
Conductivity	mS/m	48-52	56-62	65-75
TDS	ppm	332-447	485-692	402-811
COD	ppm	55-98	37-68	11-28
Chlorides	ppm	50-78	55-150	65-75
Sulphates	ppm	31-35	65-72	55-72
Phenol	ppm	0.04-0.42	0.0-0.02	0.0-0.001
Na	ppm	40-50	20-35	17-18
Zn	ppm	0.32-0.65	0.8-1.60	0.41-1.05
Hg	ppm	0.15-0.17	0.14-0.65	0.18-0.63
Pb	ppm	0.08-0.14	0.01-0.05	0.01-0.17
Ni	ppm	0.28-0.42	0.02-0.34	0.02-0.34
Cd	ppm	0.01-0.09	0.01-0.02	0.01-0.02
Cr	ppm	0.04-0.10	0.02-0.09	0.00-0.01

For determining the impact of the wastes storaged upon the soil and water environment,the following elements have been studied:

(1) Waters from a pond, from the drainage ditch as well as from the piezometers situated on the run-off direction and at a distance of some 300 meters from the yard. Results are listed in Table No.2.

(2) Chemical properties of soils around the yard. Soil samples were removed directly from the surface (0.0-0.3m); the first one in each case in a 3m distance from the yard basis. The distance between the sampling points was 50m. The soil samples have been extracted with water (0.1kg/0.3L) and the following parameters determined: conductivity, chlorides, sulphates. Figure 1 shows the plan of sampling and soil salinity. Table 3 shows the chemical characteristics of the soils around the stockyard.

(3) Plants especially growing around the landfill. These were: oat (Avena sativa L.) and pea (Pisum sativum L.).Terms of plants harvesting: overground parts in June, pods of pea in the phase of milk maturity, oat some days after coming to ear. The main elements in plants are in Table 4.

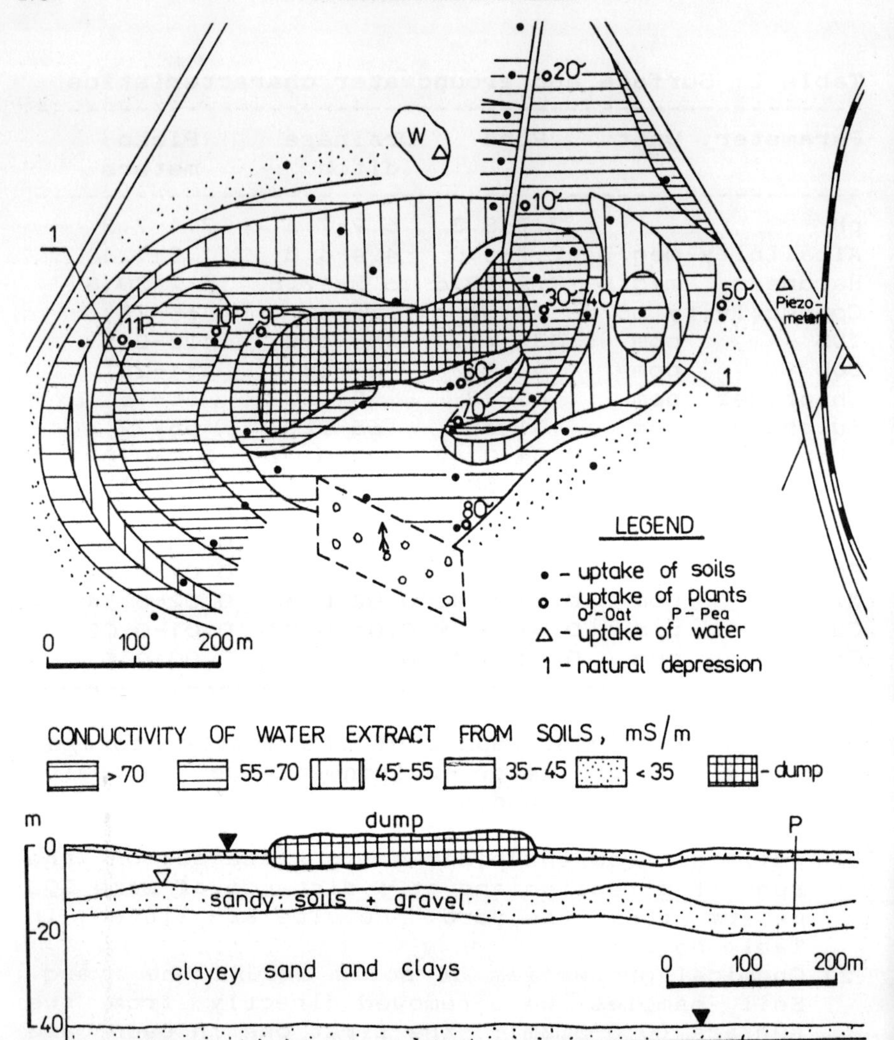

Fig. 1. Salinity of soil around the dump & soil profile.

The second stockyard

The second case study includes a landfill located in a worked out gravel heading with some 20,000 tonnes of wastes tipped of per year wherein 92.4% of precipitation sludges come from a chromium tannery (wastewater treated with FeSO4). The composition of the chrome tanning sludge is as follows: dry matter content 69.9-83.3%, organic matter 50.2-63.5% d.wt., TKN 3.3-4.2% d.wt., Phosphorus 0.13-0.28% d.wt.,oil & grease (as ether

extract) 12.1-14.9% d.wt., K 0.03-0.22% d.wt., Fe 7.71-7.8 % d.wt., total Cr 2.4-4.0 d.wt.
The rest of the solid wastes are those from the

Table 3: Soil characteristics

Parameter	Unit	Direction: N		S		E		W	
		Distance, m							
		3	300	3	150	3	200	3	200
pH (water)		7.7	6.4	5.2	5.0	6.7	5.1	5.3	4.4
TKN	ppm	672	952	840	620	672	672	784	728
P	ppm	272	309	409	240	227	270	490	360
K	ppm	682	930	930	380	980	320	1100	330
Na	ppm	150	135	190	87	120	81	140	41
Mn	ppm	218	240	97	12	190	55	210	170
Zn	ppm	25	47	18	12	23	17	13	25
Hg	ppm	7.4	17	13	7.2	14	5.8	12	7.4
Pb	ppm	30	14	21	16	24	6.2	38	15
Ni	ppm	24	8.8	12	17	13	4.7	22	2.7
Cd	ppm	2.0	0.9	0.7	0.1	0.8	0.4	2.0	0.3
Co	ppm	1.1	0.8	1.2	1.0	0.7	0.1	2.2	1.6

Table 4: Chemical components of the plants

Plant Sample*)	Total forms, % d.wt. N	P	K	Na	Microelements, ppm Mn	Zn	Cu	Ni	Cd
1O N 50	0.67	0.29	1.8	0.150	123	16	15	1.2	0.65
2O N 150	0.78	0.27	2.4	0.022	127	25	14	2.8	0.08
3O E 3	1.40	0.47	5.1	0.154	654	39	22	3.2	1.02
4O E 50	1.68	0.38	5.1	0.046	208	31	28	1.1	0.17
5O E 200	2.50	0.58	3.4	0.030	227	31	37	0.6	0.04
6O S 3	1.46	0.33	4.8	0.130	349	43	15	24.6	4.36
7O S 50	1.82	0.31	4.3	0.048	298	28	16	6.8	3.45
8O S 150	2.42	0.29	3.6	0.038	181	28	20	4.0	0.06
9P W 3	2.02	0.13	4.4	0.080	180	68	21	11.8	3.02
10P W 50	2.32	0.22	3.4	0.065	133	44	17	2.2	2.61
11P W 200	2.91	0.39	2.7	0.057	45	41	14	1.4	0.14

*) Consecution of Symbols: Sample No; type of the plant: O - oat, P - pea; direction and distance from the landfill in meters.

finishing process of skins and hides. Leachates composition the Reader can find in column 4, of table 1. At the SE side of this tip there is arable land, from the other sides the tip is surrounded by forest. The slope of the terrain is from SE to NW.

At the location of the landfill up to a depth of 0.5-1.5m occur dusty humus, aggregate muds and clay. Below this layer are present sands, gravels, sand-gravel mix, with no numerous pockets of silts and clays. The groundwater level is 1.5-2.6m below the surface. High water levels in the Bober-river are above the bottom level of the landfill. The direction of the groundwater flow is identical with the slope of the terrain.

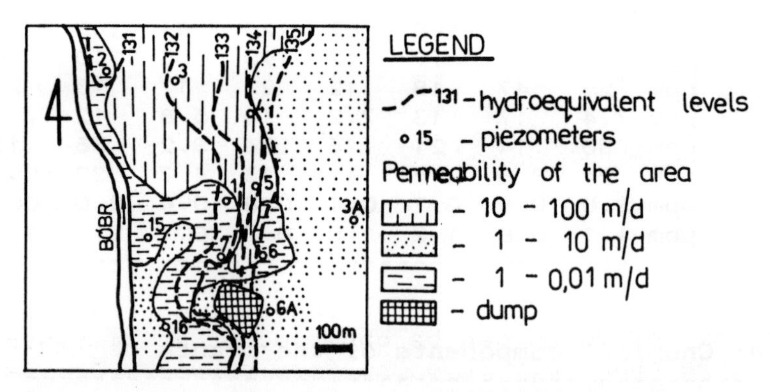

Fig. 2 Hydrogeological structure of the area.

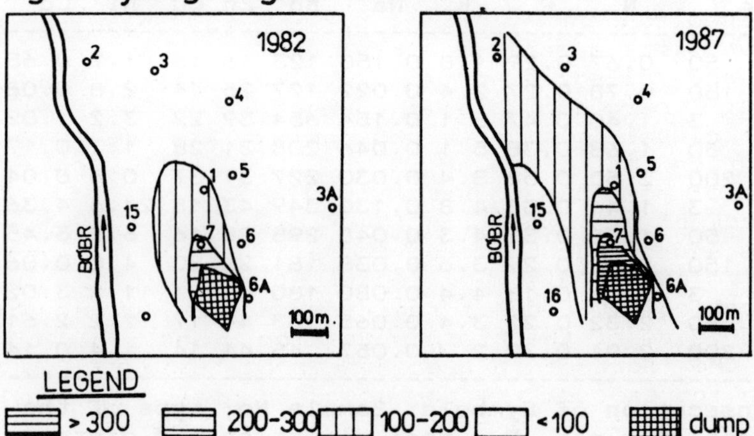

Fig. 3 Concentration of sulphates in groundwater, mg/dm^3

Starting in 1978, the quality of groundwater around the tip has been analysed systematically. The hydro-geological conditions are shown on Fig.2. On Figs. 3 & 4 concentrations of chromium and sulphates as well as their expansion in years (i.e. after 5 and 10 years of operation) are shown as examples only.

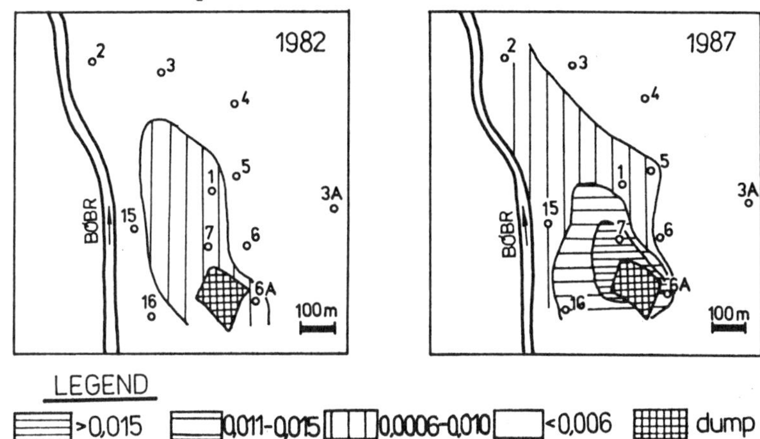

Fig.4 Concentration of chromium in groundwater, mg/dm^3

SUMMARY

A distinct impact of the first landfill on chemical properties of soils can be observed at a distance of no more than 300m from the yard's body. It is manifested particularly by higher salinity as well as by high chlorides and sulphates concentration. Increased concentrations of heavy metals have been observed in a zone varying from 30 to 50 meters only and directly adjoining to the yard. Plants growing there show higher concentrations of cadmium, nickel, manganese and sodium, and lower concentration of nitrogen which gives, in effect, yields of some 30% lower.
The lack of an insulation layer between the stocked wastes and the groundwater (Case No.2) brings about that we note a distinct impact of wastes upon the chemical composition of the groundwater. During the passed years of the tip's operation, one could observe the increase in groundwater pollution as well as an increase of the contaminated area. Although this case study gave - in terms of environment pollution - objectionable results, it gave, from the other hand, the backgrounds for forecasting the range of pollution in the coming years.

CLEANUP OF A CONTAMINATED OPERATING AREA

E. Holzmann

Introduction

The German Federal Railways own an extensive operating area in the north of Munich where rolling stock, especially steam locomotives, used to be serviced and required. As a number of repair halls are no longer used, some of the buildings concerned being over seventy years old, the German Federal Railways considered demolishing two of the larger halls with a view to altering the land use of this part of the area.

Out of the total area of 4 hectares the two former repair halls earmarked for demolition covered an area of approx. 3 ha. These halls were formerly used to repair, service and overhaul steam locomotives and rolling stock. Beside the access roads leading to the repair halls were located concrete containers for the open-air intermediate storage of waste material and scrap metal.

An inspection of the relevant documents revealed that the main harmful substances employed were oils, paints and solvents - albeit in varying concentrations, depending on the particular nature of the work in any given part of the halls. Polychlorinated Biphenyls (PCB) were presumed to have been employed in one part, where transformers were serviced.

A site inspection showed that the entire floor area of the two repair halls was superficially contaminated, although there were obvious differences in the degree of contamina-

ISWA PROCEEDINGS
VOLUME 1 ISBN 0-12-058451-4

tion according to the former utilisation of the part involved.

The competent authorities (Water Resources Department and Environmental Protection Department of the City of Munich) therefore demanded a remedial investigation of the whole operating area prior to demolition. DORSCH CONSULT was entrusted with the execution of this investigation.

One of the main difficulties which faced the company with regard to both the remedial investigation itself and any cleanup measures required thereafter was the fact that, due to external factors of overriding importance, all works - including demolition of the halls and cleanup of the area - had to be completed within a period of less than one year.

Investigation and Results

In coordination with the competent Environmental Protection Department an investigation plan was elaborated on the basis of:

- information on former utilisation and presumed harmful substances
- visual contamination
- composition of repair hall floors: wood block paving-screed-concrete-gravel.

In an initial, multi-stage work step floor material, groundwater, soil and soil gas samples were taken from borings arranged in a rough grid pattern; the grid was adjusted and compressed in accordance with the results obtained.

The results of these borings were as follows:

- There did not exist any groundwater pollution at the time

- There was no contamination worth mentioning in the open-air area either

- There did not exist any PCB contamination

- In the floors of the halls there were clear indications of Chlorinated Hydrocarbons (CHC), Aromatic Hydrocarbons (AHC) and in particular Polynuclear Aromatic Hydrocarbons (PAH) contamination

- The contamination, especially that of PAH, demonstrated varying degrees of penetration into the individual floor layers.

<u>Measurements</u>

Since, in view of the degree of contamination, the nature of the existing natural soil - gravel - and the high groundwater table, it could not be ruled out that the groundwater might become polluted, the competent authorities demanded a cleanup of the operating area. The relevant drinking water ordinance was applied as a criterion.

The following procedures were selected for the cleanup measures - taking into account the local disposal options, time and cost:

- Incineration of contaminated wood block paving from floor in a domestic waste incinerating plant

- Deposition of more heavily contaminated screed, concrete and gravel on a landfill site

- Temporary deposition of slightly contaminated/non-contaminated material for subsequent re-use, e. g. as noise protection walls.

Throughout the duration of the cleanup and demolition works soil samples were taken and analysed in order to achieve a definitive delimitation of the area to be cleaned up on the one hand and minimize the volume of material to be disposed of on the other.

In the course of these complementary investigations it was discovered that a number of small areas were more heavily contaminated than had initially been established. Furthermore, the performance of these investigations resulted, thanks to the more accurate delimitation attained, in a perceptible cost reduction.

Session 7

Computerized Solid Waste Management

THE USE OF COMPUTERS TO ASSIST IN THE PREPARATION OF STRATEGIC WASTE MANAGEMENT PLANS

P E Rushbrook
Environmental Safety Centre, Harwell Laboratory, Didcot, UK

INTRODUCTION

Planning is both an art and a science and every manager in business or public service needs to plan ahead. There are several reasons for this: i) to maintain the efficient running of current operations; ii) to assess the viability of new ventures; and iii) to predict likely future incomes, expenditures and lifetimes. Solid waste management is no exception and it is to the "waste planner" that the task of preparing plans is vested in most local waste management administrations. He (or she) needs to have not only a working knowledge of the capabilities of competing waste technologies and their possible locations, but also tact and sensitivity to balance the opposing interests and viewpoints on each.

The complexity of waste management planning has increased considerably in the last decade. Many more issues now have to be addressed than were probably appreciated when legislation was first drafted. The waste planner is frequently involved in reconciling the technically possible with the environmental desirable or the theoretically practicable. The considerable skill required to produce a workable plan is too often not fully realised by some local government administrations.

There is no one correct solution which is universally applicable for the best waste management plan. Different social outlooks and customs on the handling and disposal of solid waste, differences in the availability of disposal facilities, the quality of technical, economic and human resources, and the competing demands from other public sector services, all influence the character of a final plan. Consequently, planning at all levels of waste management is most usefully carried out locally.

It is emphasised that public authorities have the leading role in setting standards for good waste management planning. If the public sector do not have good operations and well-defined plans then no one else in the industry will.

Tactical and Strategic Planning

It is widely recognised that within the boundaries of local government policy there are two levels of planning, "tactical" and "strategic". Tactical planning is chiefly orientated towards operational matters which need to be resolved over the period of a few weeks or months. The usual upper timescale for tactical studies is 12 months. Beyond one year and perhaps up to 20 years ahead a strategic plan would be prepared instead. The principal difference is that tactical plans deal commonly with individual operations in detail (eg. collection, incineration, baling). Examples could include vehicle routing studies, design and feasibility studies, and costing exercises. Conversely, strategic plans concentrate on broader issues and are the means by which major technical and financial decisions are described. Strategic planning should, logically, precede tactical plans.

This paper will concentrate on computer-based strategic plans only.

DATA REQUIREMENTS

The most representative plans use local data. Those which instead rely heavily on nationally-derived data or values from the published literature are only second best. In general, the shorter the timescale for which a plan is being prepared then the more detailed the local data that are required. For example in tactical planning the preparation of new waste collection routes for collection vehicles or the design of a new waste incinerator would require more specific information than a strategic, pre-design review of waste treatment technologies or the preparation of a long-term plan to optimise the utilisation of existing and future disposal facilities. The type of data required for any planning exercise is not difficult to obtain, and should (or ought to be) available in some form within local government. However, it is more than likely that different parts of the data are collected by different departments. This is the first obstacle to planning a waste manager must overcome.

ROLE OF COMPUTERS

The most apparent difficulty when preparing a forward plan is the complexity of the situation, especially where many different combinations of facilities (strategies) are to be considered. Different disposal strategies may include comparing between: i) near-urban public sector landfill; ii) distant public and private sector landfill supplemented by waste transfer; iii) incineration with or without resource recovery; and iv) alternative waste handling, processing and treatment technologies such as baling, pulverisation, materials recovery, refuse-derived fuel, pyrolysis, etc.

Considerable volumes of data are amassed prior to the development of a plan which the waste planner must review and take into consideration. He can tackle this situation in two ways, either by manually sorting through the data, or alternatively using a computer to perform the same physical and financial calculations more quickly. The laborious task of preparing a plan manually is very time consuming and often the first workable combination of disposal facilities is the only one to be given serious consideration. This is not the best approach. Proper strategic planning should include the comparison of several, significantly varied, feasible combinations of facilities. Only with the use of a computer and a suitable program can this be done efficiently.

Computers are often reviewed with suspicion by waste managers and until recent years the spread of computers in waste management departments was slower than in others. However, a look around most disposal department offices now will see an encouraging increase in computer literacy.

Several computer models to assist in the strategic choice of facilities have been developed since the mid-1960's and in Table 1 a selection of recently published programs is presented. The most common approach is the use of "selection allocation" models which select suitable facilities from a group of pre-defined locations (14). The alternate, "location-allocation" models, which are designed to identify the optimum location for facilities, are less widely used since in reality a waste manager rarely has an unrestricted choice of sites.

TABLE 1

A SELECTION OF STRATEGIC WASTE MANAGEMENT PLANNING MODELS

(based on a general review of recently published literature review and supplements earlier work (eg 14))

Description	Authors	Year	Ref No.
Multi-objective decision-making model	Perlack, Willis	1985	1
Dynamic simulation model	Helferich, Hoffner, Gee	1972	2
Resource recovery planning model	Berman, Chapman, Hung	1983	3
Computational model for solid waste management	Gottinger	1986	4
Solid waste management models	Liebman	1975	5
Fixed charge planning model	Walker, Aquiline, Schur	1974	6
Multi-criteria model	Sobial, Hipel, Farquhar	1981	7
System approach to solid waste management planning	Haynes	1981	8
Parametric mixed integer programming	Henkins	1982	9
Optimal service regions	Wenger, Rhyner	1984	10
Recycling programs	Clapham	1985	11
Transportation models related to landfill design	Conrad, Hoffman	1974	12
HARBINGER, strategic waste management planning model	Rushbrook, Pugh	1987	13

There are several publications which discuss in greater detail the types of models that have been developed (eg. 5,14). However, some common features are present in most models. They all comprise an algorithm which aims to optimise the overall least cost of waste transport and/or treatment and disposal activities. The general approach is to provide a solution to allocate as cheaply as possible waste from each source-sink combination (ie. waste in each generating area to any facility). A cost function (the "objective" function) is established and is made up of separate terms to account for collection, haulage and facility operations. Various constraints can be applied to account for physical limitations (such as maximum waste inputs into facility x, or waste reduction achieved at incineration y). In some models future costs are accounted for by using discounted cash flow techniques, economies of scale can be considered and the objective function can be adjusted (weighted) to give preference to reducing costs at public or private sector operations.

The recent history of using these mathematical models has not been encouraging. Various isolated studies have been carried out in North America and Europe, though most models are still seen as academic curios and very few are being used commercially. The reasons for this are diverse though some of the principal ones are suggested here:

- many of the models have been developed by operational researchers and are not easily understandable to the waste manager. Few consider the entire waste management system or are provided in a comprehensive package for easy use.
- many contain major simplifying assumptions which negate their applicability to real-life situations. For example, some do not allow the forecast of future waste production, or operations in future years, or consider variable sizes of facilities, or the costs of waste transport;
- many do not allow several variations to the base data to be considered at once and often only very limited facilities are built in to explore the sensitivity of optimisation results to changes in the input data;
- most models are not "interactive". This is where a waste manager can make modifications at the screen and view pre- and post-optimisation data while in the program. This feature provides rapid output of results and exploration of the practicality of alternative strategies;
- some models were developed from the outset to consider one specific place or country and do not have the built-in flexibility to be directly usable elsewhere.

Waste managers need to be more aware of the expansion in the strategic planning capabilities that computer models can bring about. Modellers and computer programmers need to develop models for, and to be used by, the waste manager rather than what they think the end-user wants.

INTERACTIVE PLANNING

One example of the new generation of planning models specifically designed (packaged) to be used by waste planners is HARBINGER. This is a "selection-allocation" model in which wastes are allocated from a set of waste generating areas (ie. towns, suburbs, etc) to a selection of disposal facilities chosen by the user from a larger set previously put into the base data. It is an interactive system in which the user works directly at the computer and has an integer linear programming algorithm to perform the optimisation. The model is designed to give the planner all the information necessary to produce several costed strategies, from which a "preferred" one can be put forward as the basis of a waste management plan. The advantages of interactive use are two-fold: i) the waste manager can interject his local knowledge when required; and ii) the model has a modular structure which enables the planner to build up a waste management plan in a similar manner to doing the task manually.

The model is divided into various sub-models each performing a separate preparatory task prior to optimisation, ie. waste generation estimates, selection of facilities, waste transport haulage costs and facility operational costs.

Sensitivity analyses can be performed to vary as required most items of information. The preparatory data are then optimised in a separate sub-model. A flow sheet for the micro-computer version is given in Figure 1 and descriptive details on the structure of the objective function is in Table 2. This objective function can also be adjusted by selecting pre-defined weightings to minimise not only all costs equally, but instead preferentially public sector costs or private sector costs or all disposal costs. A further refinement is the ability to plan over a 15 or 20 year horizon in smaller time periods. Periods of 5 years duration are the most common for initial planning, though shorter periods can be used for finer, more detailed assessments.

Operational experience by waste planners using HARBINGER has already been obtained in Hong Kong (15) and three UK authorities (16).

CONCLUSION

The natural limit for waste planners to prepare manually thorough and well costed strategic plans has probably been reached. It is also apparent that strategic technical evaluations are becoming increasingly more complex. The adoption of more computer-based planning aids by waste managers is therefore only a matter of time.

The essential pre-requisites needed are the development of more easy- to-use programs which a waste manager can understand, improvement of their suitability to computers he has access to (both mainframe and micro-computers), and a willingness on the manager's part to take time out to learn how to use the model to improve his planning capabilities.

TABLE 2

COMPOSITION OF THE OBJECTIVE FUNCTION IN HARBINGER

The linear program optimises on the criterion of the minimum total cost of disposing of all wastes. This is comprised of the following elements:

Transport costs

For each waste generating area, facility and waste type:

Quantity of waste to facility X haul cost per tonne.

For each treatment or transfer facility pair and waste type:

Quantity of waste transferred to a second facility X haul cost per tonne.

For each treatment or transfer facility, landfill and waste type:

Quantity of waste to landfill X haul cost per tonne.

Facility costs

For each treatment or transfer facility:

Total waste received from all waste generating areas X cost per tonne of processing the waste.

For each landfill:

Total waste received from all waste generating areas and other treatment or transfer facilities X cost per tonne of emplacing the waste.

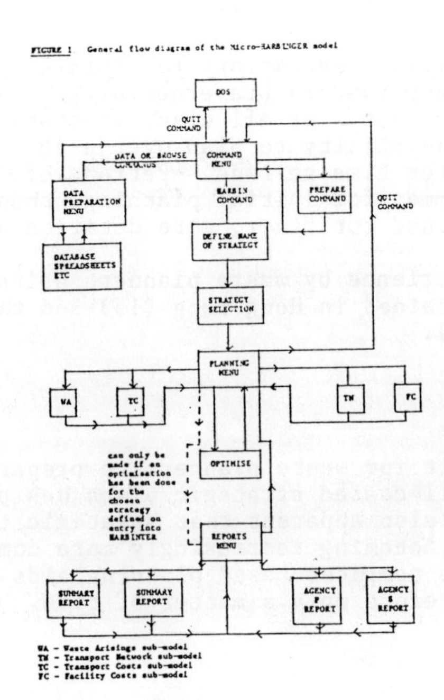

FIGURE 1 General flow diagram of the Micro-HARBINGER model

REFERENCES

1 Chang S Y (1985) reporting on Perlack R D, Willis C E (1985). "Multi-objective decision-making in waste disposal planning". J Environ. Engineering. 113(3) pp 664-667.

2 Helferich O K, Hoffner V, Gee D E (1972). "Dynamic simulation model for planning solid waste management". Int. Pollut. Contr. Mag. 1(1) pp 40-49.

3 Berman E B, Chapman R E, Hung H K (1983). "Program documentation for the resource recovery planning model". US Dept. of Commerce. Report No. NBSIR 83-2745 May 1983.

4 Gottinger H W (1986). "A computational model for solid waste management with applications". Appl. Math. Modelling 10(5) pp 330-338.

5 Liebman J (1975). "Models of solid waste management". Chapter 5 in: "Mathematical models in government planning" Ed: Goss S. Princetown, New Jersey, USA.

6 Walker W, Aquiline M, Schur D (1974). "Development and use of a fixed charge programming model for solid waste planning". Rand Corp. Paper 5307. Santa Monica, California, USA.

7 Sobial M M, Hipel K W, Farquhar E J (1981). "A multi-criteria model for solid waste management". 12(2) pp 97-110.

8 Haynes L (1981). "A systems approach to solid waste manage ment planning". Conserv. Recycl. 4(2) pp 67-68.

9 Jenkins L (1982). "Parametric mixed integer programming: An application to solid waste management. Manage. Sci. 28(11) pp 1270-1284.

10 Wenger R B, Rhyner C R (1984). "Optimal service regions for solid waste facilities". Waste Manage. Res. 2(1) pp 1-15.

11 Clapham Jr W B (1985). "Computer models help communities identify recycling programs". Resource Recycling. March/ April 1985 pp 24-25, 36.

12 Conrad R B, Hoffman EK (1974). "Transportation model applied to landfill design". Proc. Amer. Soc. Civil Eng. 100(EE3) pp 549-556.

13 Rushbrook P E, Pugh M P (1987). "An illustrated description of HARBINGER". Wastes Management 77(6) pp 348-361.

14 Wilson D C(1981). "Waste management: Planning, evaluation, technologies". Oxford University Press: Oxford.

15 Hoare R W M, Boxall H E, Wong D T W (1985). "Role of a strategic planning model in waste management in Hong Kong".

16 Gubby B, Watkins G (1986). "Installation and use of the HARBINGER Waste Management model within the West Midlands area. Proc. HARBINGER Symposium. St Catherines, Oxford. 5 February 1986.

USE OF A MICROCOMPUTER MODEL FOR SOLID WASTE DECISION MAKING

by

Samuel A. Vigil
Professor
Civil and Environmental Engineering Department
California Polytechnic State University
San Luis Obispo, California 93407

OVERVIEW OF THE SOLID WASTE FINANCIAL MODEL

Modern solid waste management practice employs waste reduction, resource recovery and recycling, energy conversion, and landfill disposal of wastes which cannot be economically recovered. The economic evaluation of various alternatives requires standardized, reliable procedures. The Solid Waste Financial Model (SWF Model) provides engineers and planners with software tools to perform such an analysis.

The SWF Model is a preprogrammed "template" which uses the Lotus 1-2-3 spreadsheet program on the IBM-PC/XT or AT or compatible microcomputers. It is composed of four parts, or modules. The first module, BASELINE, is the entry point into

ISWA PROCEEDINGS
VOLUME 1 ISBN 0-12-058451-4

the SWF Model. A menu selection system allows the user to transfer from one module to another. Data entered into the BASELINE module is automatically transferred from one module to another. The modules are briefly described below. For brevity, only the ENERGY module will be discussed in detail. The model is documented by extensive built-in HELP screens, a Users Guide (Reference 1), and a detailed Programming Guide (Reference 2).

BASELINE. This module is the normal entry point to the model. It is used to record basic data about the community including current population, future population estimates, solid waste quantities, and solid waste composition.

RECYCLE. Several types of recycling centers can be planned including: buyback, curbside, dropoff, and commercial. Use of the RECYCLE module is described in Reference 3.

ENERGY. This module uses the solid waste data from the BASELINE module to help plan waste-to-energy systems.

LANDFILL. This module helps the user evaluate landfills and make estimates of capital costs, operating expenses, closure, and postclosure costs.

THE ENERGY MODULE

The ENERGY module uses data transferred from the BASELINE module, and user input data to estimate capital and operating costs for a waste-to-energy system. The module also calculates a twenty year proforma income statement to aid in the economic feasibility evaluation of waste-to-energy systems. User selected menu options perform two basic functions, entry of required technical and financial data, and analysis of the proforma income statement.

Data Entry Suboptions

The data entry suboptions are used for the entry of financial and technical data:

TECHDES Option. The TECHDES option is used to enter basic specifications of the waste-to-energy system. TECHDES includes entries for nominal input capacity, energy output (in the form of electricity, steam, or hot water), water consumption, waste generation (wastewater and ash), annual schedule, and availability.

CAPCOST Option. The CAPCOST option is used to estimate the capital costs associated with a waste-to-energy plant. The user is prompted to enter the major capital cost components usually considered in a preliminary cost estimate. Up to ten user defined capital costs can also be entered.

OPEXP Option. The OPEXP option is used to estimate the annual operating expenses of a waste-to-energy system. Data elements include labor cost estimates, expected maintenance expenses, and utility expenses. Provision is also made for user defined annual operating expenses.

PROJREV Option. The PROJREV option is used for the input of electricity, steam, and hot water rates and calculation of annual revenues. The electrical, steam, and hot water outputs, availability, and estimated annual hours are transferred from the TECHDES option.

Data Analysis

The PROFORMA option is used to calculate a twenty year proforma income statement (also referred to as life cycle cost statement). Due to the size and complexity of a proforma income statement , PROFORMA is actually stored as a separate module. When the PROFORMA option is selected from the ENERGY menu, a new menu with seven suboptions appears.

BOND and INFL Suboptions. Additional financial data is entered into the proforma income statement via formatted data entry screens in the BOND suboption. Required data includes bond interest rate, percent equity investment and other financial parameters. Provisions are also made for projecting inflation rates for both revenue and expense elements through the INFL suboption.

XREV Suboption. After entry of the financial data, the XREV suboption computes the revenue portion of the proforma income statement. The computation utilizes data from the ENERGY option and the BOND and INFL suboptions and does not require user input.

EXP and ANAL Suboptions The EXP suboption is used to view the expense portion of the proforma income statement and the ANAL option is used to view the analysis portion of the proforma income statement. The analysis summarizes the cash flow over the lifetime of the project and calculates net present value (NPV) and residual.

SENSITIVITY ANALYSIS

Sensitivity analysis is a planning tool used to compare several options or alternatives. The Lotus 1-2-3 program has built in capabilities to simplify sensitivity analyses. These tools can be used to automatically copy data from the SWF model onto user developed Lotus worksheets and to develop spreadsheet and graphics analyses.

ACKNOWLEDGEMENT

The Solid Waste Financial Model was developed under the auspices of a grant from the California Waste Management Board. It is distributed on a non-profit basis by the California Waste Management Board at the address below:

California Waste Management Board
1020 Ninth Street, Suite 300
Sacramento, California 95814

REFERENCES

1. Vigil, S. A., and Zevely, J. A., (1985). User's Guide to the Solid Waste Financial Model. California Waste Management Board, Sacramento, California.

2. Zevely, J. A., and Vigil, S. A., (1985) Programming Guide to Solid Waste Financial Model. California Waste Management Board, Sacramento, California.

3. Vigil, S.A., Zevely, J.A., and Yeaman, R. (1987). Microcomputer Model for Solid Waste Planning, in Public Works, Vol. 118, No. 2.

AN INFORMATION SYSTEM FOR INDUSTRIAL SOLID WASTE MANAGEMENT - the case of the city of Copenhagen.

Ib Larsen, Head of The Local Office of Environmental Protection, The City of Copenhagen.

R. Victor V. Vidal, The Institute of Mathematical Statistics and Operations Research, The Technical University of Denmark.

1. Introduction. The issue of waste management and the protection of the environment is a very urgent matter and one of great public interest in Denmark. The environmental problems have only gradually been recognized over the last 15 years. There are three major problems: hazardous wastes, the impact of refuse dumps on the surrounding environment, and the pollution created by the incineration process.

During the last four years a number of initiatives have been started at The Local Office of Environmental Protection, The City of Copenhagen, both to satisfy the demands of the Danish Environmental Protection Act and the Reclamation and Recycling Acts, and the growing need to reduce the amount of waste and hereby the related environmental problems.

This paper gives a short presentation of a project concerning the development of a management information system for decision-making, supervision and control of industrial solid waste. This is one of the many activities that The Local Office of Environmental Protection has started with the objective to develop an optimal waste management policy from an environmental viewpoint subject to the economic resources available.

A main aspect to achieve this objective is the recognition of the fact that industrial waste, both in what concerns quantity and composition is susceptible to influence. That is the mere registration of industrial waste should evolve to a more active regulation/steering function to achieve the objectives of the local authorities.

Let us now look more specifically at The Local Offi-

ISWA PROCEEDINGS
VOLUME 1 ISBN 0-12-058451-4

ce of Environmental Protection's strategies for waste management in Copenhagen.

2. <u>Strategies for waste management.</u> Shortly described, the principal strategies for waste handling, reclamation and recycling are the following:

a) To sort out at the source hazardous wastes, so that hazardous wastes should be reclaimed and recycled as far as possible.
b) Reclamation and recycling initiatives for waste based on sorting out at the source, for instance some activities are going on in what concerns cardboard and paper. The main principle is to produce goods of high quality and clean raw materials. Here the economics of production and market conditions play a central role to find out what it does pay to sort out for reclamation and recyling purposes.
c) To sort out at the source wastes that reduce heating value from the rest of the waste. For instance glass and food waste use energy at incineration plants. Now-a-days these wastes are collected for reclamation purpose.
d) Power and heat production by incineration, that is incineration should be considered as a form of energy production that replaces other fuels.

The project mentioned in the last section focuses specifically on industrial solid waste management. A new area where municipalities in Denmark have not very much experience. Handling of industrial waste has usually been a private matter, but due to changes in legislation the authorities will get the necessary powers and authority for imposition of duties on industrial waste producers, for instance industries will be required to introduce a much higher degree of pre-sorting of their waste materials to enable immediate reclamation and recyling. In this connection we should mention that systems for data collection are being designed. Such data can be regularly updated as part of the regular inspection program.

Moreover, the physical (logistic) system is rather complex (it consists of production, distribution and disposal of waste) and dynamic, that is under continuous change due to economic and technological development at the different industries.

Another important aspect of the system in study is the fact that the economic and technological conditions for reclamation and recycling are also under continuous changes.

3. Problem formulation. All these factors mentioned in the last section carry us to the conclusion that the municipalities need a dynamic tool to steer/evaluate the consequences of the different strategies for industrial waste management. Such a tool should support management in both decision-making in connection with the above mentioned strategies providing the relevant information, and in the supervision/control activities of the different industries to register different types of hazardous wastes.

Due to the complexity and the dynamic aspects of the system in study it was decided to develop a computerized tool to tackle the above mentioned problems. Such a tool should provide means to collect and handle the relevant information to support decision-making at different levels.

A very important task in connection with the development of such a tool is the design of a procedure to systematize and classify industrial wastes. This procedure should give the possibility to distinguish among different type of wastes in what concerns their damage to the environment, their characteristics for reclamation and recycling, and their characteristics in relation to disposal in refuse dumps. Moreover the computerized tool should provide other facilities as for instance:

- methods to estimate the total production/disposal of solid industrial waste for different types of waste,
- methods to forecast the future development of different areas of solid industial waste,
- methods to evaluate/simulate the development of the logistic system under different changes.

4. The information system. A computer based information system is under development to support the decision and supervision/control functions described in the last section. This system should be developed in response to the needs of management for accurate, timely and meaningful data for plan-

ning, analyzing and controlling the logistic system. The system accomplishes this objective by providing means for input, processing, and output data along with a feedback decision-making capability that helps managers to respond to current and future changes both in the logistic system, legislation and priorities of the local authorities.

The main goal is that such an information system should evolve to a final decision support system with the following characteristics:

- it helps managers at the upper levels in what concerns decision-making,
- it helps managers at the middle levels in what concerns supervision and control,
- it is flexible and responds quickly to managers' questions,
- it provides "what if" scenarios,
- it takes into account the special conditions of the organization where the system is to be implemented,
- it uses features whereby non-computer-oriented people can use it in an interactive mode using a PC and eventually an electronic spreadsheet,
- it is usually aimed at both structured, semi-structured, and unstructured tasks.
- it is an adaptive system.

The information system we are constructing is a first step towards such a decision support system. The information system has 3 subsystems: a data registration level, a data handling level and a forecast/simulation level.

The data registration level will primarily be a system to record information about industrial solid waste. A central element in this subsystem will be a classification form to be able to identify both quantity of wastes for reclamation/recycling and composition of different hazardous wastes. This system should also provide relevant information for decision-making and as such it does not need to recollect huge amount of data about different industries. The system should also give the possibility of checking the received information from several sources to evaluate their reliability.

The data handling level will primarily be a system

that provides simple routines and facilities to carry out simple calculations (aggregates, averages values, standard deviations, etc.), to illustrate graphically development tendencies and it gives the possibility to use statistic methods (regression, tests, correlations, etc.).

The forecast/simulation level will primarily be a system that gives the possibility to analyze the development of industrial waste either as a forecast under a ceteris paribus assumption or as simulation to evaluate the consequences of different changes in the logistic system.

For the time being we are designing the first level. The information system requires a unique approach to systems analysis and design. The usual process of designing and implementing a system has to be interfaced with rapid and frequent user feedback to ensure that the final decision support system being built will ultimately address the decision-making needs of the managers.

The classification form is designed as a 4-level hierarchical system being consistent at each level. These levels will contain the following information:

Level 1: Total quantity of solid waste for each waste producer (factories, business branch, institutions, etc.).

Level 2: Main types of solid wastes that are classified according to the following handling and treatment, that is
- organic waste
- hazardous waste
- other inflammable waste
- other non-inflammable waste.

Level 3: Main fractions under each type (level 2) that are classified according to their relevance for the evaluation of concrete separate initiatives, for instance:
- food waste for fodder
- organic waste within individual households for biogas or accelerated compost
- garden waste for low-technological compost.

Level 4: Detailed specification of each fraction.

This hierarchical system should provide the total quantity of waste for each producer at any level. This system should be designed so that the number of levels that will be used for a given producer will depend on the need for information, the size of the fractions, the relative importance of a single producer, etc.

The lower levels (3 and 4) will change continuously as the whole solid waste system develops and changes its structure. The classification form will be constructed so that the information at the higher levels (1 and 2) should remain constant.

The higher levels should give sufficiently precise picture of the total quantity of waste. These figures should be calculated taking account of the most relevant waste producers, while estimates should be used for the other sources.

The hierarchical system will provide the quantitative information about the actual waste, this being the basic input for the decision-making system for carrying out arrangements about reclamation and recycling.

The registered information about delivered and sorted out waste quantities will be the basic input for the supervision system for carrying out the control of the producers over their observance of the political decisions on waste regulations.

We believe that it is a good idea to start building such a system before a more systematic collection of data is performed. Thus the process of system design will function as an initiator of discussions in the organization about the framework to be developed to carry through the different strategies to achieve the final goals. This is very relevant in our case where the organization has little experience with computerized management based on system analysis. In this respect it must be emphasized that no computerized system can ever take the decision making authority away from its users. The information system is just a tool to be used for making better and quicker decisions.

5. Final remarks. To design computerized problem-solving frame work is not merely a technical problem but to a high degree a problem of implementing a systematic approach in an organization with little experience in the use of system analysis and operations research. The final implementation of such an approach is a long-term learning process based on the interaction of the managers and the system designers. To develop user-friendly systems to solve first the easiest task is a good way to initiate such a process. Such a problem-solving approach should be designed before the process of final data collection because such an approach will tell us which kind of data is relevant for problem-solving. Data should be collected not just to fill up data bases but because such data is useful information for decision-making and supervision/control.

The information system is specifically tailored to the organization and problem in study, although some subsystems, as for instance classification forms, statistic methods, etc., are of such generality that they might be used by other municipalities.

6. References.

Larsen, I. (1986), Waste disposal and environmental protection, Miljøkontrollen, Copenhagen, p. 14.

Larsen, I. (1987), Affaldsbehandling i Danmark, status og tendenser, Miljøkontrollen, Copenhagen, (in Danish), p. 16.

Vidal, R.V.V. (1987), What is a decision support system?, IMSOR, The Technical University of Denmark, p. 12.

Svensson, L., and Vidal, R.V.V. (1987), Kortlægning og klassificering af erhvervsaffald, IMSOR, The Technical University of Denmark, (in Danish), p. 46.

Session 8

Thermal Treatment of Solid Waste

Keynote lecture

J. Bergström, Sweden: Energy Production with Solid Waste as Fuel
– Manuscript not received in time for publication.

Keynote lecture

J. Bergström, Sweden: Energy Production with Solid Waste as Fuel
– Manuscript not received in time for publication.

I/S RENO SYD WASTE-TO-ENERGY PLANT ANALYSIS OF FOUR YEARS' OPERATION

Niels T. Holst
B&S Miljøteknik A/S, Aarhus, Denmark

I/S Reno Syd Waste-to-Energy plant is owned and operated by 4 municipalities, Hørning, Odder, Ry, and Skanderborg. It is located in Skanderborg in mid-Jutland, Denmark. After a construction period of only 9 months the first fire was lit on Nov. 25, 1983. This paper will tell the plant's history in the 4 following years (1984-1987).

The plant is a single line 4 mt/h (106 sht/day) with a building large enough for one additional line. It is a W-400 furnace according to the BS Incinerator System, laid out to receive fuel of a calorific value of 2200 Kcal/kg. It is fitted with a fire-tube boiler, utilizing the released energy as warm water for the Skanderborg District Heating System. Emission control is an electrostatic precipitator. In front of the furnace an overhead grabble crane is feeding the fuel from a refuse bunker. This plant was the first in Scandinavia to utilize computer controlled combustion via a PC. The capital cost or total investment amounted to approx. DKK 30 million and it serves a population of approx. 60,000 in the 4 municipalities.

Economy

DKK x 1000 Year	1984	1985	1986	1987
Running costs	-3,860	- 4,894	-5,358	- 5,092
Capital costs	-2,164	- 3,958	-3,896	- 3,716
Reserve fund	- 654	- 1,322	- 371	- 940
Tipping fee, private	+ 0	+ 0	+ 0	+ 0
Tipping fee, industry	+ 176	+ 281	+ 437	+ 982
Sale of heat	+7,164	+10,357	+8,629	+10,580
Result	+ 656	+ 124	- 559	+ 1,814

Fig. 1. Economical Balance

From the Economical Balance shown in fig. 1, it can be seen that it is not necessarily a bad business to own and operate a Waste-to-Energy Facility. The most remarkable point is the line representing the tipping fee for the citizens, as they are not charged for the treatment of their refuse. The 1986 income from sale of heat was actually DKK 1.2 million larger, but after a mutual agreement it was paid back to the District Heating Works.

DKK x 1000 Year	1984	1985	1986	1987
Personnel	2,100	2,428	2,701	2,725
Building maintenance	133	235	184	174
Machinery maintenance	265	556	752	646
Electricity + water	399	384	420	341
Ash transport	147	133	145	163
Ash tipping fee	100	407	457	680
Purchase of fuel	591	669	321	131
Consultancy	74	30	249	210
Miscellaneous	51	52	129	22
Running costs	3,860	4,894	5,358	5,092

Fig. 2. Specification of Running Costs

A closer look at the running costs (fig. 2) reveals that approx. half of the costs originate from salaries to the staff. 6 shift, each of 2 men + 1 "day man" make the Operating Staff of 13 people. To this comes the administrative cost plus management. The second largest amount is ash transport and deposit costs, competing with maintenance of the electromechanical part of the plant. In a later chapter we will touch these figures again when discussing scheduled and unscheduled stops.

The most unusual figure is "purchase of fuel". This is mainly straw, purchased from neighbouring farmers. But also, at least in the two last years, it consists of purchase of paper, collected by boy scouts and sports club youth. The Plant Management decided to do so when the market price of paper for recycling went down to almost nothing, in order to maintain the voluntary organization for paper and glass bottle collection. The 4 municipalities are by the way Danish title holders in that specific discipline.

Received Refuse

At the planning stage, it was anticipated that the plant would receive 15,500 mt domestic refuse, 7,500 mt trade & shop refuse, in all 23,000 mt per year. Further, it was stipulated that up to 25% of straw could be added. The actual amounts are shown in fig. 3.

Year		1984	1985	1986	1987
Household refuse	mt	12,347	12,699	12,852	14,284
Trade & shop refuse	mt	10,276	10,497	9,591	10,914
Purchased straw	mt	2,302	2,811	1,220	2,089
Crushed refuse	mt	-	-	3,109	1,591
Paper fr. boy scouts	mt	-	-	1,133	1,167
Total	mt	24,926	26,007	27,905	30,045
Index of growth		100	104.3	112	120

Fig. 3. Received Fuel

It should be mentioned that the 1987 figures contain 3/4 year's supply of refuse from a 5th municipality (1077 mt) so that the real growth index should be 116 for that year. However, it is worth noticing that the domestic refuse when "cleaned" for foreign figures - in spite of a growing tendency to recycling - has risen 9%, and the trade & shop refuse has risen 4% over the same period. The crushed refuse figure reflects an investment in machinery, which with a very short payback period also means a prolongation of the lifetime of the controlled tip.

Residues

In the costs and income shown already, some important figures are missing. These are the income from sale of scrap iron and the sale of screened bottom ash. There is of course a reason for this, and that is because these figures belong to the accounts under the heading "Recycling", and because the income actually is mainly a tax benefit, which is a special Danish phenomenon.

Year		1984	1985	1986	1987
Bottom ash	mt	6119	5581	6301	6481
Filter ash	mt	277	263	268	324
Separated iron	mt	-	575	725	777
Total	mt	6396	6419	7294	7582
Total in % of refuse		25.7	24.7	26.1	25.2

Fig. 4. Residues

As can be seen from the figures in fig. 4, the inert material percentage of the refuse is fairly constant.

Energy Recovery

Utilizing the released energy for warm water district heating combines two important aspects, as it calls for the lowest possible investment (firetube boiler) combined with the highest possible thermal efficiency. During the efficiency test, the thermal efficiency ran as high as 81.5% based on the lower calorific value, and the indeed increased calorific value of the refuse together with modifications made on the plant makes one wonder if this record figure is not even higher today.

Year		1984	1985	1986	1987
Produced heat	Gcal	37,481	43,206	47,903	52,863
Produced heat	Gcal/mt	1.51	1.66	1.72	1.76

Fig. 5. Energy Recovery

One should pay considerable attention to the fact that during the period the amount of refuse increased by 20%, however the recovered energy increased by 41%.

If we make another calculation and add the calories during the 4 years and calculate with 0.12 mt oil per Gcal we will reach 21,774 mt oil. With Danish oil prices for heavy fuel oil this becomes the astonishing amount of DKK 67 million - or a payback time of less than 2 years. It is on occasions like these one can grow fond of the Danish energy taxes. In brackets it could be mentioned that the energy income alone amounted to DKK 36,730,000, which grossly encompasses the investment of approx. DKK 30 million.

Availability

Year		1984	1985	1986	1987
Operating	hrs %	7991 91.0	8039 91.8	7964 90.9	8028 91.6
Scheduled stops	hrs	538.5 6.1	625 7.1	597.5 6.8	496 5.7
Unscheduled stops	hrs %	254.5 2.9	96.5 1.1	198.5 2.3	236 2.7
Total stop	hrs %	793 9.0	721.5 8.2	796 9.1	732 8.4

Fig. 6. Availability

As we can see from fig. 6, the availability has been fairly constant during the years (average 91.3%) - and unusually high when it is considered that we deal with a single line plant and a firetube boiler. Subsequently, the total stop hours (average 8.7%) show the same steadiness. The balance between scheduled and unscheduled stops however varies considerably. Also the fact that the plant is restricted during the summer months, due to the fact that the District Heating Work is unable to receive all the produced energy, must be considered.

While it is understandable that planned maintenance requires planned stops and that these will vary from year to year, it is of course desirable that unscheduled stops should be kept at a minimum. As any stop besides causing work and expenditures on possible repairs also contains a loss in income from sale of heat, it is an extremely important factor to consider. Therefore a deeper analysis of the unscheduled stops is stated in fig. 7.

As will be easily seen, the two major trouble spots are the crane including grab plus slagging on the furnace walls. A conclusion from this is easy: Do not buy a cheap crane - it is going to be the most expensive solution. Another interesting story you can read is that as you intensify the performance of your unit you create a slagging problem if you are utilizing the "old" refractory technology.

Cause of Stop	Year	1984	1985	1986	1987
Crane + grab	hrs %	84.5 33.2	34 35.2	60.5 30.5	74.5 31.6
Slagging furnace	hrs %	60 23.6	1 1	35 17.6	131 55.5
Boiler cleaning	hrs %	38.5 15.1	18 18.7	12 6	2 0.8
Bottom ash pusher	hrs %	41.5 16.3	- -	14 7	- -
Feed pusher	hrs %	9 3.5	- -	41 20.7	5 2.1
PC control	hrs %	1 0.4	12.5 13	3 1.5	22 9.3
Bunker fire	hrs %	- -	22 22.8	6 3	1.5 0.6
Grate	hrs %	14 5.5	- -	9 4.5	- -
Exhauster	hrs %	1 0.4	6 6.2	4 2	- -
District heating system	hrs %	- -	- -	11 5.5	- -
ESP	hrs %	3.5 1.4	- -	- -	- -
Electrical	hrs %	- -	3 2.2	- -	- -
Water pump	hrs %	2.5 1	- -	- -	- -
Primary air system	hrs %	- -	- -	3 1.5	- -

Fig. 7. Specification of Unscheduled Stop Hours

It can also directly be seen that an improvement has b
made to the boiler cleaning system. When considering m
tenance, the expenses of crane/furnace-refractories/bo
come in as the winners, while the electrostatic precipitator follows close by without being a cause of unscheduled
stops.

Environment

During its lifetime the plant has undergone several tests to prove its ability to meet the conditions in the Permit. Unfortunately the two dioxin tests made last year are not yet available (April 88). These tests were performed by the Danish Environmental Protection Agency. The following mean values for each year are stated in fig. 8.

Particles		Year	1984	1985	1986	1987
Dust	mg/Nm^3		68	35.5	46.0	64.0
SO_2	mg/Nm^3		58	74	98	207.5
HCl	mg/Nm^3		40	52	166	131.2
HF	mg/Nm^3		-	0.9	1.2	-
CO	ppm		-	-	-	25.9
Bottom ash unburned matter			1.2%			
Fermentable (TÜV)			0.04%			

Fig. 8. Official Environmental Test Results

Taking the mean values of the results shown in fig. 8 and comparing them with the "old" permit values and the coming standards we will reach the following:

		Particles	SO_2	HCl	HF
Old standards	mg/Nm^3	150	1250	700	6
Mean values	mg/Nm^3	53.4	109.4	97.3	1.05
New standards	mg/Nm^3	40	300	100	2

Fig. 9. Environmental Comparisons

It is here one must stop and take a deep breath. The plant is close to the future demands on all counts. It will be a very difficult decision to make for the Environmental Authorities whether they are going to demand a costly acid gas cleaning device on the background of these figures.

Alterations/Additions

A trap door in the bunker wall was the first addition. Through this you can pick up undesirable bulky items with the crane grab and through the trap door deliver these to an outside container. (Refrigerators, washing machines, etc.)

Next came the magnetic separator to pick out ferrous metals from the bottom ash.

The secondary air system was altered and made larger which gave an immediate rise in thermal efficiency.

A second sonic boiler cleaning device was installed, but the single "best investment" was the combined scraper/air-blast boiler cleaning device which made it possible to have a longer period of uninterrupted operation. It is also a part of the explanation when you saw the character of undesired stop hours change during the years.

What's next

Could the plant have done everyting over again, the biggest wish would be a better crane. Secondly, a larger bunker. As this is written, new equipment for continuous measurements of particles, HCl and CO is being installed together with an advanced computerized reporting system. Also air-cooled refractories in the furnace stand high on the want list. The hope is that furnace line No. 2 will come in the not too distant future.

Acknowledgement

The Author wants to thank the Management and the Staff of I/S Reno Syd for their cooperation in preparing this paper and for making all the presented data available. Finally, it should be mentioned that the results of the plant would be impossible without a first-class well-motivated crew.

Niels T. Holst
April, 1988

THE NATIONAL INCINERATOR & EVALUATION PROGRAM (NITEP): MASS BURNING TECHNOLOGY ASSESSMENT

R.Klicius, A. Finkelstein, D.J. Hay, Urban Activities Division, Industrial Programs Branch, Environment Canada, Ottawa, Ontario KIA OH3

INTRODUCTION

This paper discusses the important design modifications and subsequent extensive combustion test program completed on a state-of-the-art mass-burning incinerator located in Quebec City, burning municipal solid waste.

One question that is raised repeatedly with both existing and proposed EFW facilities is how safe are the emissions produced. Addressing these issues and many others is the objective of the multi-faceted National Incinerator Testing and Evaluation Program (NITEP). The five-year program is mandated: to identify energy-from-waste technologies in Canada, to assess relationships among state-of-the-art designs, operations, energy benefits and emissions; to examine effectiveness of emission control systems and to develop National Guidelines for Emissions. To date, reports have been published on the extensive test programs successfully completed on a modular two-stage design in Prince Edward Island (PEI) and on two different types of air pollution control systems in Quebec City. Test results for mass burning technology is the subject of this technical paper.

PLANT BACKGROUND

The Quebec Urban Community Municipal Solid Waste Incinerator Plant is a mass-burning design developed in the early 1970's to burn as-received refuse (i.e. without any preparation) in a water-wall furnace. The plant produces steam using flue gas heat recovery boilers. The incinerator plant is owned by the Quebec Urban Community (QUC) and is located in an industrial area of Quebec City, adjacent to residential and commercial zones. It receives municipal and commercial solid waste collected by the QUC, as well as from several other municipalities and private contractors. All of the steam generated by the plant's four incinerator units is sold to a local paper company.
The incinerator plant employs the technology developed in Europe and represented a contemporary de-

ISWA PROCEEDINGS
VOLUME 1 ISBN 0-12-058451-4

sign when built in 1974. Throughout the years, a number of design changes were made to improve the operational problems in the plant, such as furnace slagging and emissions of large unburned material. However, some of the design changes compounded existing emission problems.

PLANT OPERATION

The principal elements of the Quebec Urban Community municipal waste incineration plant are shown in Figures 1 and 2, and include a refuse storage pit and crane system, four incinerators/boilers, each rated at 227 tonnes per day, electrostatic precipitators, ash quenching systems with storage pit and crane, and a common stack. Each incinerator consists of a vibrating feeder-hopper, feed chute, drying/burning/burnt-out grates, refractory-lined lower burning zone, waterwalled partially-lined upper burning zone chamber, a vertical tube mechanically-rapped waste heat recovery boiler with superheater and economizer, a two-stage electrostatic precipitator, an induced draft fan, and a wet ash quench/removal system.

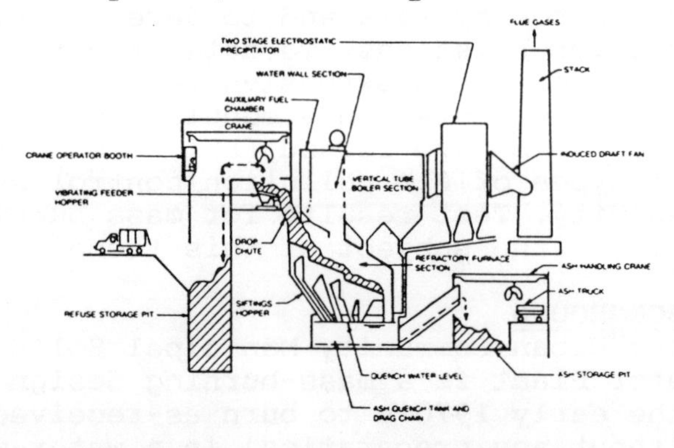

FIGURE 1
QUEBEC INCINERATOR
SCHEMATIC CROSS-SECTION

DESIGN MODERNIZATION

In May 1985, a comprehensive study was completed under NITEP on the modernization of the Quebec incinerators. The study detailed the many changes that would be required to transform the incinerators into a "state-of-the-art" mass burning design and to reduce the existing emission problems. Based on the study findings, the QUC decided to experiment by upgrading one of the incinerator units to assess the impact of the proposed changes before modifications were made to all the units.

The following important design modifications illustrated in Figure 3 were provided: (a) an improved primary air distribution system was installed, to provide independently controlled air flow to 9 zones located beneath the grates, each fitted with individual motorized and flow-controlled dampers; (b) the nozzle design and location of the secondary air system were selected to ensure suitable flue gas temperatures and the desired excess air level, while achieving sufficient mixing for completing combusting; (c) the configuration and positioning of the front and rear furnace "bull noses" were determined to optimize mixing; (d) combustion gas residence time was increased 30% by enlarging the upper furnace chamber; (e) automatic computer controls were installed.

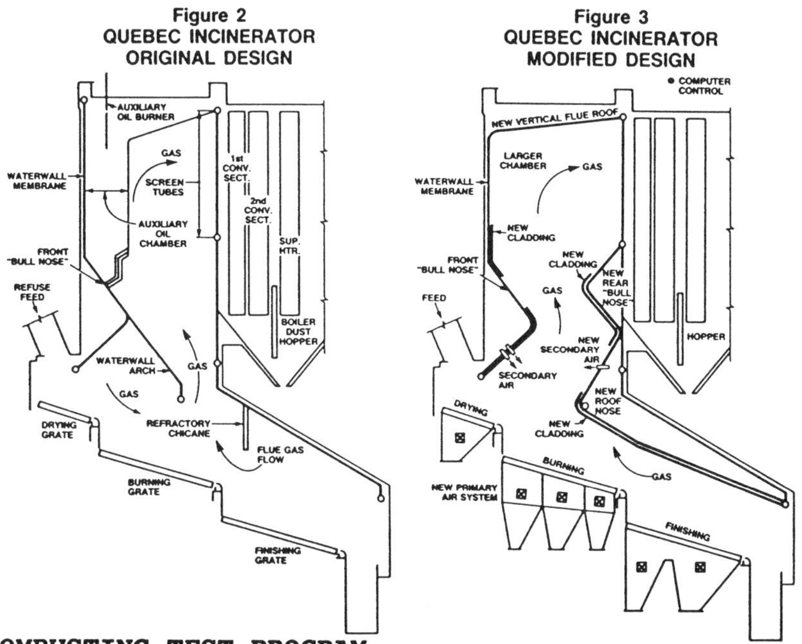

Figure 2
QUEBEC INCINERATOR
ORIGINAL DESIGN

Figure 3
QUEBEC INCINERATOR
MODIFIED DESIGN

COMBUSTING TEST PROGRAM

An extensive cumbustion test program was established to evaluate the performance and effectiveness of the new furnace design. The objective of this program was to develop correlations between various operating parameters and emissions over a wide range of different operating conditions.

Two levels of testing were involved in the combustion test program: characterization and performance. The characterization tests served as a basis for developing an understanding of the incinerator operating range, debugging of all systems, facility logistics and field crew familiarization.

Based on the characterization test results, five performance test conditions were selected, as shown in Figure 4. These represent 3 different lead conditions, as well as good and poor operations (i.e. high CO levels) at design load. These latter tests involved extensive sampling and analysis as well as process and data evaluation.

Figure 4
PERFORMANCE TEST
CONDITIONS

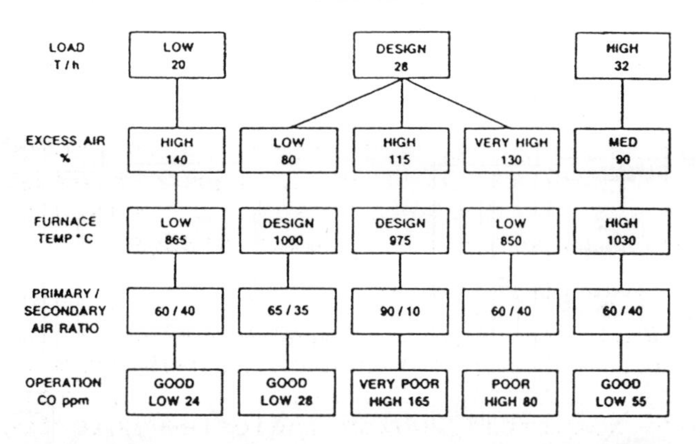

To compare input versus output, all incoming refuse and all the ashes collected were weighed and representative samples analyzed. Each sample was analyzed for two groups of compounds: organics, which consisted of dioxins (PCDD), furans (PCDF), chlorobenzene (CB), polychlorinated biphenyls (PCB), polycyclic aromatic hydrocarbons (PAH), and chlorophenols (CP); and metals, consisting of cadmium (Cd), lead (Pb), mercury (Hg) and twenty-seven others. In addition, the combusting gases leaving the stack were sampled for the compounds listed above, as well as for acid gases, such as hydrogen chloride (HCl), sulphur dioxide (SO_2) and nitrogen oxides (NO_x); and conventional products of combusting, such as carbon monoxide (CO), carbon dioxide (CO_2), oxygen (O_2) and total hydrocarbons (THC). All sampling and analyses were completed using recognized North American protocols.

Three separate micro-computer-based systems, linked into a network, were chosen to log, store and analyze all the data collected during each test day.

Each of the computer systems provided a "real time" graphic and statistical analysis of selected logged data. This provided on-the-spot monitoring of the progress of the tests and enabled project managers to make decisions quickly. Upon completion of each test day, the data generated was reviewed by a second crew overnight to verify and correct all data and produce detailed tables, as well as summary tables and graphics. This allowed the field project managers the opportunity to review previous day's results to assess the adequacy of each test and to evaluate the adequacy of each test.

An extensive amount of test data is available on the old furnace design which provided an excellent reference for the comparison and assessment of the improvements made to the furnace design.

PCDD/PCDF EMISSIONS

In 1984, Environment Quebec undertook an assessment of the dioxin and furan emissions from the Quebec Urban Community incinerator plant. Three tests were completed, which indicated relatively high concentrations of dioxins (i.e. 800-4000 ng/Sm^3) and furans, which is not uncommon for facilities of this vintage. However, after implementing the modifications as previously discussed, a significant reduction in dioxin emissions was observed (as shown in Figure 5). By employing good operating practices as defined in modern incinerator facilities, dioxin concentrations were reduced between 40 and 100 times below the 1984 test results (i.e. from 800-4000 ng/Sm^3 to 20-50 ng/Sm^3). This suggests that the new design had a significant influence on dioxin emissions; however, good operating practice was also essential to minimize dioxin formation and emission.

OTHER TRACE ORGANIC EMISSIONS

Table 1 presents a summary of the emissions of trace organic compounds for each of the five performance test conditions.

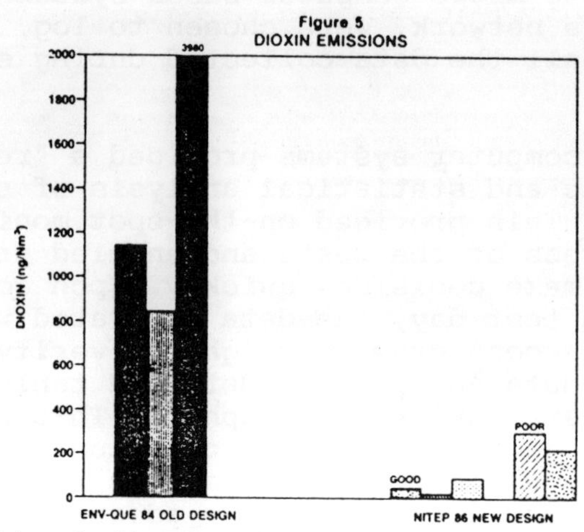

The results indicate that trace organic emissions of each major compound group were lowest under good operation at the design rate. Conversely, trace organic emissions were highest at one or both of the poor operating conditions.

TABLE 1
Trace Organic Emissions per Performance Test Mode

	Good Conditions Burning rate			Poor Conditions Design Burn Rate	
	Low	Design	High	Low Temp	Poor Dist.
PCDD (ng/Sm^3)	52.6	18.8	55.4	298.5	218.9
PCDF (ng/Sm^3)	114.5	44.5	100.7	298.3	306.4
CP (ng/Sm^3)	9.5	5.1	8.0	22.5	23.7
PAH (ng/Sm^3)	7.1	4.0	5.4	21.9	3.2
CB (ng/Sm^3)	3.5	3.3	4.4	9.9	9.5
PCB (ng/Sm^3)	4.3	3.0	4.9	7.0	1.7

Note: Values corrected to 12% CO_2

PARTICULATE/METALS EMISSIONS

Emission concentrations of particulates and nine priority metals are summarized in Table 2.

Particulate emissions were unexpectedly low (below 62 mg/Sm^3), considering that the incinerator was equipped with a two-field electrostatic precipitator. These favourable results are attributed to the low particulate load to the precipitator that resulted from modernization of the furnace design.

Metal emissions (except mercury and nickel) were generally highest under the poor operating conditions and lowest for good operation at the design rate, which corresponds to particulate emissions.

TABLE 2
Particulate/Metal Emissions per Performance Test Mode

	Good Conditions Burning Rate			Poor Conditions Design Burn Rate	
	Low	Design	High	Low Temp.	Poor Dist.
Particulate (mg/Sm^3)	26	22	36	55	62
Particle Size (% 2.5 um)	33	29	24	26	24
Cadmium (ug/Sm^3)	26	24	41	90	76
Lead (ug/Sm^3)	978	673	1599	2039	2495
Chromium (ug/Sm^3)	11	7	15	21	14
Nickel (ug/Sm^3)	9	5	8	8	7
Mercury (ug/Sm^3)	783	704	872	810	622
Antimony (ug/Sm^3)	35	36	44	112	87
Arsenic (ug/Sm^3)	2	3	5	7	6
Copper (ug/Sm^3)	39	33	54	91	89
Zinc (ug/Sm^3)	1619	1130	2061	5122	3429

Note: Values corrected to 12% CO_2.

TABLE 3
Trace Organic Concentrations (ng/g) in Ashes For Good Operation at Design Rate

	Bottom Ash	Boiler Ash	Precipitator Ash
PCDD	0.2	37	584
PCDF	1.0	31	186
CB	45	356	892
PCB	ND	ND	ND
CP	16	80	1820
PAH	538	25	111

The new furnace design resulted in a large reduction in particulate emissions compared to the old furnace design. In Figure 6, the particulate concentrations at the stack are compared between the old and new design at 28 and 32 tonnes/hr. steam flow. On the average, particulate reductions by 1 to 2 orders of magnitude were achieved with the new design concept at both of these steam rates which was comparable to the degree of reduction in dioxins. Based on current theory that dioxins are mostly bound to particulates in the combustion gases, it is postulated that good combustion conditions that reduce the carryover of particulates from the furnace, will also minimize the formation and emission of dioxines into the environment.

TRACE ORGANICS AND METALS IN ASHES

A comparison of the concentrations of trace organics in each type of ash is provided in Table 3 for the design rate and good operation. Trace organic concentrations (except PAH) were highest in the precipitator ash, followed by boiler and lowest in the incinerator (bottom) ash. Similar trends have been observed in other NITEP studies (i.e. highest concentrations in the finer ash), as well as for the more volatile metals, except nickel and copper.

No significant differences in trace organic concentrations or metal concentrations in the ashes were observed at different operating modes. However, it is relevant to note that the mass rates of boiler and precipitator ash were significantly higher under poor operation versus good operation. Accordingly, the output rate of organics and metals in ashes was higher under poor operating conditions.

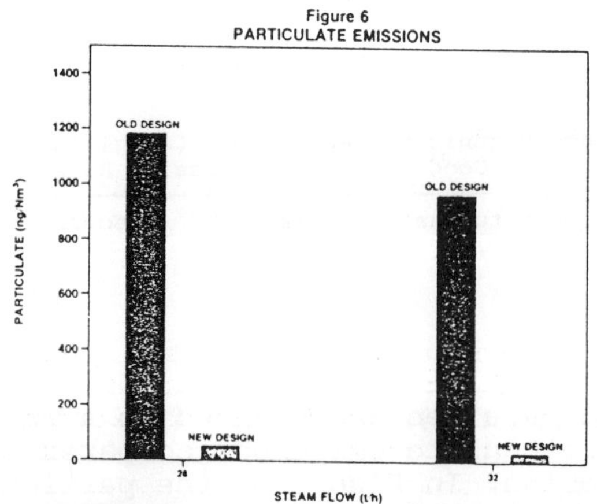

Figure 6
PARTICULATE EMISSIONS

STATISTICAL ANALYSES

A key objective of the tests was to identify relevant relationships between the emissions of selected trace organics and inorganics versus process operating variables and versus continuous emission data. Simple and multiple linear regression analyses were completed to identify these relationships by calculating correlation coefficients between selected dependent and independent variables. These relationships were used to generate "prediction models", which predict trace organic emissions based on readily measurable variables (surrogates), and "control models", which determine how to control the incinerator to minimize emissions.

Based om simple linear correlations, the following relationships are important to note:

1. Dioxin and furan exhaust emissions showed strong correlations with the following parameters: (a) CO emissions, (b) particulate emissions, (c) exhaust gas flow and primary air flow, (d) chlorobenzene (CB) and chlorophenol (CP) emissions, and (e) copper emissions.
2. No correlation was found between dioxin or furan emissions and dioxin/furan concentrations in the refuse.
3. Particulate emissions correlated with the following: (a) CO emissions; (b) exhaust gas flow and primary air flow; (c) some of the trace organics emissions, namely dioxin, furan, and CB; (d) most of the priority metal emissions, such as cadmium, copper and lead.
4. Emissions of selected metals correlated well with CB, CP, dioxin and furan emissions.
5. Lower radiation chamber temperature correlated well with combustion parameters such as O_2 and CO_2 concentrations.

Although simple regression analyses provided a good first-order screening, as indicated by the above results, multiple linear regression analyses resulted in much stronger prediction models and control models:
1. Multiple linear regression analyses produced prediction and control models with coefficients of determination (r^2), which are indicators of the model's strength, ranging from 0.74 to 0.90 for the best fit models.
2. The prediction models which best characterized trace organic emissions of dioxins (Figure 7), furans, CB and CP used two to three of the following monitoring parameters: (a) CO, (b) NO_x, (c) O_2, (d) H_2O in flue gas, and (e) lower radiation chamber temperature.
3. The best control models to minimize trace organic emissions (Figure 8 for PCDD) used three of the following operational settings: (a) total air flow, (b) primary/secondary air ratio, (c) steam rate or refuse rate, and (d) secondary air front/rear ratio.
4. Carbon monixide was determined to be the best single surrogate in the one variable (simple) prediction model for most of the trace organics, with the exception of PAH and PCB.
5. Although NO_x was the best second variable to improve the prediction capabilities of the models,

its significance is not clearly understood at this time.

6. Primary air flow was the most influential operational setting for the one-variable control models for dioxins, furans, CB and CP.

7. The PCB and PAH models examined contained either a relatively low r^2 value or the data scatter implied a poor predictive model. Thus, no useful models were found for these two groups of trace organics.

8. The models developed for dioxins, furans, CB and CP are consistent with the fact that poor incinerator operating conditions resulted in higher emission concentrations.

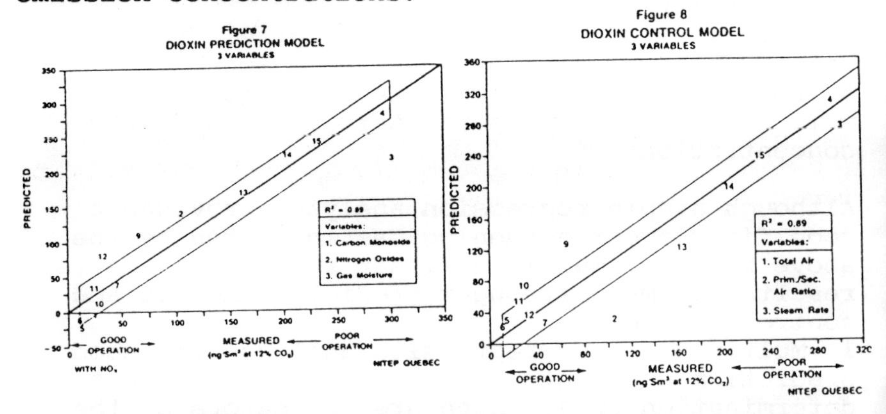

Carbonaceous Particles in the Fly Ash - A Source for the Formation of PCDD/PCDF in Incineration Processes

L. Stieglitz, H. Vogg
Kernforschungszentrum Karlsruhe, Postfach 36 40, D-7500 Karlsruhe 1

INTRODUCTION

In previous investigations of the thermal behavior of PCDD/PCDF in fly ash it was shown that the formation of chlorodioxins and -furans takes place within the fly ash itself at temperatures around 300° C by gas-solid reactions, rather than by high temperature chemistry of the gas phase. Oxygen and certain transition metals are important prerequisites for the reaction 1, 2 . In order to answer the question about origin, nature and role of organic precursor compounds for the PCDD/PCDF formation a series of experiments with specially treated fly ash of municipal waste incinerators and with synthetic mixtures was performed.

RESULTS

Experiments with Fly Ash

In the study of PCDD/PCDF formation in fly ash samples of different origin and composition it was indicated that the potential for the low temperature synthesis of dioxins and furans may be correlated to the carbon content of the fly ash 3 . Carbon free fly ash was prepared by heating the ash in air at 500° C for 6 hrs. To this material (carbon content < 0.2 % C) charcoal (purified by exhaustive extraction and vacuum drying at 400° C, 4 hrs., and 500° C, 2 hrs.) was added in amounts to give concentrations of 0.7, 1.2, 2.2 and 4.2 % (wt) carbon. These mixtures were heated in an air stream at 300° C for 2 hrs. The PCDD/PCDF concentrations resulting after the thermal treatment are shown in table 1. In samples without addition of charcoal, containing 0.2 % residual carbon only minor concentrations of 1.3 ng/g total PCDD and 18 ng/g PCDF were produced. Upon addition of charcoal a concentration increase of the PCDD/PCDF congeners is noted which is roughly proportional to the carbon content, a concentration of 1 % carbon yielding roughly 700 ppb PCDD and 2500 -3000 ppb PCDF. The PCDD/PCDF formation is accompained by a decrease of the carbon content by 40 - 60 %, as also shown in table 1 (bottom), whereas the chloride concentration remains constant at 6.95 % to 7.05 %. A rough calculation indicates that during the 2 hrs. annealing a fraction of the carbon of $5.2 \cdot 10^{-5}$ is converted to PCDD and of $2.9 \cdot 10^{-4}$ to PCDF.

ISWA PROCEEDINGS
VOLUME 1 ISBN 0-12-058451-4

Experiments with Synthetic Mixtures

Preliminary results from other investigations led to the assumption that the chlorides of certain metals contained in the fly ash, especially of copper (II), play a dominant role in the processes observed 3 . In order to differentiate the contribution of individual inorganic compounds of the fly ash to the PCDD/PCDF formation, the thermal experiments were performed with synthetic mixtures, in which defined concentrations of inorganic reagents were added to a Mg-Al-silicate matrix. The use of these mixtures allows the investigation of the influence of varying concentrations of carbon, inorganic chloride and additives of selected metal chlorides to simulate the composition of the fly ash.

Table 1: Formation of PCDD/PCDF in Fly Ash as a Function of Carbon Content

CONGENER	CONCENTRATION PCDD / PCDF (ng/g)				
	PERCENT CHARCOAL (INITIAL)				
	0.2	0.7	1.2	2.2	4.2
T4CDD	0.2	114	260	380	690
P5CDD	0.5	210	435	680	900
H6CDD	0.01	140	310	540	770
H7CDD	<0.01	25	50	110	150
O8CDD	0.6	0.04	3.7	9	14
PCDD	1.3	485	1060	1717	2524
T4CDF	2.5	880	2000	2920	5810
P5CDF	6.9	990	2600	3960	8300
H6CDF	5.8	280	680	1100	2470
H7CDF	3	30	55	110	200
O8CDF	<0.01	3	2	6	50
PCDF	18	2183	5337	7998	16650
% *CARBON	<0.2	0.44	0.49	0.92	1.66
% *CHLORIDE	7.02	7.05	7.03	6.96	6.96

*AFTER ANNEALING

a) Influence of the Addition of Metal Chlorides. For the experiments Mg-Al-silicate was mixed in a stainless steel mill with 1 % active charcoal, 1 % KCl and with 1 % of selected metal chlorides as listed in table 2. The mixtures were annealed at 300° C for 2 hrs. in an air stream containing 100 mg/l water vapor. The PCDD/PCDF concentrations after the experiments are presented in table 2. Obviously the system responds quite differently to the addition of the individual metal chlorides. Without additive the total PCDD/PCDF concentration is below 0.5 ng/g for each compound class. Similarly low are the concentrations achieved by the addition of the chlorides of Ca, Fe(II) and Cd. A stronger influence is noted with the chlorides of Mn, Ni, Hg, Pb, Fe(III), with concentrations of 3 - 8 ppb PCDD and 5 - 50 ppb PCDF being

Table 2: Influence of the Addition of Metal Chlorides on the Formation of PCDD/PCDF

METAL CHLORIDE ADDED (1%)	TOTAL CONCENTRATION (ng/g) PCDD	PCDF
NO ADDITION	<0.5	<0.5
$MgCl_2 * 6H_2O$	2.0	3.6
$CaCl_2$	<0.5	<0.5
$ZnCl_2 * 2H_2O$	8.1	10.4
$SnCl_2 * 2H_2O$	1.2	7.4
$FeCl_2 * 4H_2O$	<0.5	<0.5
$FeCl_3$, anhydr.	7.6	53.0
$MnCl_2 * 4H_2O$	3.0	4.9
$NiCl_2 * 6H_2O$	0.4	5.5
$CdCl_2 * H_2O$	<0.5	<0.5
$HgCl_2$	0.5	7.4
$PbCl_2$	5.0	20.0
$CuCl_2 * 2H_2O$	679.0	4340.0

Table 3: Influence of Copper Concentration on the Formation of PCDD/PCDF

CONGENER	CONCENTRATION PCDD / PCDF (ng/g) PERCENT Cu^{2+} (ADDED) 0	0.08	0.24	0.4
T4CDD	1.3	8	20	13
P5CDD	1.0	20	80	65
H6CDD	0.9	37	240	400
H7CDD	0.3	24	230	860
O8CDD	1.0	12	200	110
PCDD	4.50	101	770	1448
T4CDF	7	100	310	260
P5CDF	11	310	1290	1550
H6CDF	3	230	1150	3100
H7CDF	1.6	100	690	2730
O8CDF	0.07	20	200	840
PCDF	22.6	760	3640	8480

formed. In contrast to this, the addition of 1 % $CuCl_2 \cdot 2H_2O$ gives rise to 679 ng/g PCDD and 4340 ng/g PCDF. Apparently the role of copper is not restricted to catalytic activities in the degradation and hydrogenation of PCDD/PCDF 4 , but is also inducing reactions in carbonaceous particles leading to an oxidation of the carbon and to the formation of dioxins/furans. The influence of copper concentration on the PCDD/PCDF synthesis in a standard system (Mg-Al-silicate, 1 % charcoal, 1 % KCl) with copper concentrations varying between 0 and 0.4 % was investigated. The results are shown in table 3. The addition of 0,08 % of Cu(II) as chloride promotes the formation reaction and results in concentrations of 100 ng/g PCDD and 760 ng/g PCDF. Increase of the copper concentration to 0.24 % causes an overproportional rise of the dioxin- and furan content, indicating the sensible response of the system to the presence of copper.

b) Influence of Temperature. Experiments with reaction times of 2 hrs. were performed with synthetic mixtures (Mg-Al-silicate, 1 % charcoal, 1 % KCl and 1 % $CuCl_2 \cdot 2H_2O$ (ca. 0,4 % Cu) in an air stream at 250°, 300°, and 350° C. As in previous studies with fly ash 1 , maximum concentrations were obtained at 300° C.

c) Influence of Annealing Time. The kinetics of dioxin/furan formation was studied in a system of Mg-Al-silicate, 4 % charcoal, 3 % chloride (KCl) and 0.4 % Cu at 300° C with annealing times of 0.25, 0.5, 2, 4, 6 hrs. After a fast increase a practically constant concentration level is obtained after 2 hrs. From the time dependence a quasi-first order reaction is indicated.

d) The Role of the Nature of Carbon. Carbon of different origin (sugar coal, charcoal, soot) was added in quantities of 1 % to a Mg-Al-silicate matrix containing 1 % KCl and 0,4 % Cu(II). The dioxin/furan concentrations after annealing at 300° C(2 hrs.) in air (100 mg water/l) depend on the nature of the carbon added. The total dioxin concentrations obtained are 212 ng/g with soot, 944 ng/g with sugar coal and 2439 ng/g with charcoal and are probably a function of the surface area. For all three carbon specimens the profile of the congeners is similar showing maximum concentrations with the octa chlorodioxins. The amounts of furans formed are higher and the difference in yield is less pronounced, ranging from 5050 ng/g (sugar coal) to 7750 ng/g (charcoal). The congener profiles show maximum concentrations for hexachlorofurans.

e) Influence of Carbon Content. In the system Mg-Al-silicate (1 % KCl, O,4 % Cu(II)) the influence of carbon concentration was investigated in the range between 1 % to 8 % carbon. Conditions: 300° C, 2 hrs., airstream with 150 ng/l water. As reported above for the experiments with fly ash also in these experiments a good linear relationship between carbon content and dioxin/furan concentrations is obtained.

f) Formation of other Chloroaromatics. In the experiments reported apart from PCDD/PCDF a variety of other chlorocompounds was formed. In a system with Mg-Al-silicate (1 % KCl, 0.4 % Cu(II)) and 2 % charcoal annealed at 300°C (2 hrs.) the following compounds were newly formed: 1, 3, 5 trichlorobenzen 50 ng/g, 1, 2, 4 trichlorobenzene 450 ng/g, 1, 2, 3 trichlorobenzene 610 ng/g; the concentrations of the tetra chlorobenzenes were for the 1, 2, 3, 5 isomer 1030 ng/g, for the 1, 2, 4, 5 isomer 650 ng/g, and for the 1, 2, 3, 4 isomer 270 ng/g. The corresponding values for the pentachloro- and hexachlorobenzene are 11650 ng/g and 8800 ng/g. Further compounds identified are polychloronaphthalene (n Cl = 4 - 6) and polychlorobiphenyls (Cl = 7, 8).

CONCLUSIONS

Results from laboratory experiments prove that in incineration of municipal waste PCDD/PCDF, as well as other organochlorocompounds, are formed at moderate temperatures from carbonaceous particles of the fly ash by gas-solid reactions with oxygen and inorganic chlorides, influenced and activated by copper(II) chloride.

From the correlations between the formation of PCDD/PCDF and the content of particulate carbon and of copper(II) in the fly ash it is concluded that minimization of dioxins/furans may be achieved in incineration plants by
- high burn-out of the fly ash, resulting in low residual organic carbon
- low residence times for fly ash and particulates in the critical temperature zones of 300° C
- removal/recycling of metallic uncombustible material, especially of copper prior to incineration.

We acknowledge the financial support by the project "Wasser, Abfall, Boden" of the State of Baden-Württemberg.

REFERENCES

/1/ Vogg, H., Stieglitz, L. (1986). Thermal Behavior of PCDD/PCDF in Fly Ash from Municipal Waste Incinerators, Chemosphere 15, 1373-1378.

/2/ Stieglitz, L., Vogg, H. (1987). On Formation Conditions of PCDD/PCDF in Fly Ash from Municipal Waste Incinerators, Chemosphere 16, 1917-1922.

/3/ Vogg, H., Metzger, M., Stieglitz, L. (1987). Recent Findings on the Formation and Decomposition of PCDD/PCDF in Municipal Solid Waste Incineration, Waste Management & Research 21, 285-294.

/4/ Hagenmaier, H., Kraft, M., Brunner, H., Haag, R. (1987). Catalytic Effects of Fly Ash from Waste Incineration Facilities on the Formation and Decomposition of Polychlorinated Dibenzo-p-dioxins and Polychlorinated Dibenzofurans, Environ. Sci. Technol. 21, 1080-1084.

BASIC RESEARCH ON THE EMISSIONS OF HAZARDOUS SUBSTANCES CAUSED BY WASTE INCINERATION

K.Yasuda, M.kaneko
Environmental Research Center of Kanagawa Prefecture
Yokohama, Japan

1. Introduction

The emissions of hazardous substances caused by municipal waste incineration vary according to waste composition and operating parameters such as furnace temperature, excess air, and carbon monoxide. However, to obtain a sample sufficient to measure the emissions of hazardous substances at trace levels, it is necessary to operate the incinerator for many hours. Since during these lengthy periods it has not always been possible to maintain stable conditions, it is very difficult to determine the relationship between these emissions and waste composition.

In our basic research, therefore, we used municipal waste with an artificially regulated composition for our combustion experiments, and by using an experimental incinerator we examined the emission behavior of hazardous substances (heavy metals, PAH) with respect to changes in waste composition and combustion conditions.

2. Experimental method

2.1 Experimental incinerator

As shown in Figure 1, the experimental incinerator is a fluidized bed incinerator. The total waste firing capacity of this incinerator is about 400 kg/h. The mean gas residence time was estimated at about one second.

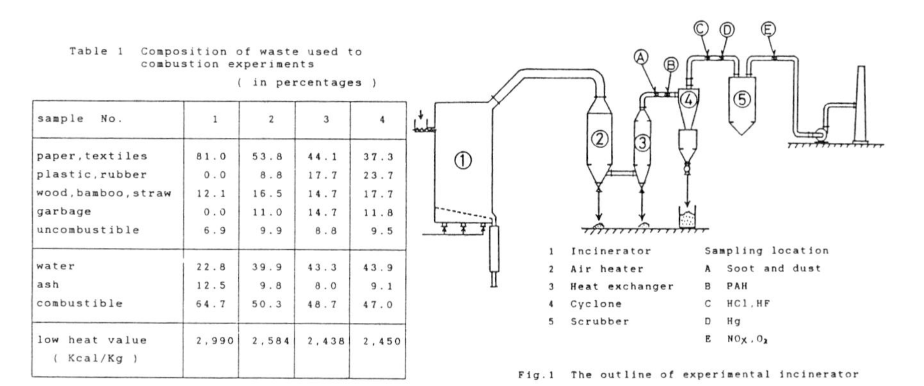

Table 1 Composition of waste used to combustion experiments
(in percentages)

sample No.	1	2	3	4
paper, textiles	81.0	53.8	44.1	37.3
plastic, rubber	0.0	8.8	17.7	23.7
wood, bamboo, straw	12.1	16.5	14.7	17.7
garbage	0.0	11.0	14.7	11.8
uncombustible	6.9	9.9	8.8	9.5
water	22.8	39.9	43.3	43.9
ash	12.5	9.8	8.0	9.1
combustible	64.7	50.3	48.7	47.0
low heat value (Kcal/Kg)	2,990	2,584	2,438	2,450

Fig.1 The outline of experimental incinerator

ISWA PROCEEDINGS
VOLUME 1 ISBN 0-12-058451-4

2.2 Combustion experiments

The municipal waste used in combustion experiments, shown in Table 1, was sorted into five categories: paper and textiles; plastics, rubber and leather; wood, bamboo and straw; garbage (food waste); and ferrous metal.

This municipal waste was used in combustion experiments which were conducted at two different combustion temperatures of 700 °C and 900 °C.

3. Methods of sampling and analysis

3.1 Dust concentration and heavy metals contained in dust (except Hg)

Dust sampling was performed by isokinetic extraction from the flue gas in the duct. The dust was collected in a silica thimble (whatman 88RH).

Concentrations of heavy metals in dust from the silica thimble were determined by leaching with HNO_3 and H_2O_2, after which the solutions were analyzed by atomic absorption.

3.2 Mercury

Almost all the mercury in the flue gas is in the vapor phase after passing through the cyclone. The sampling train for on-line monitoring of vapor-form mercury is presented in Figure 2.

Samples were sucked directly from the heated (150 °C) sampling probe (5, 13) at a partial flow rate of 1 l/min. through two wash bottles (7,8) connected in series, after which vapor-form mercury was continuously analyzed by cold vapor atomic absorption (11).

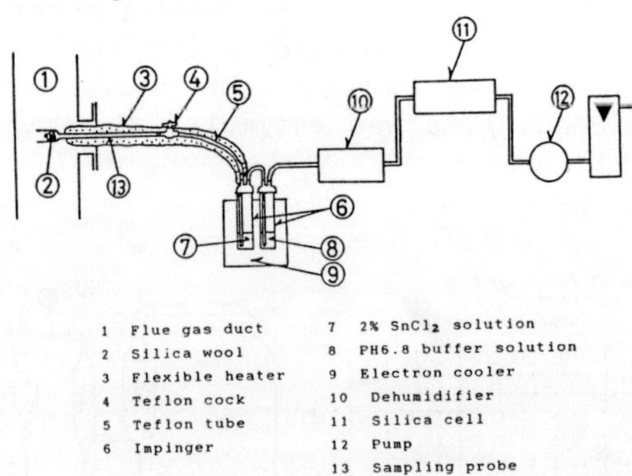

Fig.2 Scheme of the sampling train (vapor-form Hg)

3.3 Polycyclic aromatic hydrocarbons (PAH)

Using the sampling system shown in Figure 3, gas and particulate matter were collected from the incinerator by isokinetic conditions. To avoid absorption onto the particles of compounds present in the gas phase, the silica thimble (whatman 88RH, 2) was placed on the top of the sampling probe inside of the stack.

The gas first passed through a Liebig condenser (5) and then into a short neck Kjeldall flask (7) cooled with ice to collect the condensate, after which it passed through a glass fiber filter (GF, whatman, 9) and two polyurethane foam plugs (PUFP, 10). Typical sampling periods were 30 minutes and the sampling volume was 200 to 260 l. During sampling the temperature of the sampling probes was kept at approximately 150 °C by a flexible heater.

The silica thimble and GF+PUFP were subjected to extraction separately for 4 hours in a Soxhlet apparatus using acetonitrill. The condensate was extracted 3 times using a separate funnel with 250 ml cyclohexane per liter of sample. These extracts were reduced in volume with a Kuderna Danish apparatus to a concentration of 0.5 ml. Each of these fractions was then analyzed by high-performance liquid chromatography (HPLC) on a 5 µm polymeric C_{18} column using an acetonitrill/water (8/2) mobil phase at a flow rate of 1.5 ml/min. and fluorescence detection.

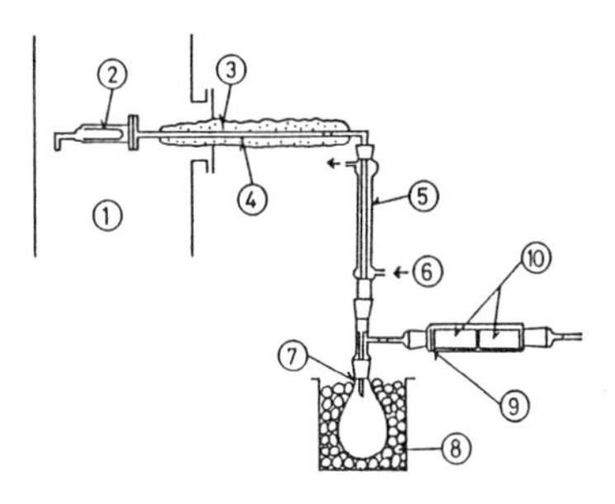

1 Flue gas duct
2 Silica thimble
3 Flexible heater
4 Sampling probe
5 Liebig condenser
6 Cooling water(Icy-cold water)
7 Short neck kjeldall
8 Ice
9 Whatman fiber filter
10 Polyurethane foam plug

Fig. 3 Scheme of the sampling train (PAH)

4. Results and discussion

4.1 The emission of heavy metals caused by the incineration of dry cells

Table 2 indicates that dry cells contain a large number of heavy metals. Therefore, when municipal waste containing used dry cells is incinerated, there is a possibility that heavy metals will be discharged. For this reason we examined the emission behavior of heavy metals (Hg, Cd, Pb, Mn, Zn and Ni) caused by the incineration of dry cells.

(1) Background level of mercury

When incinerating municipal waste containing no used dry cells or fluorescent tubes, the concentration of Hg in the flue gas was 0.03 to 0.05 mg/m^3, as seen in Figure 4. This concentration corresponds to 1/5 to 1/10 of the concentrations which have been measured at full-scale municipal incinerators in Japan.

Table 2 Contents of heavy metals in a dry cell

kind of dry cell	Hg mg/cell	Cd mg/cell	Pb mg/cell	Mn g/cell	Zn g/cell	Ni g/cell
maganese cell(R-20)	3.0	10.0	30.0	12.6	25.0	——
alkaline-manganese cell(LR-20)	750	10.0	30.0	30.0	18.0	——
(LR-6)	140	2.7	8.1	4.1	2.8	——
nickel-cadmium cell(KR-20)	——	100g	——	——	——	70.0

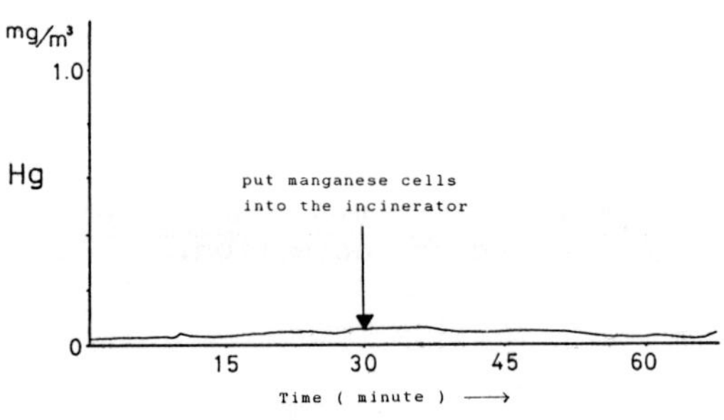

Fig.4 Measuring chart of Hg (1)

(2) Incineration of alkalin-manganese cells (experiment Nos. 5 and 6)

When alkalin-manganese cells (LR-6, 5 cells) were put into the incinerator as shown in Figure 5, the concentration of Hg in the flue gas became 20 to 100 times that of before the cells were placed in the incinerator. At times the concentration was over 1 mg/m^3. But mercury was not always discharged rapidly, there were many cases in which it was discharged gradually inside the furnace. The concentration of Mn in the flue gas was little changed before and after the dry cells were put into the incinerator, as shown in Figure 6 and 7.

Figure 7 also shows that the concentration of Zn in the flue gas was high when the furnace temperature was over 850 °C. The emission of Zn corresponded to 22 to 32 % of the total zinc within these cells.

(3) Incineration of manganese cells (experiment Nos. 2, 3 and 4)

When manganese cells (R20, 15 cells) were put into the incinerator, which is shown in Figure 4, the concentration of Hg in the flue gas was little changed before and after they were put in.

On the contrary, the concentration of Mn and Zn in the flue gas were 2 to 6 times and 2 to 2.2 times, respectively, that of before, as seen from Figure 6 and 7. The emissions of Mn and Zn were 22 to 30 %, and 10 %, respectively, of the total within these cells. The emission rates of these metals are generally high when the furnace temperature is high.

Figure 7 also shows that the concentrations of Cd and Pb were high when the furnace temperature was over 850 °C.

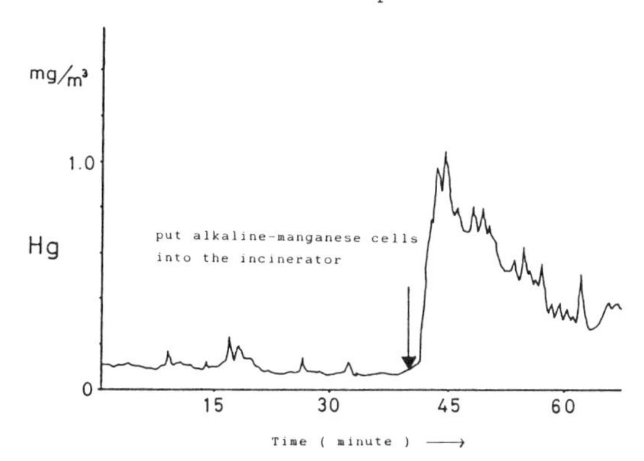

Fig.5 Measuring chart of Hg (2)

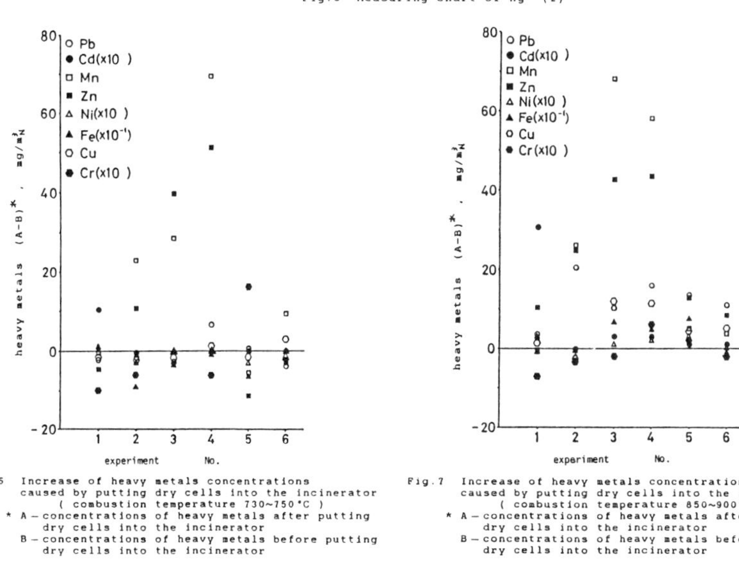

Fig.6 Increase of heavy metals concentrations caused by putting dry cells into the incinerator (combustion temperature 730~750 °C)
* A – concentrations of heavy metals after putting dry cells into the incinerator
B – concentrations of heavy metals before putting dry cells into the incinerator

Fig.7 Increase of heavy metals concentrations caused by putting dry cells into the incinerator (combustion temperature 850~900 °C)
* A – concentrations of heavy metals after putting dry cells into the incinerator
B – concentrations of heavy metals before putting dry cells into the incinerator

(4) <u>Incineration of nickel-cadmium cells (experiment No. 1)</u>

When nickel-cadmium cells (KR20, 10 cells) were put into the incinerator, the concentration of Cd in the flue gas was, as in Figure 6 and 7, 6 to 10 times that of before they were put in. The concentration of Ni in the flue gas was 1.2 to 1.4 times that of before. The concentrations of other metals (Hg, Pb, Mn and Zn) were unchanged after the batteries were put into the incinerator.

According to these experiments, the emission of heavy metals is clearly dependent upon the proportion of used dry cells. It seems, therefore, that the elimination of these cells from municipal waste by separate collection prior to incineration can effectively diminisch the emission of heavy metals.

4.2 The emission of PAH

(1) We measured the emission forms of PAH caused by municipal waste incineration. The results are shown in Table 3. The PAHs found in the flue gas were almost all vapor-form compounds.

Table 3 The emission forms of PAH

sample No.	PAH particle-form quartz thimble	%	$\mu g/m^3_N$ vapor - form condensed water	%	GF+PUFP a)	%	Total
1	0.08	1.8	1.21	27.8	3.07	70.4	4.36
2	N.D	0.0	4.52	13.1	29.9	86.9	34.4
3	0.11	0.2	10.6	17.8	49.0	82.0	59.7
4	N.D	0.0	5.79	6.6	81.5	93.4	87.3

note: a) glass fiber filter + polyurethan foam plugs
combustion temperature 840~900 °C
flue gas temperature of sampling location 340~370 °C

(2) When municipal waste was incinerated at 850 to 950 °C, as shown in Figure 8, the PAH concentration in the flue gas increased rapidly as the proportion of plastics in the waste increased from 0 to 24 %. And it appears that amounts of PAH tend to increase at high furnace temperatures.

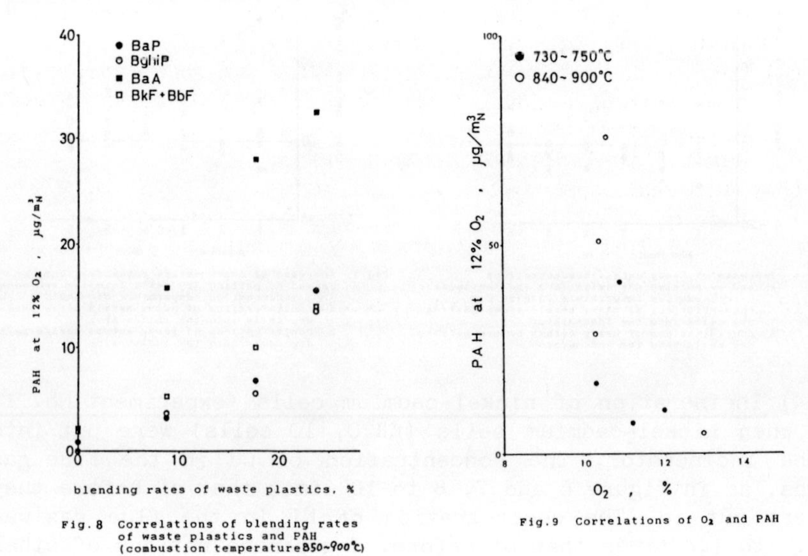

Fig. 8 Correlations of blending rates of waste plastics and PAH (combustion temperature 850~900 °C)

Fig. 9 Correlations of O_2 and PAH

These data indicate that the possibility of imcomplete combustion becomes greater with an increasing proportion of plastics in the waste at higher temperatures.
(3) As shown in Figure 9, there is a negative correlation between the PAH and O_2 concentration in the flue gas.
(4) A mass balance showed that detected PAH emissions were as much as 100 times greater than the amount put into the furnace. These data indicate that substantial amounts of these compounds can be created between the feed inlet and the stack under the rather poor combustion conditions of these diagnostic experiments.

It is clear from these experiments that the elimination of plastics from municipal waste by separate collection and the improvement of combustion conditions can effectively diminisch the emission of PAH.

5. Conclusions

The following key conclusions have been drawn from the extensive test data on the combustion experiments.

(1) When incinerating municipal waste containing no used dry cells or fluorescent tubes, the concentration of Hg in the flue gas was 0.03 to 0.05 mg/m^3.
(2) When alkalin-manganese cells were put into the incinerator, the concentration of Hg in the flue gas was 20 to 100 times that of before the cells were placed in the incinerator.
(3) When manganese cells were put into the incinerator, the concentrations of Mn and Zn in the flue gas were 2 to 6 times and 2 to 2.2 times, respectively, that of before they were put in.
(4) When nickel-cadmium cells were put into the incinerator, the concentration of Cd was 6 to 10 times that of before they were put in. The concentration of Ni was 1.2 to 1.4 times that of before.
(5) When municipal waste was incinerated at 850 to 900 °C, the concentration of PAH in the flue gas increased rapidly as the proportion of plastics in the waste increased from 0 to 24 %.
(6) There is a negative correlation between the PAH and O_2 concentration in the flue gas.

Semi-technical Testing of the 3R-Process

H. Braun[+], K. Horch[++], J. Vehlow[+], H. Vogg[+]

[+] Kernforschungszentrum Karlsruhe GmbH, Karlsruhe,
[++] Deutsche Babcock Anlagen AG, Krefeld,
F. R. Germany

1. Introduction

The safe disposal of residues from the flue gas cleaning process is an increasing problem in operating a municipal solid waste incinerator (MSWI) in Germany. Special attention is being paid to the fly ashes because of predominance of their flow and their content of mobile heavy metals.

If fly ashes are leached at a constant pH-value of 4, nearly 90 % of Cd, 50 % of Zn and smaller quantities of other metals are dissolved. This pH is reached by today`s acid rain and this leaching process may occur on landfills after a long period of time, when all alkaline salts are removed from the material.

MSWIs with wet scrubbing systems offer the opportunity to let this long-term leaching process take place in the incineration plant itself. At the Kernforschungszentrum Karlsruhe (KfK) this type of treatment of residues, called "3R-Process" was proposed some years ago [1-3] (3R = Rauchgas-Reinigung mit Rückstandsbehandlung, i.e. flue gas cleaning with treatment of residues).

2. The 3R-Process

Figure 1 gives a flow diagram of the 3R-process. The HCl containing solution of the first scrubber is used to leach the fly ash in a dissolving reactor. Mobile heavy metals like Cd are transferred from the solid into the liquid. The residue is fed back into the incinerator to bind it into the ash and to destroy adherent organic compounds like PCDDs.

ISWA PROCEEDINGS
VOLUME 1 ISBN 0-12-058451-4

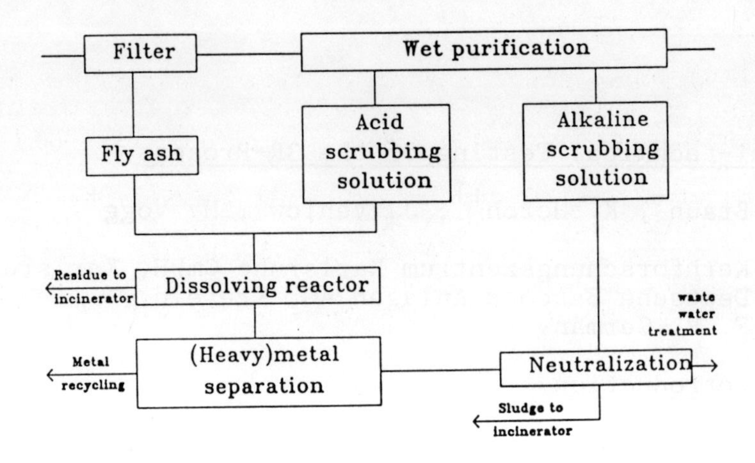

Fig. 1: Flow diagram of the 3R-process

The dissolved heavy metals can be separated from the liquid by precipitation or ion exchange. Precipitation gives an amount of about 1.5 kg toxic residue from 1 metric ton of burnt refuse. Ion exchange offers a better chance of recycling the metals.

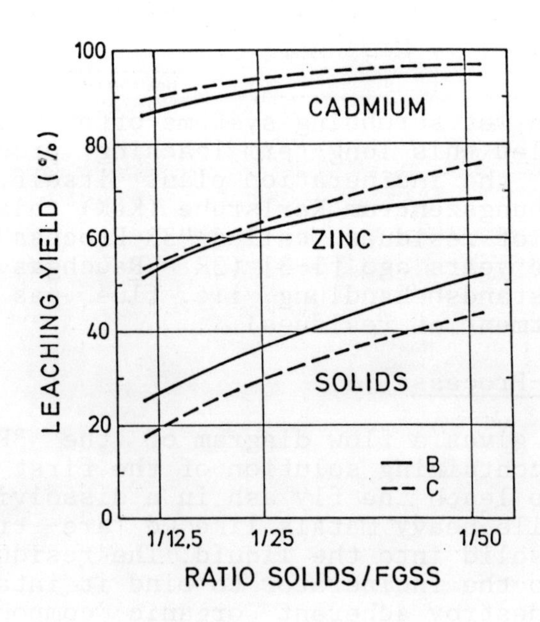

Fig. 2: Leaching yield after a time of 5 minutes

3. Laboratory-scale tests

In laboratory scale leaching tests original acid scrubbing solutions (pH=1) and fly ash from a full-scale incinerator were used. The pH of the mixture reaches values between 3 and 4 within leaching times of about 20 minutes. Figure 2 shows the 5 minutes leaching yields of Cd and Zn as a function of the liquid-solid ratio. At liquid-solid ratios of 25:1 more than 90 % of Cd and 60 % of Zn are extracted. Between 30 and 40 % of solids are also dissolved.

4. DORA pilot plant

To test the 3R-process on a semi-technical scale a demonstration plant called "DORA" was erected at the MSWI Oberhausen by the Deutsche Babcock Anlagen AG (DBA) and KfK (DORA = Demonstrationsanlage Oberhausen zur Rückstandsbehandlung bei der Abfallverbrennung, i.e. demonstration plant Oberhausen for treatment of residues from waste incineration).

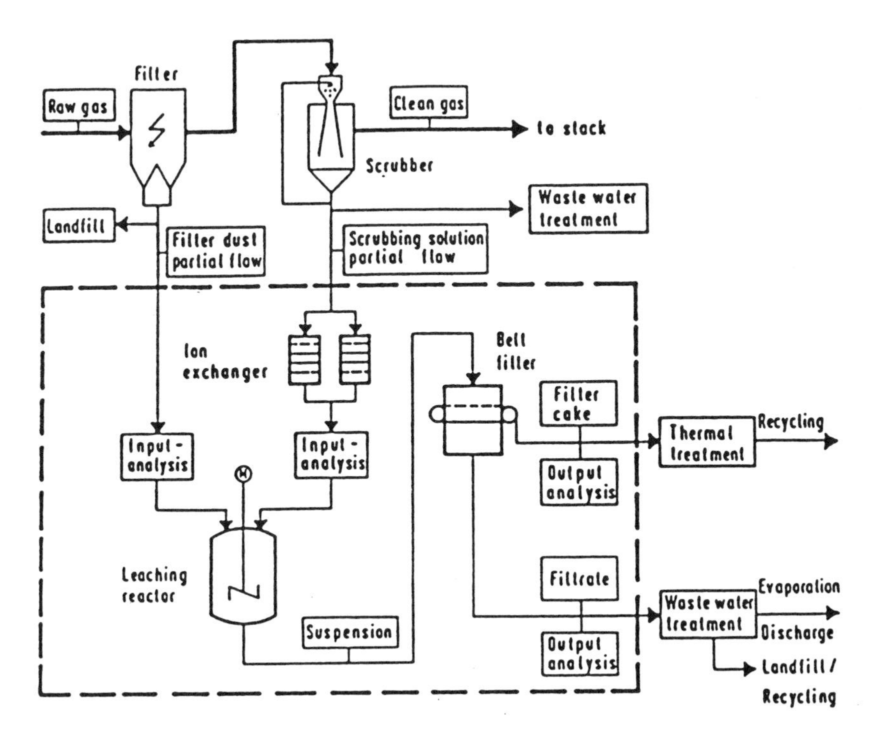

Fig. 3: DORA pilot plant

This pilot plant is operated as a bypass to the flue gas cleaning system of the MSWI. It has a capacity corresponding to a refuse throughput of 2 t/h. The flow diagram of the plant is given in figure 3. The partial flows of fly ash and acid scrubbing solution are directly discharged from the corresponding aggregates of the MSWI. An ion exchange column is inserted between the scrubber and the leaching reactor to separate Hg. The Hg contained in the flue gas is absorbed in the acid scrubber and would otherwise be strongly adsorbed by the carbon particles of the fly ash [4].

The residence time in the leaching reactor is about 1 hour. A vacuum belt filter is used for liquid-solid separation. The fly ash feed system contains an automatic sampling device. From all other inputs and outputs two samples per hour are taken for analysis. The solids are immediately analyzed for their metal content by means of a fast XRF-system [5].

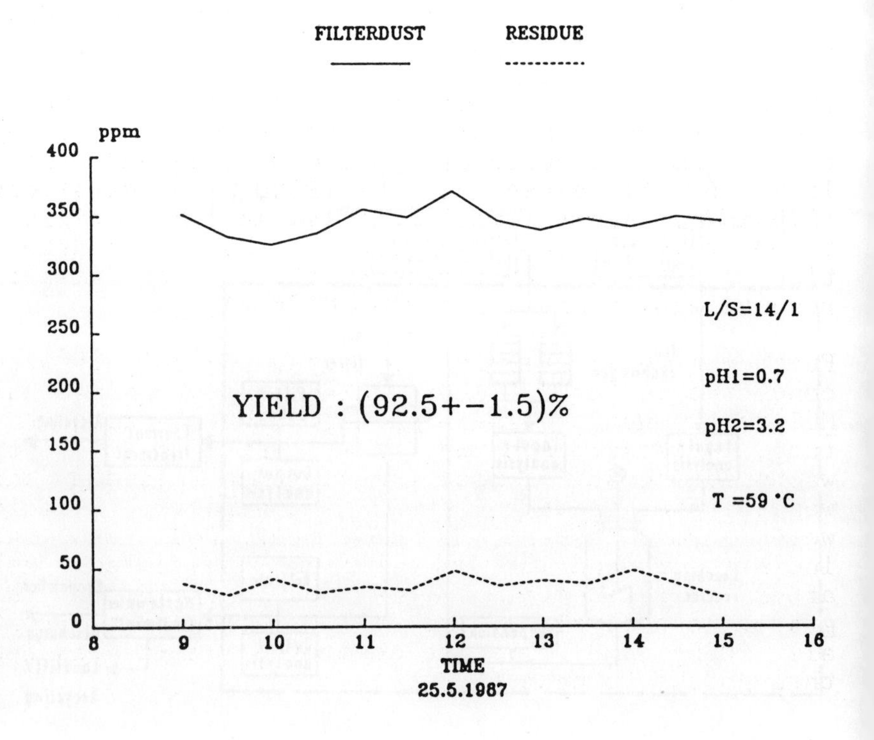

Fig. 4: Concentration of Cd in solids

5. First results of DORA

In the first test run in 1987 DORA was operated with 700 l/h scrubbing solution (pH=0.7), 50 kg/h fly ash, a residence time of 1 h and a leaching temperature of 60 °C. After leaching the slurry showed a pH-value of about 3.2.

In figure 4 the concentrations of Cd in the fly ash and in the 3R-residue are to be seen. The calculated mean leaching yield of (92.5 ± 1.5) % is in good agreement with our laboratory tests. The results for Zn were also in good accord with the preliminary investigations. A leaching yield of (62 ± 7) % was reached.

6. Decomposition of organic compounds

The second step in the 3R-process is feeding 3R-residues back into the incinerator to destroy organic pollutants and to bind the residues into the bottom ash. These experiments will be carried out in 1988, but some preliminary tests have been done on the thermal decomposition of PCDDs and PCDFs.

In the laboratory the thermal behavior of 3R-residues was investigated in air atmosphere according to the same procedure as applied by Vogg and Stieglitz for fly ashes [6]. The PCDD concentrations (300 ng/g) show a small increase at the critical temperature of 300 °C and decrease below the detection limit at a temperature of 500 °C. Similar results are obtained for the PCDFs [7].

Further experiments with compacted 3R-residues were conducted in the "TAMARA" (TAMARA = Test-Anlage zur Müllverbrennung, Abgasreinigung, Rückstandsverwertung, Abwasserbehandlung, i.e. test facility for waste incineration, flue gas cleaning, residue and waste water treatment) test incinerator. Pellets were formed by addition of cement, fed into the incinerator and singled out from the bottom ash after a residence time of about 30 min on the grate. An analysis of the pellets demonstrates that all PCDDs and PCDFs are destroyed while passing through the furnace [8,9].

7. Conclusions

A new process, the 3R-process, was developed to inertise fly ashes from MSWIs. It has been demonstrated on laboratory and semitechnical scale that mobile heavy metals are extracted from fly ashes by leaching them with the scrubbing solution from the acid stage of a wet gas purification system. The process includes the removal of Hg from the flue gas cleaning solution by means of ion exchange. Recycling of metals (Hg, Zn, Cd) is possible.

Experiments on a laboratory scale and in a test incinerator have made certain that organic pollutants are decomposed when leached ashes are fed back into the furnace.

Further experiments will be conducted to ensure binding of 3R-residues into the bottom ash of an MSWI. This product should have a quality which allows to reuse it, for instance in road construction.

8. Literature

[1] Vogg, H. (1984). Chem.-Ing.-Tech. 56, 740.
[2] Vogg, H. (1987). Int. Chem. Eng. 27, 177.
[3] Vogg, H., Christmann, A. and Wiese, K. (1987). In Recycling von Haushaltsabfällen (Ed. K.J. Thomé-Kozmiensky), 471. EF-Verlag für Energie- und Umwelttechnik, Berlin.
[4] Braun, H., Metzger, M. and Vogg, H. (1986). Müll und Abfall 18, 62 and 89.
[5] Lubecki, A. and Walk, H. (1986). KfK-4079.
[6] Vogg, H. and Stieglitz, L. (1986). Chemosphere 15, 1135.
[7] Vogg, H., Vehlow, J. and Stieglitz, L. (1987). VDI-Report No. 634, 541.
[8] Horch, K., Herden, A., Vehlow, J., Vogg, H. and Braun, H. (1987). In Müllverbrennung und Umwelt (Ed. K.J. Thomé-Kozmiensky), 756. EF-Verlag für Energie- und Umwelttechnik, Berlin.
[9] Vehlow, J. and Vogg, H. (1987). Int. Municipal Waste Incineration Workshop, Montreal.

PULSED AIR CLASSIFICATION FOR RESOURCE RECOVERY AND ENERGY PRODUCTION

J.Jeffrey Peirce

I. INTRODUCTION

Historically resource recovery and waste-to-waste production facilities have met with mixed success in communities throughout the world. Operating difficulties and unanticipated expenses have acted to preclude the utilization of a range of technologies as decision makers face costly waste utilization and disposal issues. Basic research conducted at Duke University over the past seven years has, however, led to the development of a promising alternative method for processing waste for material recovery or prior to energy production. [1,2,3,4,5] Pulsed flow air classification is a concept successfully demonstrated in the laboratory in its ability to separate particles for resource recovery and energy production. The objectives of this research thus include: (1) Document the theory of pulsed air classification, (2) Design and construct pulsing and non-pulsing classifier systems, (3) Test the abilities of the classifiers to separate: wheat/chaff; peanuts/shells; and combustible/non-combustible fractions of municipal solid waste (MSW). Each objective is addressed below.

II. THEORY OF PULSED AIR CLASSIFICATION

The premise of air classification is that a mixture of

ISWA PROCEEDINGS
VOLUME 1 ISBN 0-12-058451-4

particles entering a rising air column can be separated into constituents by adjustment of the velocity of the stream. Examples of the currently utilized non-pulsing designs are the straight and zig-zag air classifiers. That portion of the feed which is drawn upward with the air stream is called extracts, in waste-to-energy production facilities ideally all of the combustibles in MSW, while the portion which falls with gravity counter to the flow is termed rejects, in waste-to-energy production facilities ideally all of the non-combustibles in MSW. These current designs are also applied with mixed results in various materials recovery applications in mining and agriculture.

Pulsed-flow air classification was developed to differentiate between separation by aerodynamic characteristics and separation by particle densities. [1,2,3,5] The elimination of such aerodynamic characteristics as those involving airfoil behavior and tumbling leads to the isolation of mass and overall size as defining particle behavior, thereby isolating particle density as the particle characteristic defining the separation. In the ideal air classifier, density would define the separation. In passive pulsing systems discussed below the pulse varies with vertical position in the classifier, while in active pulsing systems the pulse varies with time. The mathematical expression for the forces acting of such particles is well documented. [1,2,3,4]

III. LABORATORY EQUIPMENT AND PROCEDURES

Complete pulsing and non-pulsing classifier systems were constructed in the laboratory to study the separation of wheat/chaff, peanuts/shells, and combustible/non-combustible portions of MSW. The components of the active-pulsing system are illustrated in Exhibit 1.

Test samples were injected into the classifier through the air lock chamber: wheat/chaff into small classifier throats ($\approx 100\ cm^2$ cross section, 1 meter tall) - 15 grams per sample, 5 replicates per mean rising air velocity; peanuts/shells into small classifier throats ($\approx 100\ cm^2$, 1 meter tall) - 20 grams per sample, five replicates per mean rising air velocity; combustible/non-combustible portion of MSW into large classifier throat ($\approx 1000\ cm^2$, 6 meters tall) - 4 kg per sample, 3 replicates per mean rising air velocity. Once a complete sample had exited the classifier system to either of the settling chambers, the efficiency of separation could be determined as:

Exhibit 1

Active Pulsing Classifier System

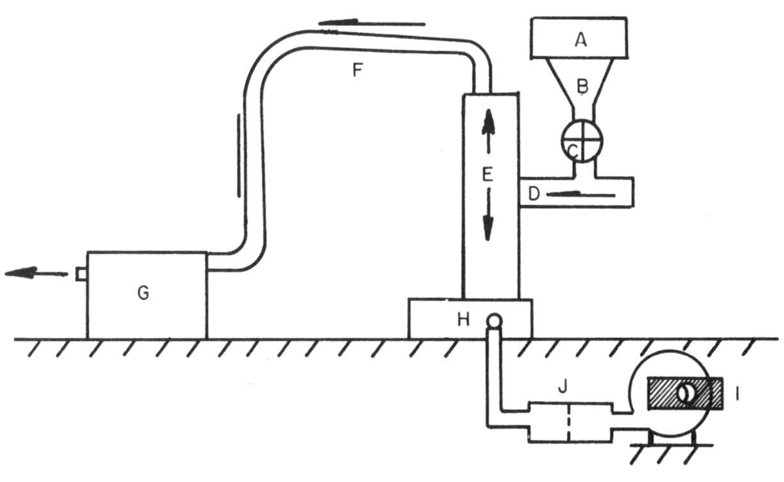

A = STAGING TABLE
B = SURGE BIN
C = AIRLOCK VALVE
D = MINI CONVEYOR
E = CLASSIFIER THROAT
F = DUCT WORK
G = SETTLING CHAMBER
H = SETTLING CHAMBER
I = FAN WITH SLIDING GATE VALVE
J = PULSING VALVE

$$E = \sqrt{\frac{X_e}{X_o} * \frac{Y_r}{Y_o}} * 100 \qquad (1)$$

where: E = efficiency (%)

X_e = mass of less dense particles (chaff, shells, combustibles) exiting top of classifier (gms)

X_o = mass of less dense particles input to the classifier (gms)

Y_r = mass of more dense particles (wheat, peanuts, non-combustibles) exiting the bottom of the classifier (gms)

Y_o = mass of more dense particles input to the classifier (gms)

This E valve is used to represent the efficiency of a particular classifier in separating the components of a sample at a specific mean rising air velocity. R_{90} is defined to the operating range of mean rising air velocity (M/S) over which at least 90% of efficiency is attained.

IV. RESULTS

Laboratory results are reported in terms of the maximum separation attained, E_{max}, as well as the R_{90} achieved, for each combination of classifier and particle feed tested at Duke University; E_{max} and R_{90} are presented in Exhibit 2 and 3 respectively. Exhibit 4, which focuses on MSW separation, is a graphical example of data collected for each classifier.

Exhibit 2

Maximum Efficiency E_{max} (%)

Particles Separated	Classifier: Zig-Zag	Passive Pulse	Active Pulse
Wheat/Chaff	94	100	no data
Peanuts/Shells	100	100	no data
Combustibles/ Non-Combustibles	98	99	99

Exhibit 3

Operating Range (M/S) Where Efficiency Exceeds 90% (R_{90})

Particles Separated	Classifier: Zig-Zag	Passive Pulse	Active Pulse
Wheat/Chaff	9.2 to 10.7 Δ = 1.5	9.2 to 13.8 Δ = 4.6	no data
Peanuts/Shells	5.6 to 15.8 Δ = 10.2	7.7 to 24.0 Δ = 16.3	no data
Combustibles/ Non-combustibles	0.5 to 2.9 Δ = 2.4	1.9 to 7.1 Δ = 5.2	0.9 to 4.9 Δ = 4.0

V. CONCLUSIONS

1. New pulsed air classifiers achieve a higher separation efficiency than existing non-pulsing classifiers. In all cases, the pulsing classifiers were seen to be capable of operating at an E_{max} of 99% efficiency or above.
2. New pulsed air classifiers achieve a higher separation efficiency over a much larger operating range than existing non-pulsing classifiers. The pulsing classifiers were seen to be consistently capable of operating above 90% efficiency over a range of air velocities of 1.5 to 2 times the range of existing non-pulsing classifiers.

Exhibit 4

Separation of Combustible/Non-combustible MSW Components

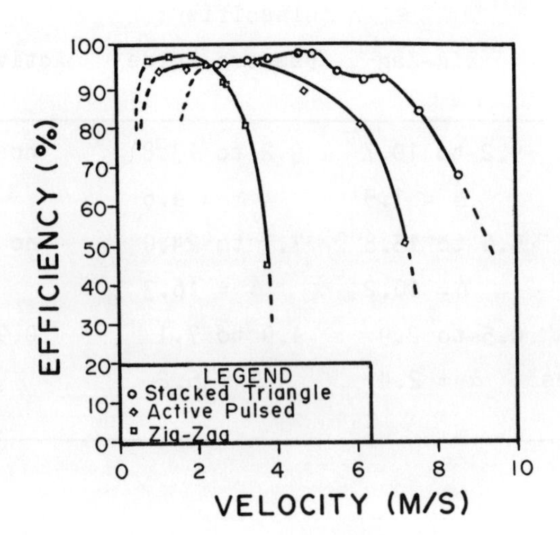

REFERENCES

1. Peirce, J. J. and Stessel, R. Particle Separation in Pulsed Airflow, Journal of Fluids Engineering of the American Society of Mechanical Engineers, New York.
2. Peirce, J. J. and Stessel, R. (1986). Comparing Pulsing Classifiers for Waste-to-Energy, Journal of the Energy Engineering Division of the American Society of Civil Engineeris. New York.
3. Peirce, J. J. and Stessel, R. (1986). Passive Pulsing Air Classifiers for Energy Production. Journal of the Energy Engineering Division of the American Society of Civil Engineers. New York.
4. Peirce, J. J. and Stessel, R. (1986). Separation of Solid Waste with Pulsed Airflow, Journal of the Environmental Engineering Division of the American Society of Civil Engineers. New York.
5. Peirce, J. J. and Wittenberg, N. (1984). Zig-zag Configurations and Air Classifier Performance, Journal of the Energy Engineering Division of the American Society of Civil Engineers. New York.

Simultaneous Combustion of Municipal Waste and Sewage Sludge in an AFBC Unit

P. Steinrück; F. Knoll
Research & Development
SIMMERING-GRAZ-PAUKER AG, Austria

1. Introduction

For decades, society's discarded materials have been disposed of as waste. However, facing increasing problems with landfill a growing number of people have come to question this practice and ask for alternatives.

All suggested technologies of resource recovery - source separation and recovery process systems - yield fractions of refuse derived fuel (RDF) with a high heating value.

The combustion of mechanically dewatered sewage sludge has been put into practice in many plants using the fluidized bed technology. High equipment costs and the need for auxiliary fuels to sustain stable combustion cause economic problems, especially for small units.

Up to now it is common practice to coburn sewage sludge in grate fired municipal waste incinerators. However, for some reasons the allowable quantity is limited. It has been shown that the mixing ratio can be varied within much larger ranges using fluidized bed technology.

But the FBC-systems installed to burn RDF so far suffer from poor reliability, high fuel preparation requirements and unsatisfactory combustion quality.

It is the objective of this paper to present a new unique FBC technology to overcome these troubles.

Figure 1 Fast Internally Circulating Bed

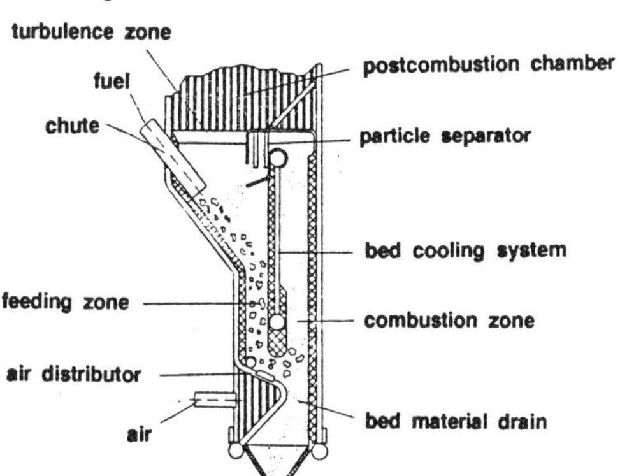

ISWA PROCEEDINGS
VOLUME 1 ISBN 0-12-058451-4

2. SGP's Fast Internally Circulating Bed (FICB) Process

This new **FICB** process is a development of Simmering-Graz-Pauker AG in collaboration with the Technical University of Vienna. It has been designed specifically to utilize sewage sludge, refuse and hazardous waste for thermal energy conversion. In addition, it also permits efficient and low-pollutant burning of coal and biomass.

The specific advantages of the FICB combustion technology are

- enhanced fuel flexibility
- simple fuel treatment
- easy removal of bulky incombustibles
- lower emission rates
- drastically reduced size
- significantly lower investment costs

These advantages are based on the principles incorporated into the design of the FICB process (fig. 1). An essential characteristic of the FICB process is its combination of circulating fluidized bed and post combustion chamber. The circulating bed is formed basically by two zones. The bed in one zone gets only small quantities of combustion air (bubbling bed), and the fuel is fed into this zone (feeding zone). Owing to the low gas velocity in this zone, small particles which occur frequently in waste fuels do not escape.

Extremely large quantities of air are added to the second zone (combustion zone) to produce a turbulent fluidized bed. This fluidized bed expands considerably and bed particles are swept out. A separator positioned above the outlet of this fast-moving zone separates the solids from the gas flow and drops them onto the slowly fluidizing feeding zone below. A shower of sand hits the incoming fuel particles, causing them to mix well with the bed below.

With both zones of the combustion chamber linked by a gap, the material fed into the slow zone can spill over into the fast zone, so that circulation of the bed material is achieved.

The fuel fed into the feeding zone moves with the bed material. In the course of this movement it is first dried, then deaerated and gasified before it finally passes into the turbulent combustion zone to be burned.

The gasification products released in the feeding zone float up from the feeding zone and in the mixing zone (turbulence zone) they meet the oxygen-containing flue gases leaving the particle separator. The flue gases have an extremely high flow velocity, causing heavy turbulences and thus create ideal conditions to achieve reactions.

The combustion zone has heat removal surfaces integrated in the chamber walls. This design effectively solves the problem of heavy wear of heating surfaces encountered in conventional fluidized bed systems. The arrangement also accounts for the high rate of partial load operation (40 %) possible with this system.

Movement of the bed material also ensures transport of any coarse particles that may have been added with the fuel to the bed drain, located below the combustion zone, from where they can be removed easily.

Both the ash removed at this point and the fly ash caught in the downstream dust collectors are completely burned out.

3. Combustion Tests with RDF and Sewage Sludge

To prove the claimed advantages, numerous combustion tests have been performed using the 1 MW_{th} pilot plant installed at the research center of Simmering-Graz-Pauker AG in Vienna. Test series with RDF originating from the Viennese Municipal Waste Separation plant (composition see fig. 2) revealed that FICB is capable of handling both fines and fluff. Due to the sand falling onto the incoming fuel, fluff is entirely mixed into the fluidized bed. Fines are kept down by the extremely low fluidization within the feeding zone (fluidization velocity $\sim$ 0.4 m/s).

Tests with different mixtures of dewatered sewage sludge originating from the Vienna Sewage Purification Plant and RDF (fig. 3) demonstrated that the emissions of SO_2, CO, NO_x and HCl (fig. 4) can be kept below the standards set by the new environmental protection laws using a bag-house filter for fly-ash collection and a one-stage scrubber for the elimination of HCl. Complete burnout of organic compounds has been achieved by the combination of high temperatures and turbulence in the secondary zone.

Figure 2 Composition of RDF in weight %

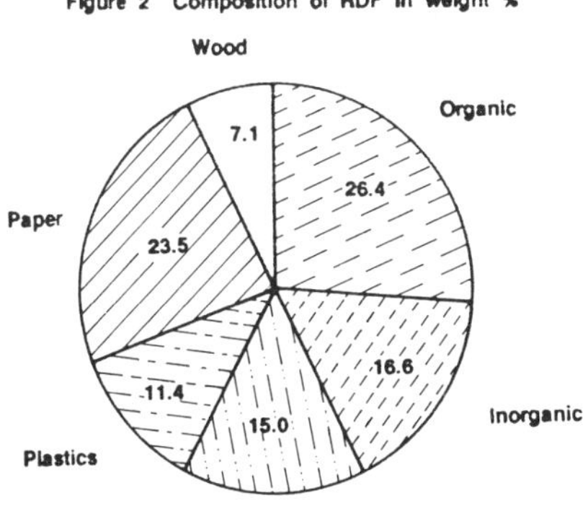

Figure 3 Different Mixtures of RDF and Sewage Sludge

Figure 4 Emission Profile at Inlet of Flue Gas Purification Plant

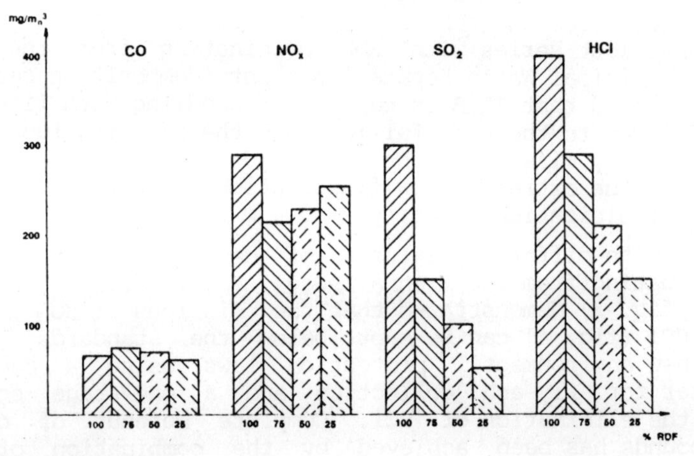

4. Case Study

Up to now (lit. 1) FBC combustion of municipal wastes (RDF, sewage sludge) has been considered to be advantageous in regard to costs since flue-gas cleaning facilities can be dispensed with. Although FBC in general and specifically the FICB process features excellent emission profiles, recent requirements set up by some municipalities, e.g. Vienna (table 1), cannot be met without use of sophisticated flue-gas purification equipment.

Thus it should be ascertained whether FICB can offer an economically feasible solution even with this new aspect.

Table 1 Plant Specification

Emission regulations	
NO_x	80 mg/m_n^3
SO_2	30 mg/m_n^3
HCl	20 mg/m_n^3
HF	0.5 mg/m_n^3
Capacity	
Refuse derived fuel (RDF)	25,000 tons/yr
Sewage sludge	8,000 tons/yr
Cogeneration cycle	
Hot water	8.0 MW_{th}
Electricity	1.7 MW_{el}
Revenues	
Hot water	US$ 12.5 /MWh
Electricity	US$ 66.7 /MWh
Landfill fee savings	US$ 47.5 /MWh

It is the goal of this case study to pose this question for a comparatively small installation. Moreover, the plant presented here demonstrates new technologies enhancing the environmental soundness of the entire process, i.e. a unique ash treatment process featuring both substitution of caustic material needed for operation of the flue-gas scrubbers and removal of mobile inorganic toxic compounds.

4.1 Design criteria and plant specification

The basic data of the specific plant are given in table 1. In order to meet local regulations (NOx < 80 mg/mn^3) SCR (selective catalytic reduction) technology has to be applied. It is worth mentioning that SGP's experience with both gas purification of municipal waste incinerations and SCR enables the design of a high dust system, avoiding auxiliary heat exchangers and consequently a lot of costs.

The different components of a composite plant capable of meeting the requirements are listed in table 2. The calculation of capital costs lead to a total investment of US$ 9,600,000.--.

Table 2 Modules and Capital Costs of a 10 MWth plant

Component	Investment	
Fuel Handling	US$	475,000
FICB module and boiler	US$	2,710,000
Flue gas purification, fly ash treatment	US$	2,230,000
High dust SCR	US$	665,000
Cogeneration cycle	US$	1,560,000
Process control	US$	1,040,000
Building, auxiliaries	US$	920,000
Total	US$	9,600,000

Table 3 Cost-Benefit-Analysis

Costs		
Total annual debt service [1)]	US$	1,090,000
Labor [2)]	US$	710,000
Residue disposal [3)]	US$	140,000
Electricity [4)]	US$	200,000
Supplies and chemicals	US$	300,000
Total	US$	2,440,000

Revenues		
Electricity [5)]	US$	910,000
Hot water [6)]	US$	800,000
Landfill-fee-savings [7)]	US$	1,570,000
Total	US$	3,280,000

[1)] Assumes a bond rate of 7.5% and a 15 year pay-back period
[2)] 1 plant manager, 15 plant operators, 1 maintenance mechanic
[3)] calculated as follows: 4800 tons/yr inert material x 12.5 US$/ton landfill fee +
1600 tons/yr fly ash x 47.5 US$/ton landfill fee
[4)] Calculated as follows: 8000 h/yr x 66.7 US$/MWh x 0.375 MW
[5)] Calculated as follows: 8000 h/yr x 66.7 US$/MWh x 1.7 MW
[6)] Calculated as follows: 8000 h/yr x 12.5 US$/MWh x 8.0 MW
[7)] Calculated as follows: 33.000 tons/yr x 47.5 US$/ton

4.2 Economics

A cost-benefit-analysis (table 3) reveals the economic benefits of the presented FICB-system in combination with other innovative technologies. The residue disposal costs entered in that calculation are comparatively low due to the good quality of the products of the ash treatment plant (landfill category II according to recommendation of Nord-Rhein-Westfalen).

5. Conclusions

Coburning of sewage sludge and RDF in a combined waste incineration plant is a practical alternative to landfilling. The use of innovative technologies - FICB-combustion, fly ash treatment and high dust SCR - yields excellent emission profiles, necessary to meet even very stringent emission regulations without incurring excessive costs.

6. Literature

(1) Bilitewski, B. (1987). Wirtschaftlichkeitsvergleich verschiedener Entsorgungssysteme, p. O1 - O22. Grazer Seminar Regionale Abfallwirtschaft.

LEACHING EVALUATION OF SOLID RESIDUE FROM DRY AND WET-DRY FLUE GAS TREATMENT IN THE COMBUSTION OF COAL AND MUNICIPAL SOLID WASTES (MSW)

Giugliano M., Cernuschi S., Bonomo L., De Paoli I., Istituto di Ingegneria Sanitaria del Politecnico di Milano, Italy

INTRODUCTION

A typical pulverized coal-fired plant produces 20-40 kg of bottom ash and 80-160 kg of fly ash, collected by the precipitators, per ton of coal burned. An MSW incinerator produces 225-315 kg of bottom ash and 25-35 kg of fly ash per ton of MSW burned. The current emission limits normally require the treatment of acid flue gases with cleaning processes which give rise to a further type of waste product. This investigation has examined 4 different residues from dry and wet-dry cleaning systems, which produce 10-40 kg of residue per ton of waste burned and 50-70 kg per ton of coal burned, depending on the dosage of dry and wet lime sprayed into the flue gases. The present study evaluates the chemical properties of the residues and their behaviour to leaching out with different leaching media.

EXPERIMENTAL

A representative sample of the following residues was used for the chemical composition and leaching test:

(1) A: flue gas desulphurization residue (without fly ash) of a coal-fired district heating power station. Wet-dry system.

(2) B: flue gas cleaning residue (with fly ash) of a MSW incineration plant (200 t d^{-1}). Wet-dry system.

(3) C: flue gas cleaning residue (with fly ash) of a MSW incineration plant (150 t d^{-1}). Dry system.

(4) D: flue gas cleaning residue (with fly ash) of a MSW incineration plant (36 t d^{-1}). Wet-dry system.

ISWA PROCEEDINGS
VOLUME 1 ISBN 0-12-058451-4

Tab. 1 - Content of trace elements and leachates characterization of a flue gas desulphurization residue of a pulverized coal-fired plant (A).

Parameter	Residue composition	Leachates characterization ($mg\ l^{-1}$)			MSW leachate filtered characterization
		Leaching media			
	Dry basis ($mg\ kg^{-1}$)	Acetic acid 0,5 N (pH ≃ 5)	Water continually saturated with CO_2	MSW leachate	($mg\ l^{-1}$)
pH final	-	5.1	5.8	7.75	8.30
Chloride	-	2510	2595	3510	2680
Sulphate	-	1079	343	373.6	197
Sulphite	-	< 0.05	< 0.05	642	10.5
Tot.Diss.Sol.	-	15535	8079	20083	15710
Arsenic	< 5	0.063	0.01	0.036	0.063
Bromine	320	22.4	22	24	7.6
Cadmium	< 30	0.07	< 0.05	0.07	0.05
Chromium (tot.)	< 14	< 0.10	< 0.10	0.29	0.67
Copper	< 20	0.09	0.05	0.12	0.10
Iron	1000	0.59	0.37	4.85	13.9
Lead	< 10	0.9	0.60	0.62	0.57
Manganese	50	4.1	3.4	0.09	0.05
Mercury	< 0.5	< 0.005	< 0.005	< 0.005	< 0.005
Nickel	20	0.53	0.38	1.69	1.10
Selenium	6	0.25	0.06	0.11	0.05
Zinc	14	0.42	0.17	0.62	1.06

The leaching behaviour of the residues was evaluated by shaking tests with a solid phase/leachate ratio of 1:16 (CNR-IRSA, 1985) and with the following leaching media: distillate water with acetic acid 0,5 N (pH=5), distillate water saturated with CO_2, and MSW leachate. The composition of the residues was determined by PIXE (proton induced X-ray emission), the leachates **analytical** characterization was conducted with atomic absorption and UV-VIS spectrophotometry.

Tab. 2 - Content of trace elements and leachates characterization of a flue gas cleaning residue of a MSW incineration plant (B).

Parameter	Residue composition Dry basis (mg kg^{-1})	Leachates characterization (mg l^{-1}) — Leaching media: Acetic acid 0,5 N (pH ≃ 5)	Water saturated with CO_2	MSW leachate	MSW leachate filtered characterization (mg l^{-1})
pH final	-	5.0	12.0	11.9	8.5
Chloride	-	7992	7785	13009	3650
Sulphate	-	438	799	993	219
Sulphite	-	2.5	0.3	3.0	10.5
Tot.Diss.Sol.	-	63650	18494	30346	16472
Arsenic	< 30	0.005	< 0.003	0.018	0.027
Cadmium	108	6.0	0.05	0.07	0.03
Chromium (tot.)	80	2.10	0.27	0.65	1.21
Copper	460	21.1	0.07	1.85	0.12
Iron	1800	11.7	0.60	0.56	30.3
Lead	3600	195	50.0	8.9	0.4
Manganese	210	2.12	0.06	0.08	0.42
Mercury	< 0.5	< 0.003	< 0.003	< 0.003	< 0.003
Nickel	60	1.28	0.43	0.75	0.81
Selenium	< 20	< 0.003	< 0.003	< 0.003	< 0.003
Zinc	4400	235	3.7	0.83	0.46

RESULTS AND CONCLUSIONS

The results of the investigation are reported in Tab. 1 to 4. With the exception of manganese, the toxic metals leachated out from the coal combustion residue are very low, according with the composition of the waste product (Tab. 1). Among the salts dissolved in the leachates the behaviour of the sulphites (toxic anions) is particularly interesting: substantially stable with acid acetic and CO_2, they are mobilized in large quantities with MSW leachate, probably in consequence of complexation effects developed by ligands, both organic and inorganic, which com-

Tab. 3 - Content of trace elements and leachates characterization of flue gas cleaning residue of a MSW incineration plant (C).

Parameter	Residue composition	Leachates characterization ($mg\ l^{-1}$)			MSW leachate filtered characterization
		Leaching media			
	Dry basis ($mg\ kg^{-1}$)	Acetic acid 0,5 N (pH ≃ 5)	Water saturated with CO_2	MSW leachate	($mg\ l^{-1}$)
pH final	-	5.0	12.3	12.0	8.5
Chloride	-	7266	6678	10518	3650
Sulphate	-	1109	1230	1432	219
Sulphite	-	2.25	14.8	9.0	10.5
Tot.Diss.Sol.	-	75896	17892	27056	16472
Arsenic	0.2	0.007	< 0.003	0.015	0.027
Cadmium	140	6.8	0.05	0.07	0.03
Chromium (tot.)	148	3.70	0.34	0.51	1.21
Copper	452	20.1	0.06	1.10	0.12
Iron	8000	50.2	0.59	0.70	30.3
Lead	4400	205	26.2	7.4	0.4
Manganese	212	2.89	0.07	0.09	0.42
Mercury	< 0.5	< 0.003	< 0.003	< 0.003	< 0.003
Nickel	56	1.32	0.40	0.71	0.81
Selenium	-	< 0.003	< 0.003	< 0.003	< 0.003
Zinc	5480	255	1.2	0.41	0.46

plex the cations and leach out the sulphite from the residue (Stanforth et. al., 1979). This effect was confirmed with **synthetic MSW leachate, containing typical complexing agents** as pyrogallol (1,2,3 - trihydroxybenzene) and EDTA (ethylenediaminetetraacetic acid).

The leaching tests conducted on MSW flue gas cleaning residues show that the leachability of salts and heavy metals is highly dependent on leaching media. The acetic acid solution, in consequence of the low pH reached in the test, is much more aggressive in the extraction of toxic metals than the other leaching media. The MSW leachate appears to be the

Tab. 4 - Content of trace elements and leachates characterization of a flue gas cleaning residue of a MSW incineration plant (D).

Parameter	Residue composition Dry basis ($mg\ kg^{-1}$)	Leachates characterization ($mg\ l^{-1}$) Leaching media: Acetic acid 0,5 N ($pH \simeq 5$)	Water saturated with CO_2	MSW leachate	MSW leachate filtered characterization ($mg\ l^{-1}$)
pH final	-	5.1	11.3	8.1	8.5
Chloride	-	13321	13667	17265	3650
Sulphate	-	1350	1669	2047	219
Sulphite	-	0.6	< 0.05	7.5	10.5
Tot.Diss.Sol.	-	47604	27778	36016	16472
Arsenic	-	0.006	< 0.003	0.016	0.027
Cadmium	440	26	0.07	1.44	0.03
Chromium (tot.)	192	2.95	0.30	0.48	1.21
Copper	2160	116	0.18	38.3	0.12
Iron	9200	8.2	0.96	1.30	30.3
Lead	11000	380	24.3	5.8	0.4
Manganese	-	4.54	0.07	0.16	0.42
Mercury	< 0.5	< 0.003	< 0.003	< 0.003	< 0.003
Nickel	-	1.64	0.62	0.95	0.81
Selenium	-	< 0.003	< 0.003	< 0.003	< 0.003
Zinc	22640	1120	1.3	26	0.46

least aggressive for all the subtances with the exception of copper (residue D). The comparison with leaching tests of pure incinerator fly ash shows a significantly enhanced mobilization of toxic metals from flue gas cleaning residue (Francis et al.,1987; Cernuschi et al., 1986). The binding of trace element, therefore, seems to be much more weak in flue gas cleaning residue than in pure fly ash. In the case of acid acetic test all the cadmium, copper, lead and zinc are leached out. The observed behaviours should make more difficult the landfill disposal and the co-disposal of the residues with the MSW.

ACKNOWLEDGEMENTS

The research was supported by the MPI 40% fund. The authors would like to thank Mrs. M. Casati and Mr. E. Gelmi for their technical assistance.

REFERENCES

Cernuschi, S., Giugliano, M., Marforio, R. (1985). Short term environmental behaviour of fly ash and slag from municipal solid waste incineration. Proceedings of "International Recycling Congress and Exhibition". 343-349. Berlin

CNR-Irsa (1985). Metodi Analitici dei Fanghi. Quaderni 64. Roma

Francis, C.W., White, G.H. (1987). Leaching of toxic metals from incinerator ashes. Journal W.P.C.F., 59, 579-586.

Stanforth, R., Ham, R. (1979). Development of a synthetic municipal landfill leachate. Journal W.P.C.F., 51, 1965-1975.

LONG-TERM BEHAVIOR OF BOTTOM ASH LANDFILLS

Jürg Krebs, Hasan Belevi and Peter Baccini
Swiss Federal Institute for Water Resources and Water Pollution Control, CH-8600 Dübendorf, Switzerland

1. Introduction

The long-term behavior of bottom ash from municipal solid waste incinerators in landfills is not known. Bottom ash can react in many different ways with rain water and various newly formed compounds in the leachate. Microbiologically mediated reactions of the organic carbon in the bottom ash and the formation of many degradation products, e.g. carbon dioxide, methane, organic acids, are probable. Existing reducing zones in the bottom ash can favor redox reactions. The course and the duration of these reactions are not yet known. Therefore it is not possible to predict correctly the element and compound fluxes from a bottom ash landfill into the environment. The objective of this study is 1) to elucidate the main reactions of bottom ash with water and to assess its long-term mobilization potential with respect to a few "indicator elements", 2) to discuss the "final storage quality" of a monolandfill for bottom ashes, i.e. a physical and chemical state of the landfill with residual fluxes compatible with the environment of the landfill site.

2. Methods

2.1. Principles

A bottom ash landfill is a "chemical and biological fixed bed reactor". There are different chemical zones in the reactor (e.g. oxidative or reductive). Dry parts in the reactor exist because the water flow through the reactor is not homogeneous. It follows that laboratory experiments can not simulate the complex reaction combinations in a real landfill. They are designed to give basic informations on reaction types with respect to the master variables and the order of magnitudes of reaction rates. In a first approach both systems (laboratory and landfill) are taken as two-phase systems (water/solid). The solid consists of two main parts: a first part which reacts with water and is well soluble or "mobile", and a second part which is practically "inert" after the reaction with water. From the laboratory experiments the "mobile" part of the bottom ash is to be assessed.

2.2. Sampling

The raw bottom ash (having passed a slag bath, residence time about 30-40') is collected, mixed thoroughly and screened with a sieve (> 5 cm) in the field. In the laboratory the Bottom ash is dried for 24 h at 105°C, ground for 2 h in a steel

ISWA PROCEEDINGS
VOLUME 1 ISBN 0-12-058451-4

bowl mill and screened again with a sieve. The resulting fraction (< 0.5 mm) is analyzed and used for all experiments.

2.3. Reaction with water

A batch procedure was chosen. The relation water/solid was 10:1. 30 g of bottom ash (grain size < 0.5 mm) were mixed in a Erlenmeyer flask with 300 ml distilled water, the flask covered with parafilm to guarantee a closed system and rotary-type extracted (with 150 rpm) for 0.5, 5, 50 or 500 hours respectively. The samples were filtered (membrane filter < 0.45 micron) and the electrolytic solutions analyzed. To characterize the solution, the following parameters were analyzed: Ca, Na, K, $Al_x(OH)_y^{3x-y}$, OH^-, Cl^- and SO_4^{2-} were considered as main elements in the solution. Furthermore organic carbon (as DOC), P (as phosphate), N (as nitrate and ammonium), Zn, Pb, Cd and Cu, which are enriched in bottom ash compared with granite [1], were determined. The additional reaction was carried out analogously with a bottom ash which was leached once. The alkalinity, defined operationally by the amount of acid to reach pH 7, was determined in eight parallel experiments. The pH of the suspension of 2 g bottom ash in 100 ml HNO_3-solution was measured after 48 hours of rotary-type shaking (150 rpm) in a closed system. As "reference solid" concrete was chosen.

3. Results and discussion

3.1. Composition and alkalinity of bottom ash

Table 1: Composition of bottom ash, granite and cement
a) Main elements (> 1%, calculated as oxides*, in g/kg)

	Na,K	Mg	Al	Si	Ca	Fe	C	S	Total
Bottom ash	60	25	107	513	153	116	16	8	998
Granite[2]	82	6	142	706	18	32	1	1	988
Cement [3]	12	15	55	225	640	25	<1	23	995

*: assuming that the total Al and Fe are in an oxidized form

b) Trace elements (< 1%, in g/kg)

	Cu	Zn	Cd	Hg	Pb	F	Cl
Bottom ash	1.200	3.50	0.0080	0.00030	1.300	0.29	3.0
Granite [2]	0.012	0.05	0.0001	0.00002	0.020	0.80	0.2
Cement [3]	0.120	0.12	0.0005	0.00050	0.017	n.d.	0.1

n.d. : not determined
Alkalinity of bottom ash : 1.5 mol base / kg bottom ash.

The main elements (> 1%) in bottom ash are Si, Ca ,Fe , Al , Na, K and Mg. As oxides they are responsible for more than 95% of the total mass. The non-metals carbon (1.5%) and sulfur (about 1%) belong also to this group. Compared with a common rock like granite the composition of bottom ash is very similar: Only Ca and Fe are slightly enriched (enrichment factors EF < 10). The non-metals carbon and sulfur are strongly enriched (EF > 10). Compared with cement all metals are slightly enriched (EF < 10) in bottom ash except Ca which is the main element (64%) in cement.
Important trace elements in bottom ash are Cu, Zn, Cd, Hg, Pb, F and Cl. All these elements except F are enriched with factors > 10 compared with granite. Compared with cement the concentration of these elements in bottom ash are at least an order of magnitude higher. The only exception is Hg.

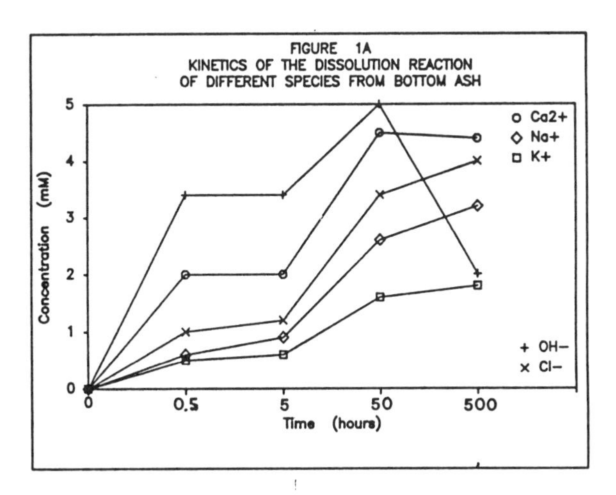

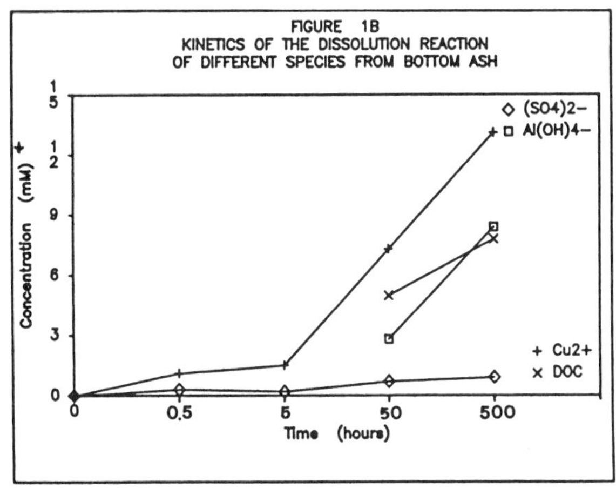

+ Cu^{2+}-concentration : (μM)

3.2. Reaction with water

3.2.1. Kinetics

The concentrations of the main cations Ca^{2+}, Na^{+} and K^{+} and of the main anion Cl^{-} are increasing in the electrolytic solution from 0.5 to 50 hours reaction time (Fig.1A and B). From 50 to 500 hours the concentrations of these parameters are either constant or only slightly increasing (Fig. 1A). The hydroxide ion concentration is the only one which decreases from 50 to 500 hours (about 3 mmol). This might be due to the reaction of $Al(OH)_3$ (s) with OH^{-}, increasing the $Al(OH)_4^{-}$ concentration from 3 to 8 mmol (Fig. 1B). In the same time period copper and DOC are also increasing (Fig. 1B). Cu^{2+} forms stable soluble complexes with organic ligands at this pH.

3.2.2. Ion balances in the electrolyte solutions

In a first reaction of bottom ash with water the resulting electrolyte solution consists mainly of Ca^{2+} as cation and of OH^{-}, Cl^{-} and $Al(OH)_4^{-}$ as anions. The pH of the solution is about 11.5. The solution is not saturated for any species. In the same experiment the solution of the reaction of the reference material concrete is practically a $Ca(OH)_2$-solution (about 90%) which is slightly oversaturated and gives a pH-value about 12.5.

Table 2: Ion balance in the electrolyte solution

	Bottom ash				Concrete			
	Experiment 1		Experiment 2		Experiment 1		Experiment 2	
	meq/l	%	meq/l	%	meq/l	%	meq/l	%
Ca^{2+}	9	68	9	80	36.9	90	5.2	84
Na^{+}	2.6	20	1.7	15	0.7	2	0.3	5
K^{+}	1.6	12	0.5	5	3.3	8	0.7	11
Total Error	13.2 (+/-0.5)	100	11.2 (+/-0.7)	100	40.9 (+/-0.5)	100	6.2 (+/-0.7)	100
OH^{-}	5	39	< 0.1	0	38.9	99	2.6	43
Cl^{-}	3.4	26	1.4	12	0.4	1	0.2	3
$(SO_4)^{2-}$	1.5	12	8.8	77	< 0.1	0	2.1	35
$Al(OH)_4^{-}$	3	23	1.3	11	< 0.1	0	< 0.1	0
CO_3^{2-}	< 0.1	0	< 0.1	0	< 0.1	0	1.1	18
Total Error	12.9 (+/-1.9)	100	11.5 (+/-1.0)	100	39.3 (+/-1.9)	100	6.0 (+/-1.0)	100

Experiment 1: reaction of bottom ash with water (relation 1:10), reaction time 50 h

Experiment 2: additional reaction of bottom ash with new water (relation 1:10), reaction time 50 h.

In the additional reaction of bottom ash Ca^{2+} is again the main cation in the solution. The main anion (about 80%) is now the sulfate, mixed with little amounts (about 10%) of Cl^- and $Al(OH)_4^-$. It supports the first observation (Fig.1) that the dissolution of sulfates is slower than the formation of hydroxides. The pH-value of the solution decreases to 9.5. The additional reaction of concrete with water also results in a decrease of the pH-value (from 12.5 to 11.5) and an increase of the sulfate concentration. The solution consists mainly of Ca^{2+}, OH^- and sulfate. A minor compound is the carbonate (18%).

3.2.3. Assessment of processes in monolandfills

Organic carbon concentration (TOC) in the bottom ash is about 10 g C/kg [4]. Provided that 50% of organic carbon would be microbiologically degraded to CO_2, nearly 1 mol base could be neutralized per kg bottom ash, i.e. about equivalent to the alkalinity determined (1.5 mol/kg, Table 1). The proton flux by precipitation (1 m^3/m^2&y rain with a pH of 4 on a landfill prism of about 10m in depth) is about 10^{-5} mol per kg bottom ash and year. In a time period of 1000 years this proton source (the air/landfill mixing with additional CO_2-fluxes is not known) would be still two orders of magnitude lower than the proton potential of the bottom ash itself. It follows that TOC is a master variable for the long-term behavior of a bottom ash landfill [4]. During the first decades after disposal leaching should occur under alkaline conditions.

Table 3: Assessment of the element concentrations in leachates from landfills for bottom ash

Element	m [mg/kg]	t [y]	c_E [mg/l]	c_L [mg/l]	c_{QS} [mg/l]
Cl	1600	10	3200	3200	100
SO_4-S	1500	100	300	80	30
DOC	1000	100	200	400	2-20
Pb	2	1000	0.04	0.02	0.05-0.5
Cu	15	1000	0.3	0.01	0.01-0.1
Zn	6	1000	0.1	0.2	0.2-2
Cd	0.03	1000	0.0006	n.d.	0.005-0.05

m : amount in bottom ash mobilized in lab. experiments
t : chosen leaching time
c_E : calculated concentration expected in leachate
c_L : mean concentration in leachate from a landfill (age ≈ 1 y) for bottom ash mixed with fly ash [5]
c_{QS} : quality standards [6,7]
n.d.: not determined

The measured total amounts m of seven elements mobilized in two consecutive laboratory experiments are given in Table 3. According to their mobilities [7] it is arbitrarily chosen that chlorine will be leached in 10^1 y, sulfate and organic carbon in 10^2 y and metals in 10^3 y with a zero order leaching rate. A landfill with 10 m in depth, a precipitation of 1000 mm/year and a 50% yield in leachate generation would produce 0.05 l leachate per kg bottom ash and year. The calculated concentrations c_E in leachates expected for such a landfill are in the same order of magnitude as the concentrations c_L in leachates from a landfill for bottom ash mixed with fly ash [5] (Table 3). The quality standards c_{QS} for leachates compatible with the environment are chosen on the basis of the Swiss quality standards for running waters ($\leqslant$ 10 x standard values) [6,7]. The leachate concentrations are higher than the quality standards for nonmetals and in the same order of magnitude as the quality standards for metals.

4. Conclusions

Under laboratory conditions the important reactions between water and bottom ash (for a first assessment of their long-term behavior) last at least days or even weeks. Consequently, only one test method carried out in hours to days is insufficient to assess the behavior of a material in a landfill. Additional properties, e.g. chemical composition, alkalinity, mobile fraction, have to be also considered.

The organic carbon may be the most important proton producer due to microbiological degradation to CO_2. Contents > 1% could suffice to acidify at least parts of the bottom ash and thereby increase the leaching of heavy metals strongly. Laboratory experiments and field assessments support the hypothesis that bottom ash cannot be considered as a material with final storage quality. Mobilization of nonmetals would cause high concentrations in leachates first. Consequently bottom ashes have to be pretreated to achieve this quality. Otherwise the leachate has to be treated during some decades.

Acknowledgment

This work was supported financially by the Swiss Federal Office Environmental Protection

References

[1] Baccini P.,Brunner P.H. (1985),Gas-Wasser-Abwasser 65,403.
[2] Wedepohl F. (1969), Handbook of Geochemistry 1.
[3] Lang Th. (1982), VDZ 43.
[4] Brunner P. et al. (1987), WMR 5, 355-365.
[5] Grabner E. et al. (1979), Bottom Ash, VGL.
[6] Swiss Ordinance on Waste Waters (1975).
[7] Belevi H., Baccini P. (1987), ISWA Symposium, Cagliari.

CHARACTERIZATION AND LEACHABILITY OF RAW AND SOLIDIFIED U.S.A. MUNICIPAL SOLID WASTE COMBUSTOR RESIDUES

Carlton C. Wiles, USEPA, Cincinnati, Ohio, U.S.A.

The United States, in common with most industrialized nations, faces the prospect of having fewer and fewer land sites in which to dispose of ever increasing amounts of municipal solid waste (MSW). Combustion processes (whether strictly for incineration or for energy recovery) offer the advantages of destroying the organics and reducing the volume of material remaining to be disposed of. In the United States and elsewhere, however, there is concern about the potential toxicity of materials and the potential harm to human health and the environment from leachates of municipal waste combustor residues disposed in the land. The United States Environmental Protection Agency (USEPA) is presently seeking to determine the true probable risk resulting from MSW combustor ashes along with the best regulatory approach to ensure the elimination of unacceptable risks.

Toxic Constituents in Ash and their Leachability

Results from past United States studies concerned with MSW combustion ashes have varied considerably because study objectives differed. Early studies principally focused on evaluating those ashes from the combustor designer or heat-recovery-boiler designer's point-of-view. Slagging characteristics, ash fusion temperatures, and similar measures were tested and analyzed with almost no information given to help define the quantity or leachability of the toxic heavy metals.

ISWA PROCEEDINGS
VOLUME 1 ISBN 0-12-058451-4

In the early 1970's, studies identified and quantified recoverable and recyclable metals in MSW combustor residue streams. Although some studies indicated the quantity of potentially marketable metals such as copper and nickel, the more toxic metals, which commonly exist at much lower concentrations, were historically ignored. In the late 1970's and 1980's, studies attempted to define the total and leachable quantity of toxic heavy metals in MSW ash streams. Here, however, different leaching tests were used to determine the quantities of leachable constituents, and authorities have disagreed about the validity of the different tests to predict the quantities of metals that will leach under field disposal conditions; thus, comparing the data is difficult.

In the United States, the present test to legally classify a waste as hazardous or nonhazardous, based on its toxicity characteristics, is the Extraction Procedure Toxicity Characteristic (EP_{Tox}) test which uses a weak acid leachant. This test may be replaced by the Toxicity Characteristics Leaching Procedure (TCLP), which was designed to determine the leachability of metals, pesticides, semivolatile organic compounds, and volatile organic compounds. The leachant used is one of two solutions of acetic acid diluted in deionized water, with highly alkaline wastes being extracted with the more concentrated acetic acid solution. Results from using these two tests in MSW combustor ashes is a source of controversy in the United States. Are the ashes hazardous or nonhazardous? Are human health and the environment threatened? The issue is raised because ash will, at times, fail the EP_{Tox} test, primarily for lead and cadmium.

The available data characterizing the toxic constituents in MSW incinerator residues can be very confusing. The reported values of total metals in the ash streams as well as the leachable metal content of those ashes vary considerably (Table 1). Because insufficient information exists about the manner of sampling and testing, as well as the control conditions on the combustor systems being tested, the reasons for this wide variability are undefined.

Table 1. Ranges of Total and Leachable Toxic Metal in United States MSW Combustor Ash Values Found by Researchers

Compound	Bottom Ash mg/kg	Bottom Ash Leachate mg/l	Fly Ash mg/kg	Fly Ash Leachate mg/l
Pb	31 - 36,600	0.02 - 34	2.0 - 26,600	0.019 - 53.35
Cd	0.18 - 100	0.018 - 3.94	5 - 2,210	0.025 - 100
As	0.8 - 50	ND(0.001) - 0.122	4.8 - 750	ND(0.001 - 0.858)
Cr	13 - 1,500	ND(0.007) - 0.46	21 - 1,900	0.006 - 0.135
Ba	47 - 2000	0.27 - 6.3	88 - 9000	0.67 - 22.8
Ni	ND(1.5) - 12,910	0.241 - 2.03	ND(1.5) - 3,600	0.09 - 2.90
Cu	40 - 10,700	0.039 - 1.19	187 - 2,300	0.033 - 10.6

ND = Not detectable; ( ) = Detection limit

Possibly this high variability is entirely real and is, in fact, unpredictable. Or possibly, the high variability results from shortcomings or inconsistencies in the sampling and analytical procedures, or from the inherent variability of metallic constituents in MSW from load to load.

The quantity of metallic constituents in the bottom ash will vary from those of the fly ash depending on the combustion system and the air pollution controls and the manner in which they are operated. These latter two factors also affect the chemical form (i.e., the actual salts of the various metals) present in the fly ash and bottom ash. These factors may be part of the reason for the high variability seen in the metals in EP_{Tox} and TCLP leachates from various combustor systems. The different solubilities of these different salts account for the fact that there appears to be no correlation between total metals present in the ash and the metals extracted in the leaching procedures. In other words, ashes with the same concentration of individual metals will not result in the same concentrations of those metals in the leachate from any of the leaching test procedures.

The conclusion is, then, that no one consistent statement can be made about the potential for leaching toxic metals from MSW combustor ashes. This is true in general as well as for specific installations. On any one day at any particular facility, the ash produced and tested can fail any of the leach test procedures for one or more of the metals of concern. Lead and cadmium are the metals which most often fail. No predictive procedure is available to identify the cause of such failure from such readily identifiable factors as location, season, combustor type, or combustor operating procedures.

Solidification/Stabilization as a Treatment Method

Because of potential harm to the environment from leachates of MSW combustor ashes, studies have been initiated on methods to control, minimize, or eliminate such threats--primarily studies of procedures for the careful disposal and long-term monitoring of MSW ashes disposed in "monofills," i.e., landfills which contain a single type of waste such as MSW ashes, and procedures for the solidification and stabilization of these ashes. One study's purpose is to find a product that will be resistant to, or relatively impervious to, the leaching of metal salts. Another is to develop solidification/stabilization processes whereby these ashes could be used as a major constituent in construction and building materials, obviously with no risk to the environment.

In the United States, initial studies indicate this solidification/stabilization technology does work. Most of the solidification processes tested indicate that leachability of toxic constituents from the final product is greatly reduced and that such final products will meet the leaching test procedures used to date (Table 2). This success does not come without cost, however: an indirect cost of 5 to 50 percent volume increase of the final product over the volume of the ash being solidified and a direct cost of $50 to $100/ton for reagent materials and processing the ash to a solidified product. If MSW combustor ashes are classified as hazardous, however, these cost factors may be insignificant when compared with the potential liability associated with the escape of leached toxic metals.

Table 2. Results of Stabilization Processes on Leachability of Metals from MSW Combined Bottom/Fly Ash; Concentration (mg/l) in extracts from EP_{Tox}

Compound	State	A	B	C	D
Pb	Raw Ash	0.44 - 0.58	0.014	0.011	70
	Stabilized	< 0.1	0.003	0.002	0.2
Cd	Raw Ash	< 0.05	0.042	0.022	0.05
	Stabilized	< 0.05	< 0.001	0.006	0.005
As	Raw Ash	-	0.092	< 0.004	0.17
	Stabilized	-	0.004	0.013	< 0.01
Cr	Raw Ash	< 0.05	1.5	< 0.009	0.66
	Stabilized	< 0.05	0.37	0.057	0.01
Ni	Raw Ash	-	0.25	< 0.02	-
	Stabilized	-	< 0.02	0.04	-
Cu	Raw Ash	-	0.04	0.54	-
	Stabilized	-	< 0.01	0.28	-

These cost factors may be controllable with more research. For example, studies indicate that the leachable metal fraction is inversely proportional to the size of the particles in the ash. Thus, ash may be screened so that the fraction of waste composed of large-sized particles would prove to be nontoxic and the much smaller fraction of small-sized material (which would fail the leach tests) would then be the candidate for solidification or other treatment to prevent leaching.

One concern for solidification processes performed primarily to produce building materials is the exposure of these materials to biota. Samples of such solidified concrete block used for breakwaters and sea walls were examined for any attached motile forms of marine life. The blocks were scraped and the organisms tested for heavy metal content. No toxic metal uptake was found. Although preliminary, the study is promising and shows potential for success in the use of fly ash as a constituent in building materials.

Current United States Studies and Regulatory Strategy

The USEPA supports research of MSW combustor ash in several areas: establishing a data base of information of the characteristics and management options for the ashes, further characterization of the ashes, and investigation of treatment and use alternatives. The USEPA is also field testing different combustor designs and air pollution control devices and studying the effect of waste stream characteristics on emissions. On an interim basis, regulatory guidance being formulated is expected to emphasize monofilling of MSW ashes. Long-term management strategy emphasizes utilization of the ashes.

LEACHATE FROM MUNICIPAL INCINERATOR ASH

by

Ole Hjelmar
WATER QUALITY INSTITUTE
11 Agern Allé, DK-2970 Hørsholm
Denmark

INTRODUCTION

The solid residues from a municipal solid waste incinerator (MSWI) consists of 80-90% (by weight) of bottom ash from the furnace bottom and 10-20% of fly ash from the dust collection system (electrostatic precipitators, fabric filters, cyclones).

If the incinerator is equipped with a flue gas cleaning system, a flue gas cleaning residue (FGCR) will also be produced. Depending on the system used, the FGCR may be a dry powder or a wet slurry, and it may contain various amounts of fly ash. Wastewater may or may not be produced from bottom ash quenching and flue gas cleaning.

Although some of the bottom ash is often sorted and utilized, a major part of the residues from the incineration and gas cleaning processes must be disposed of, usually by landfilling.

During recent years, growing attention has been paid to the potential hazard to groundwater and surface water bodies constituted by leachate from landfills containing incineration products, particularly fly ash and FGCR.

In order to improve the planning of new landfills for incinerator residues it is important to develop and verify methods of predicting quality and quantity of leachate.

This paper presents some results of laboratory leaching tests on fly ash and mixtures of fly ash and bottom ash from two large Danish MSWIs not (yet) equipped with scrubbers. The results are compared with data on the composition of leachate from an incinerator ash monofill.

ISWA PROCEEDINGS
VOLUME 1 ISBN 0-12-058451-4

ASH LEACHING TESTS

Ten column leaching tests were conducted on fly ash and mixtures of fly ash and bottom ash from two large MSWIs, A and B. Both plants receive household waste and non-hazardous industrial waste.

The chemical composition of the test materials is described in tables 1 and 2 and compare reasonably well with results from a study of residues from 6 Swedish MSWIs (Hartlén & Elander, 1986).

ELEMENT		UNIT	BOTTOM ASH		FLY ASH FROM ESP		NATURAL SOIL
			INC.A	INC.B	INC.A	INC.B	SANDY CLAY
Silicon	Si	g/kg	271	285	172	186	298
Aluminum	Al	g/kg	49	47	78	77	37
Chlorine	Cl	g/kg	-	-	-	-	-
Iron	Fe	g/kg	27	31	18	19	16
Calcium	Ca	g/kg	68	65	116	102	83
Potassium	K	g/kg	12	13	30	44	17
Magnesium	Mg	g/kg	8.4	7.7	19	17	6.9
Sodium	Na	g/kg	37	35	23	29	6.6
Sulfur	S	g/kg	2.2	1.3	11	18	1.0
Phosphorus	P	g/kg	3.7	4.1	8.3	9.6	0.7
Titanium	Ti	g/kg	3.2	4.2	9.5	9.3	2.2
Manganese	Mn	g/kg	0.9	1.0	1.0	1.1	0.4
Silver	Ag	mg/kg	7.1	14	-	-	0.048
Arsenic	As	mg/kg	24	19	49	71	13
Barium	Ba	mg/kg	1900	1700	1600	1000	-
Cadmium	Cd	mg/kg	1.4	3.2	250	250	0.28
Cobalt	Co	mg/kg	30	23	61	29	-
Chromium	Cr	mg/kg	230	230	180	220	24
Copper	Cu	mg/kg	2400	4300	890	1300	8.5
Mercury	Hg	mg/kg	0.09	0.14	1.4	2.4	0.11
Molybdenum	Mo	mg/kg	2.5	5.4	26	29	<4
Nickel	Ni	mg/kg	68	97	120	110	8.3
Lead	Pb	mg/kg	3100	2300	7400	7800	11
Selenium	Se	mg/kg	6	8	6.1	11	<1
Strontium	Sr	mg/kg	250	170	250	250	300
Vanadium	V	mg/kg	36	36	52	32	34
Zinc	Zn	mg/kg	2300	2500	25000	22000	37

Table 1 Analyses of bottom ash and fly ash from incinerators A and B and of a natural soil. Part of the ferromagnetic material and material larger than 40 mm (approx 30% of total) was removed from bottom ash prior to analysis.

PARAMETER	UNIT	BOTTOM ASH		FLY ASH	
		INC.A	INC.B	INC.A	INC.B
pH *	-	10.7	10.7	10.4	7.0
Alkalinity **	eqv./kg	1.8	1.8	2.5	2.6
TOC	gC/kg	4.8	5.6	17	7.1

* 1% slurry suspended in demineralized water for 30 minutes.

** Titration to pH = 7.

Table 2 Properties of bottom ash and fly ash from incinerators A and B.

The fly ash and fly ash/bottom ash mixtures were packed in acrylic plastic columns (diameter = 14.5 cm) to a height of 60-80 cm (8-10 kg on a dry weight basis), and artificial rainwater or artificial brackish seawater (salinity, S = $10^{o}/oo$) was passed through an upflow at a rate corresponding to 18 mm/24 hrs for an empty column. 6 consecutive fractions of leachate were collected from each column (L/S = 0-0.1, 0.1-0.2, 0.2-0.5, 0.5-1.0, 1.0-2.0, 2.0-5.0). L/S in the liquid/solid ratio.

A more detailed description of sampling procedures, analytical methods, and test design has been given elsewhere (Hjelmar, 1987). Some of the results of the leaching study are presented below.

Figure 1 shows concentrations of major components of the leachate (Na^+, K^+, Ca^{2+}, Cl^-, and SO_4^{2-}) vs. L/S for the rainwater leaching test on a 87%/13% (w/w) bottom ash/fly ash mixture from incinerator B. The corresponding concentrations of some trace elements (As, Cd, Cu, Mo, Pb, and Se) are shown in Figure 2.

The highest concentrations of most major and trace elements occur in the first fractions of leachate. This is not true for some of the solubility controlled elements, e.g. As (and V). The concentrations of As will often reach a maximum at relatively high L/S values.

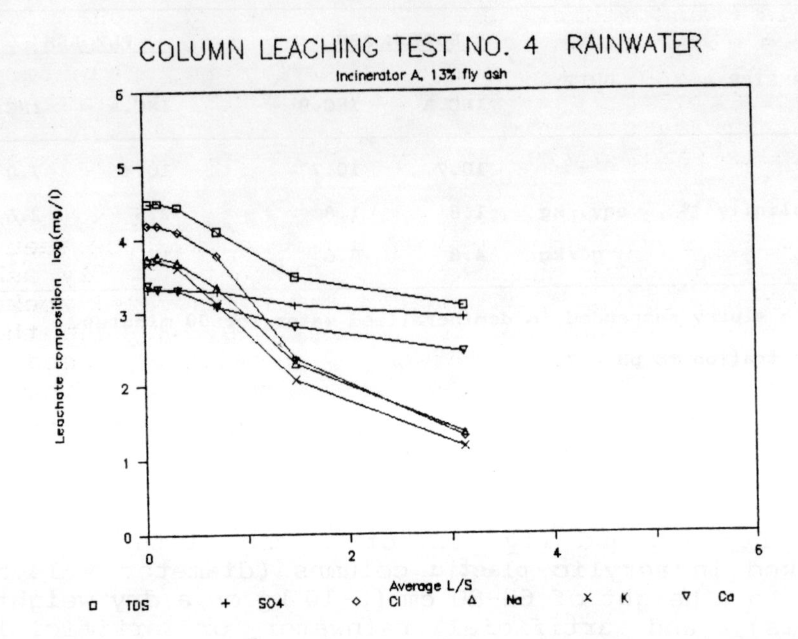

Figure 1 Concentration of major leachate components vs. L/S.

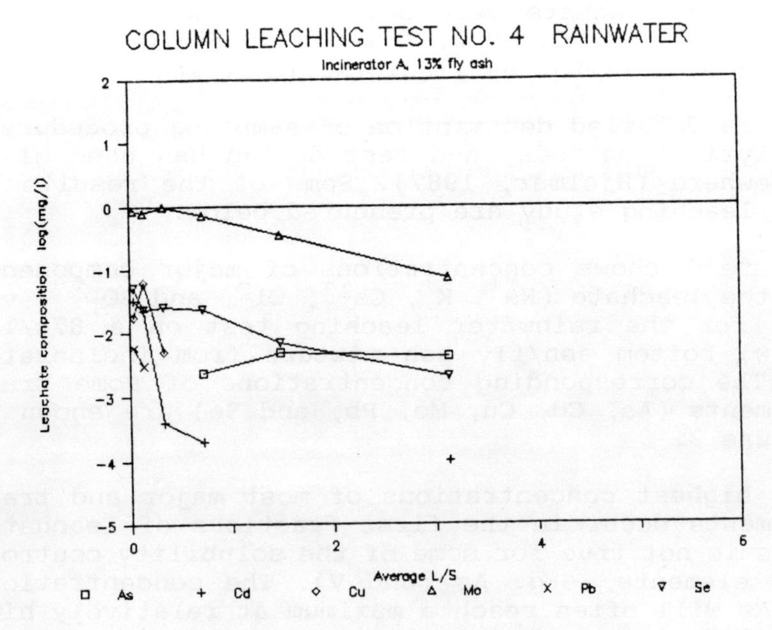

Figure 2 Concentration of trace elements in leachate vs. L/S.

Table 3 and Table 4 summarize the total amounts of trace elements (As, Cd, Mo, Pb, Se, and Zn) leached out for L/S = 0-5 with rainwater and seawater, respectively. It is clearly seen that for most of these trace elements, the amounts leached increase with increasing fly ash content in the ash mixture.

Large differences in leaching of Cd, Pb, and Zn from ash from incinerators A and B can also be seen. This is probably due to the fact that the fly ash from incinerator A had been moist for several weeks when it was packed into the columns, whereas the fly ash from incinerator B was dry. This caused a lower initial pH in the columns with fly ash from

RAINWATER Leaching of trace elements (mg/kg ash)							
INCI-NERATOR	%FLY ASH	As	Cd	Mo	Pb	Se	Zn
A	100	0.025	0.019	8.8	<0.1	0.20	<0.1
A	49	0.019	0.51	3.7	0.05	0.11	<0.1
A	13	<0.017	0.0054	1.3	<0.005	0.037	<0.09
B	100	<0.04	7.1	10	0.41	0.48	87
B	51	0.034	1.4	6.0	1.3	0.22	<0.3
B	15	0.016	0.012	1.9	<0.03	0.057	<0.1

Table 3 Total amounts of trace elements leached in column experiments with rainwater for L/S = 0-5.

SEAWATER, S = 10 $^{o}/oo$ Leaching of trace elements (mg/kg ash)							
INCI-NERATOR	%FLY ASH	As	Cd	Mo	Pb	Se	Zn
A	100	<0.02	0.035	10	<0.01	0.24	<0.1
A	13	<0.03	0.06	6.0	<0.015	0.044	<0.1
B	100	<0.055	6.6	11	<0.8	0.51	20
B	15	0.040	0.008	2.3	<0.02	0.065	<0.1

Table 4 Total amounts of trace elements leached in column experiments with brackish seawater for L/S = 0-5.

incinerator B, thus favouring the initial release of Cd, Pb, and Zn. Tests have shown that moistening of dry fly ash from incinerator B within a few days changes pH on reaction with water from an initial value of 7.0 to 10.0. The reason for this must be that the alkaline fly ash particles initially are coated with acidic condensation products.

Leaching of trace elements from incinerator ash with rainwater and brackish seawater, respectively, are seen to be of the same order of magnitude.

COMPARISON OF LEACHATE FROM LABORATORY LEACHING TESTS TO REAL LEACHATE

When the concentrations of contaminants in the leachate are expressed as functions of the liquid/solid ratio (L/S) it is possible to compare and extrapolate the results of column leaching tests to leachate monitoring data from full scale ash disposal sites.

In Table 5 the range of variation of composition of leachate from a Danish incinerator ash monofill (disposal site 1) which has been monitored regularly throughout its 15 years of existence is compared to the corresponding range of variation of composition of the 3 first fractions of leachate from column experiments No. 3 and No. 4 (leaching with rainwater). The content of fly ash in the monofill is assumed to be 10-40%. The design of the monofill and the leachate monitoring data have been described elsewhere (Hjelmar, 1987).

Despite all changes that may have occurred in waste composition between 1973 and 1985, there is good agreement between the results of the laboratory column experiments and the results of leachate monitoring at an actual landfill site. Some of the differences, e.g. in SO_4^{2-} and Ca^{2+} concentrations, may be explained by super-saturation of leachate from the accelerated column experiments. With a few exceptions, the predicted values are on the safe side (i.e. they are higher than those found in the actual leachate).

PARAMETER	UNIT	DISPOSAL SITE 1	COLUMN EXPERIMENTS		DANISH DRINKING WATER STANDARDS
			NO. 3 INC. A	NO. 4 INC. A	
TDS	mg/l	(15800-51200)	36100-72000	27000-30000	1500
pH	-	8.7-10.5	8.5-8.9	8.7-9.6	7.0-8.5
Alkalinity	meqv/l	1.4-5.1	1.2-2.3	1.4-2.1	-
Conductivity	mS/m	1650-3340	5100-9570	3980-4440	30-100
SO_4^{2-}	mg/l	1950-4900	2000-2560	2060-2130	250
Cl^-	mg/l	3700-11400	17500-38900	12400-15500	300
NH_3-N	mg/l	6-47	2.4-3.8	3.4-5.7	0.5
Na	mg/l	2900-7300	5800-12300	4800-5500	175
K	mg/l	1300-4300	8100-16000	4100-5100	10
Ca	mg/l	300-1000	2200-4690	2000-2400	200
Mg	mg/l	0.09-41	<3-61	<3-82	30
As	mg/l	0.006-0.025	0.0024-0.0028	<0.002	0.050
B	mg/l	(0.47-0.76)	1.1-3.1	0.98-1.1	1
Cd	mg/l	<0.0001-0.001	0.058-3.5	0.0004-0.032	0.005
Co	mg/l	(<0.005-<0.05)	<0.0001-0.0013	<0.0001-0.0029	-
Cr	mg/l	<0.002-0.08	<0.002-0.035	<0.002-0.012	0.050
Cu	mg/l	<0.0005-0.21	0.014-0.036	0.005-0.061	0.1
Hg	mg/l	<0.00005-0.0004	<0.0003-0.0019	<0.0003-0.001	0.001
Ni	mg/l	(<0.005-0.002)	<0.0005-0.001	0.0007-0.0016	0.050
Pb	mg/l	<0.0005-0.019	0.009-0.077	<0.001-0.006	0.050
Zn	mg/l	<0.01-0.32	0.02-0.091	<0.02-0.032	5
L/S range:		0-0.41	0-0.53	0-0.47	-
% fly ash:		Unknown	49	13	-

Table 5 Range of variation of composition of leachate from incinerator ash monofill (disposal site 1) and the comparable fractions of leachate from column experiments No. 3 and No. 4.

As can be seen, the drinking water standards are exceeded excessively by the concentrations of major ions (inorganic salts) in the leachate. For trace elements in the leachate from disposal site 1, however, the drinking water standards are exceeded slightly only for Cr and Cu. The high concentration of Cd in the initial leachate from column test No. 3 indicates that Cd (and Pb, Zn, and others) may leach out in considerable quantities under unfavourable conditions (especially at low pH). pH and alkalinity of the ash are key factors in determining short and long term leachability of a number of trace elements from incinerator ash.

CONCLUSION

Comparison of the results of column leaching tests on incinerator ash with leachate monitoring data from an incinerator ash monofill indicates that laboratory leaching tests which simulate the leaching conditions in a landfill may be useful tools for predicting the composition of leachate produced at incinerator ash landfill sites.

REFERENCES

Hartlen, J. and Elander, P. (1986). Residues from Waste Incineration - chemical and physical properties, SGI Varia 172, Swedish Geotechnical Institute, Linköping.

Hjelmar, O. (1987). Leachate from incinerator ash disposal sites. In Proceedings from the international Workshop on Municipal Waste Incineration, Montreal, October 1-2, 1987, NITEP, Environment Canada. (In press).

Session 9

Solid Waste Managment in Developing Countries

SOLID WASTE MANAGEMENT IN HOT CLIMATE: ISSUES,CONSTRAINTS,PROBLEMS AND MANAGEMENT STRATEGIES IN DEVELOPING COUNTRIES IN ASIA

B.N.Lohani

INTRODUCTION

Population densities are higher than any developed country, but Asia does not have adequate financial resources needed to provide equivalent solutions. Many changes have been taking place.The character of the solid wastes has changed in line with rising standards of living, retail distribution methods and fuel technology. The volume has increased; storage methods have evolved from open heaps through portable and disposable containers; transport has improved from horses to motor vehicles and from open trucks to compactor vehicles. The advent of high-rise buildings, supermarkets, institutions, hospitals and industries have given rise to new problems and efficient solutions have been devised. The quality of the urban environment is a matter of growing concern and the importance of efficient solid waste utilization and management is increasingly being recognized.

THE ASIAN SCENARIO

Nature of Solid Wastes

Tables 1 and 2 presents a comparative data on solid wastes situation of some countries. A quick run-through of the above data indicates qualitative and quantitative information of urban solid waste constituents. Differences in composition by weight percentage for each constituent varies, but a common trend is seen. With rare exceptions, the major constituents are paper and vegetable/putrescible matter, accounting about 50 percent by weight in industrial countries,and up to 90 per cent in developing countries.

The refuse analysis as presented in Table 2 is particularly useful for considering various stages of solid wastes management, viz: handling, storage, collection, disposal and recovery. These are briefly outlined in the following paragraphs.

Handling and Storage

On-Site Handling. In Asia, on-site handling of solid wastes are carried out for residential and commercial wastes. The residents or tenants of low-rise dwellings are in general responsible for placing solid wastes that are generated and accumulated at various locations in and around their dwellings in the storage containers. In some cases, wastes are also collected from door to door. Household compactors are not used to reduce the volume of wastes to be collected. Separated collection is not practiced by the household but a collector may go from house to house buying reusable items like glass bottles to sell them to companies for recycling. Separation of usable plastic and to some extent paper are also common in some countries of Asia as a source of income.

Storage Methods. Storage methods which are commonly used in Asia can be divided into two major groups, namely: the household storage methods and the communal storage methods. Household storage is in fact used to supplement the function of communal storage bins in most Asian Cities. Each family is asked to use containers to store their daily generated wastes which are dumped to nearest communal storage by private cleaners or directly brought to disposal sites by collection crew. Some of the households storage containers that are available in most Asian Countries are:

- plastic bucket with lids having capacities from 7 -11 litres which are suitable for a family of six with a daily collection service.
- plastic bins with lids having capacities from 20 to 30 litres and suitable for twice a week collection system.
- galvanized steel or plastic bins with lids having capacities of 50-70 litres and mainly used when there is twice a week collection from a high income group or daily collection from shops.
- expendable plastic sacks having different sizes but relatively less in use due to the cost of sacks.

In the communal storage method, wastes are stored in communal containers provided, in most cases, by government officials. Some of the typical communal storage methods in Asia are:

- Depots: These consist of a single storage building about the size of a large garage on the ground floor of a multipurpose building and are commonly adopted for the storage of domestic and trade wastes in a number of Asian cities. Depots are not highly recommendable because of the difficulty in acquiring sites and the high cost associated with such sites. Furthermore depots in Asia are badly operated.
- Enclosure: An enclosure is probably the most common communal storage method; its essential feature is a wall of timber, corrugated iron, brick or concrete which screens and contains the wastes. They have been observed with capacities from 1 to 10 m . The objectives to this type of storage are: (i) rain, animals, flies and scavengers have free access; (ii) the collection process is dirty and unhygenic; (iii) wastes tend to be thrown just inside the entrance; and (iv) enclosures are sometimes used for urination and defecation, increasing the risk to the health of workers.
- Fixed Storage Bins: These are also commonly used in Asia. This type of container is usually built from masonry wall or concrete blocks and it differs from the enclosure by having no entrance. The average capacity is usually not more than 2m and the walls are about 1.2-1.5 metre high. In one side of the wall, an opening, covered by a flap is provided through which the wastes are removed. The real objections to these containers are: (i) the collectors have to climb inside and thus they come in contact with the wastes; and (ii) the flaps covering the bottom opening tend to break off and disappear so that the contents overflow at that point.
- Concrete Pipe Sections: These are used in some Asian Cities with a low population density. The most common size is about 1 metre diameter, the length being of similar or shorter dimension. Although such a container is almost indestructible, it fails because: (i) the wastes are exposed to view; (ii) they are accessible to flies,rats,domestic animals and scavengers; and (iii) they are difficult to be racked away and provide the worker an intimate contact with the wastes.
- 200 Litre Drums: A few cities in Asia have demonstrated the possible use of 200-litre drums with reasonable success and in all these cases

the standard of management by the local authority has been very high.

Portable Steel Bins: The traditional steel (or plastic) bin of 70-100 litres used in the industrialized countries for domestic storage is also in use in few cities in Asia for communal use where generation is low and collection frequency high. Galvanized bins of about 100 litres have also been in use in some cities in Asia.

Basic Collection Systems in Asian Cities

Four collection systems have been employed in Asia and they are communal collection, block collection, kerbside collection and door to door collection.

Collection from Communal Site. In this system, the collectors just collect the waste from the communal storage which may thus require delivery of the wastes by the householder over a considerable distance. This collection system is used in Bangkok, Bangalore, Delhi, Madras, Manila, Rangoon and is introduced recently in Kathmandu.

Block Collection. In this system, a collection vehicle travels regular routes at predetermined intervals and the householders deliver the wastes to the vehicle at the time of collection; usually collection is done twice or thrice in a week and the residents are informed of the arrival of the vehicle by ringing a bell or playing a familiar musical tone.

Kerbside Collection. In this system, the residents place their bins on the footway in advance of the collection time and remove them after they are emptied. Kerbside collection is never entirely satisfactory in Asia because: (i) the bins are sorted through by scavengers; (ii) bins are stolen; (iii) traffic accidents are caused by bins rolling on the road; (iv) bins are sometimes turned over by animals; and (v) the householder fails to retrieve the bins quickly.

Transportation of Solid Waste in Asian Cities.

Some of the typical means of transportation of solid waste in Asian cities are mentioned below:

Handcarts. Handcarts are used in parts of Asia for street sweeping as well as for daily house to house collection especially in narrow streets and alleys. Most handcarts in Asia are often open boxes and the means of transporting waste is by dumping on the ground. This is wasteful of labour and increases vehicle standing time.

Pedal Tricycle. Pedal tri-cycles with a box carrier in front are common in Asia but their volumetric capacity is less than a handcart.

Motor-Tricycles. The two store, three-wheel motor cycle is in common use in several cities in Asia and West Asia. Its relative high speed gives it an operating radius of about 10 kms.

Tractor & Trailers. Agricultural tractors are almost universally available in countries of Asia. It offers the cheapest method of transport for solid wastes up to 6m .

Tonne Truck. This type of vehicle is used in Asia for transport of wastes from communal storage.

Container-hoist. The container-hoist is an alternative to tractor-trailer unit and is primarily aimed at handling wastes which will go more than a tonne to cubic meter. This is also in use in several Asian cities.

Disposal Methods

About 90 per cent of the waste in Asia is disposed of by crude dumping, although sanitary landfilling or controlled tipping is being practised in Korea, Singapore,Taiwan and Hong Kong. Non-mechanized or manual composting is successfully operated in India but is rare elsewhere. In Taiwan, provincial governments have erected semi-mechanical refuse composting treatment plants and a mechanized composting plant is also being operated in Bangkok. All these are not commercially successful because much of the product remains unsold.

Few incinerators operate in the region and to a very limited extent. An incinerator is being operated in Sri Lanka mainly for the destruction of foodstuff condemned as unfit for human consumption and disposal of animal carcasses. Incineration remains the preferred method of refuse treatment in Hong Kong and Singapore due to land constraints. A small incinerator plant in the Philippines is at present non-operational due to its high operating cost.

Labour-intensive recycling operates on a vast scale. Householders have newspapers, tin cans, scrap metals, plastic recovered and sell them. At every stage of storage, collection and disposal, saleable items are salvaged so that recovery is negligible when the wastes reach the disposal site.

Constraints and Problems for Effective Solid Waste Management Climate and Weather Conditions. A common denominator to all collection systems is climate and weather conditions. Heavy rains increase leachate problems and run-off causing surface water pollution. Accumulation of solid waste causes blockage of ditches, streams and canals after heavy rains and flooding which increases vehicular traffic - all helps to hinder waste collection. Moisture content and the density of wastes vary with the season and thus impose problems on both collection and disposal methods. Wet weather impedes plant operations which result in higher maintenance costs and easier wear and tear.

Obnoxious Odors and Air Pollution. Open communal storages and dump sites and open collection trucks produce obnoxious odors and wastes and litter are accidentally blown out into the streets or scattered by stray dogs and animals and illegal scavengers. Temporary storage, burning of refuse at street corners and roadsides also causes air pollution.

Rodents and Insect Vectors. The very large numbers of open communal storage and unofficial dump sites, together with the universal failure to cover landfills properly, encourages the breeding of flies and rodents.

Direct Health Hazards. Aside from the health risk brought about by these insect vectors, the methods of collection allow workers to be exposed directly by regular skin contact with wastes which sometimes contain fecal matter and offensive materials. This accounts for the increased incidence of certain diseases and accidents among refuse collectors.

Water Pollution. Leachate from dumps, static water in dump sites, the use of streams and canals for solid waste disposal and decomposition of solid wastes are potential sources of water pollution. Water drainage systems are frequently impaired due to blockage of ditches, gullies and sometimes major waterways by solid wastes.

Land for Waste Disposal. Almost everywhere, finding suitable landfill sites is a problem for governments due to the unavailability of and the high cost involved in acquiring the land. Often the landfill sites are outside the city in which transfer and transport of collected refuse add to the cost of operation.

Dust. In countries where controlled tipping or sanitary landfilling is practised, the process of excavation, hauling and placing of earth cover materials, under dry weather conditions, present a major dust problem. In Singapore where there is an extensive silting of roads caused by earth littering from constructional activities, the sweeping of heavily-silted areas by the use of the mechanized sweeping units generates dust and occasionally causes accidents as a result of dust obliteration.

Accidental Fires and Gas Leaks. Accidental fire at a landfill is not surprising and may arise due to several causes: a lighted cigarette thrown down by a careless worker, hot ashes in a vehicle delivering waste, the sun's rays through a fragment of glass. Fires have often been seen on the surface of landfills in most Asian countries due to high density, high vegetable components on the waste unlike in some European countries where serious underground fires have caused the collapse of the surface into voids caused by fire. Toxic gases like carbon monoxide, carbon dioxide, methane and hydrogen sulphide are produced as a result of decomposition. These gases produce an extremely offensive odor which at certain concentrations may cause unconsciousness.

Lack of Motor Vehicles and Equipment. Almost everywhere in Asia, motorized collection vehicles are not only too old but too few in number. Poor maintenance and the lack of a vehicle replacement policy aggravates the problem. A considerable amount of equipment used is no longer in good working condition. In most cities, regular checkups of the vehicles and spare parts replacement are ignored or not considered seriously due to lack of funds.

Lack of Cooperation from the Public. Public dissatisfaction with cleansing standards is often aired through the radio, press, letters and other media in the same way that most government officials complain of the lack of cooperation from the public. It is evident that services have not reached that stage which commands public respect and cooperation, with the exception of a few countries (e.g. Singapore), where more stringent laws and regulations are enforced and people are quite satisfied with the services that they are getting.

Delayed and Inefficient Refuse Collection. Improper scheduling of collection vehicles causes delays in collection. There are times when the collections are held up for such along time that households have to dispose of their wastes in their backyards,vacant lots or sometimes into streams or canals. Often, garbage collectors are too careless in transferring refuse from the containers to the dump trucks causing the scattering of litter and refuse on the streets.

Dense Traffic. In most large cities in Asia, the work of collection vehicles is impeded by dense traffic. Road congestion and street obstructions caused by the increasing number of roadside parked cars is the most notable problem in Singapore where mechanized street cleaners are being used. This problem is somewhat minimized by the collection of refuse at night or dawn.

Lack of Funds. The most pressing factor is the economic stability of the country, the availability of sufficient funds to enable the government to provide a good standard of service and the ability of the community to pay for such a good standard of service. Very often, solid waste problems are being ignored and given minimum priority both by the government and the public sectors. An average of about US$0.60 is spent in most Asian cities compared to about US$10/head/year in developed countries, inspite of a greater total workload due to access problems, the need for daily collection,more complex housing and inefficient traffic systems.

Limited Foreign Exchange. Shortage of foreign exchange for most countries in Asia means the purchase of foreign equipment is often blocked as preference is given to other essential items.

Social and Religious Factors. Although there is no wellfounded information of these matters, some strict religion do not tolerate the storage of kitchen wastes in the house, and they must be removed from the premises immediately after each meal. The manner of rearing domestic animals somehow depends on certain social and religious norms. The social and religious attitude towards cleanliness plays an important role too.

TECHNICAL OPTIONS FOR SOLID WASTE RECOVERY AND RECYCLING.

Several technological options for resource recovery and recycling exists, some of which are:

Separation or Initial Recovery

Known as the "front and system", separation or initial recovery can be done at the source manually and through physical and mechanical means. But for it to be a success in urban areas, dedication, interest, careful planning and administration and cooperation from the citizens is required. Residents have to be educated which ultimately means a change in their daily habits. Source separation is not a "technical option" in the truest sense, but it greatly facilitates the use of subsequent conversation processed which require sophisticated technology.

Conversation Process or Further Treatment

Incineration (with energy recovery), composting, pyrolysis, single cell protein (SCP) production, anaerobic digestion and landfill gasrecovery are some of the methods used for further treatment of the waste.

Incineration

In many developing countries, the calorific value of refuse is so low that a fuel supplement may be necessary as part of the incineration process. Most Indian wastes are unable to sustain their own combustion. The high moisture content of the Philippines´wastes also makes them unsuitable for incineration without the addition of auxiliary fuel or pre-treatment by sorting to remove the combustible material. All these additional steps would result in higher incineration costs.

There are a few countries like Korea which are considering incineration with heat recovery on a long-term basis. In Taiwan, Republic of China, the use of central incinerators is preferred as an efficient and sanitary method, but this requires large capital investments and operational costs. The mixed refuse of low heat value is removed first from a salvage or open dump by use of separate collectors,leaving only the combustible material.

There are three existing incineration plants in Hong Kong, each comprising four units, each unit having a maximum capacity of 10.5 tonnes per hour.The variable moisture content of the waste posed problems. To incinerate the waste to an acceptable degree, the basic design was modified. Furthermore, there are difficulties in public acceptance of incineration as a disposal process due to high capital and recurrent operating costs. In Jakarta, Indonesia incineration was regarded as not appropriate as about 30% of the volume of waste is in the form of dust, hard matter or metal and other materials.

Incineration was found to be the most cost-effective method in Singapore. Incineration cannot be applied to most urban wastes in Asia because of the high organic and putrificable fraction, high ash and non-combustibles, high moisture content and rather low calorific values.

Composting. There are several arguments which speak in favor of composting as a resource recovery option particularly suitable for developing countries, and these could be: (i) simple technology; (ii) process is readily adaptable to local conditions; (iii) only a small amount of residue is left to be disposed of; (iv) compost may sell well; (v) compost has a low toxic substance burden; (vi) composting meets requirements concerning hygiene; and(vii) refuse composition is appropriate for composting.

There are at least five pre-conditions which are necessary for the successful operation of composting projects in developing countries. These are: (i) suitability of waste; (ii) a market for the product near the compost source; (iii) support from government authorities, particularly those responsible for agriculture; (iv) price of the product which is applicable to most farmers; (v) net disposal cost (plant costs less income from sales) which can be sustained by the local authority.

Marketing of the compost products has been a problem and furthermore composting cannot recover certain valuable materials. Wood, rubber, textiles,metals and plastics are either impossible or very difficult to compost.

Pyrolysis. The results of pyrolysis developmental work abroad demonstrate that the process is self-sufficient and regenerative. The gas obtained is sufficient for heating the retort, the by-product tar-oil can be used as a substitute for furnace oil and the coke can be used as domestic fuel. Investigation of pyrolysis on a pilot or semi-commercial scale in the Indian cities of Calcutta, Delhi, Bombay and Madras estimated that 2000 tonnes day of glass and 3o tonnes day of coke, 70,000 to 90,000 1 day of oil, 30 tonnes day of glass and 30 tonnes day of metals could be obtained.

Single Cell Protein Production (SCP). Urban solid wastes that are rich in organics and cellulose can be utilized for SCP production in which a "controlled fermentation process where selective use of aerobic thermophilic organisms are used for the formation of protein from cellulose and lignin wastes".

Anaerobic Digestions. About70% methane, 30% carbon dioxide, traces of oxygen and hydrogen sulfide are the end products of an anaerobic digestion thus forming a valuable fuel gas. However, the constant requirement of carbon to nitrogen ratio 30:1 cannot usually be met in Asia due to the changing nature of the urban wastes.

Landfill Gas Recovery. This is rapidly attracting interest in many countries. Traditionally regarded as a nuisance, landfill gas is not only a potential source but its utilization helps alleviate problems of gasmigration from landfills and improves air quality adjacent to these sites.

SCAVENGING OF SOLID WASTES

Recycling in developing countries is higly labor-intensive. This is possibly only because the rapid rate of population growth in the urban sector, combined with the slow growth of the economy has led to a high level of unemployment and hence, low wages. Under these conditions, hand-picking of refuse becomes a viable economic proposition. Crude dumping is almost universal in these countries and often supports a large army of scavengers who extract valuable materials. It is paradoxical that the poorest countries are in this way achieving a high level of recycling despite the small proportion of saleable matter in the waste.

The grim realities of child labor and public health hazards must not be overlooked. As the whole family is usually employed in scavenging, even the very young and the elderly are exposed to a wide variety of pollution effects, obnoxious odors and most specially disease vectors which may seriously endanger the health of all workers who come in direct contact with the waste.

Common Recycling Methods

Household or Internal Waste Sorting. Some materials in collected refuse are noticeably low as they are already sorted out at the household level, used and/or sold. Cinders, coal, coconut shells are extracted for fuel, metal cans for domestic vessels and vegetable wastes for animal feed. Others,e.g. paper (for repulping), tin cans (for remelting), glass (for re-use, remelting or abrasives manufacture) and plastics (for re-use or inferior grade production), are sold to small merchants who operate collection and sorting

Recycling Potentials through Organized Scavenging

Collection from Disposal Sites of Municipalities. This is practised to a high degree in developing countries and scavengers perform efficiently as soon as collectors empty their loads at the disposal sites. Sometimes whole families of scavengers support themselves living on these dump sites and little of value is thrown away.

Scavenging is also highly organized at major dump sites in Manila. The system in effect is self-contained through the action of middlemen and subsequent interaction with buyers of secondary materials. Although this provides a method of resource recovery, the scavenger, who is at the heart of the system, must work under deplorable and hazardous conditions.

In Bangkok, salvage of materials by garbage collectors and other workers is performed during the collection stage. The volume of retrieved materials is equal to about 5 bamboo baskets; paper and vinyl bags or the like are retrieved. The salvaged materials are purchased by junk dealers. Monthly earnings from this source amount to an average of US$80-90.

Salvaging operations are largely uncoordinated; scavengers are allowed to operate at all stages of collection, storage and waste disposal. While some capital-intensive, high-technology solutions alleviate the human costs of scavenging, these ignore the potential for efficient waste recovery and hence, the source of employment and resource base for a variety of small enterprises. Existing scavenging practices provide employment for new arrivals and household income ranging from base survival in exploitative entrepreneurial recovery. Sorting and marketing of recyclable materials is ordinarily in the informal sector and small enterprises return them to the formal sector.

Organized Scavenging. The potential of developing organized scavenging in Asia should be emphasized. In the Philippines, a prototype solid waste project called "Resource Recovery" was launched. This is aimed at demonstrating a more systematic way of recycling starting from education at the household level to training and fielding of "ecology aides" or Eco-Aides to the identification of end-buyers. The establishment of such a working system will provide a more decent means of living for scavengers and minimize their exposure to filthy conditions.
The project plan is to hire as many of those currently engaged in scavenging as Eco-Aides. They will make the rounds of households and be authorized to buy at predetermined prices those recyclable materials which have been previously sorted by the household into wet organnic or dry re-usable items. The materials will be redeemed at redemption centers at about 20-25% more than what they paid for them, and where the materials can later be sold to junk dealers. As high as 60% of the garbage in Metro Manila may be recycled, 20% consisting of "wet garbage" can be digested for methane production, 10% for composting and the rest to be landfilled. Difficulties in administering the project has been identified.
In refuse disposal sites in Bangkok groups of 12-30 scavengers demarcate the site radially into sectors by means of wooden logs and separate the valuable materials into baskets. The disposal trucks come to the center of the plot and dump the solid waste. This form of scavenging is so profitable that the leader of a group even gives cash or liquor to the driver so that he does the favor of dumping the waste more often into his zone. All materials are collected and sold to middlemen who visit the site.
If emphasis is put on a better working standard at the site, vector control measures to prevent

diseases, and basic amenities, the scavengers would be a happier group of people. Furthermore, if the system is controlled to avoid middlemen cheating these hard workers, organized scavenging would be a very prosperous livelihood for many, an economic gain for the city and an achievement in the field of solid waste management.

A recent study at Bangkok on the health conditions of the scavengers revealed that generally speaking scavengers had poor health; colds were very common as they were exposed to sun and rain for a long duration;the prevalence of muscle and tendon disease was very high, particularly muscle ligaments affecting the back; skin disease was common and difficult to cure; hand and foot injuries were very common; and depression and "moody" feelings crept in among the scavengers due to long, continuous working hours.

It is very difficult to get a picture of the exact nature of the trade at the wholesaler level. They are very secretive about divulging any information regarding the trade. Overall reaction from the wholesalers was that their profits were large. The wholesaler has an establishment including a shop, a godown, 4-5 laborers. Besides the overheads, he can still make a profit which shows that by establishing an organized scavenging system, the scavengers could benefit more by cutting out the middlemen.

SOCIO-ECONOMIC CONSIDERATIONS IN RECOVERY AND RECYCLING OF URBAN SOLID WASTE

In the decision-making of waste management the economic analysis is concerned with the assessment of the costs and benefits that are attached to any particular course of action is required.

Cost and Benefits of Resource Recovery and Recycling. The first and most fundamental step in using economic analysis in waste management is to identify all the relevant costs and benefits that need to be taken into account in the particular decision being examined. The attractiveness of resource recovery and recycling as a method of handling waste depends on how the costs and benefits of the available resource recovery and recycling methods compare with the costs of conventional methods of waste collection and disposal that would otherwise be employed. From the point of view of the responsible authority

recovery should be of interest if the cost of collecting the material or product plus any cost of sorting or separation of the material or by-product generation less any saving in overall waste collection and disposal costs is less than the revenue obtained from the sale of recovered materials, energy or other by-products.
From the point of view of the buyer of the reclaimed material it will be worthwhile if the cost of buying the reclaimed material/energy is less than the cost of purchasing or producing virgin material/energy less any extra cost involved in processing and using reclaimed material/energy less any additional transport cost involved in obtaining the reclaimed rather than the virgin material/energy.
In general the scope for using recovered rather than the virgin material/energy and the level of extra costs for the user to be set against any saving in price depends in practice on a number of factors such as:
- the technical ability to turn reclaimed material or by-product into an acceptable substitute material.
- the market´s attitude to final products made with recovered material.
- the supply conditions for reclaimed as compared with virgin material.
- the location of production facilities in relation to sources of materials.

Social Aspects. The attractiveness of resource recovery and recycling should also be looked at from the society as a whole since the views of the user of the recovered materials or the waste handling authority may be different. Some of the issues of interest which should be considered in the assessment of costs and benefits of resource recovery may be:
- the value of the society of the waste material not be recovered.
- the value of the materials in terms of saving of imports and scarce raw materials.
- the impact on pollution of the natural environment.
- the effect on public health and safety
- the acceptance of the recovered materials by the users.

SOLID WASTE MANAGEMENT PLAN

The objectives of solid waste management plan is to ensure efficient and economic collection, handling, utilization and disposal of solid wastes with minimum acceptable environmental impacts. Since such a plan has to be investigated on a longer time framework, it is often desirable to include solid waste management in the urban development masterplan. The structure of the solid waste plan consists of the following stages. (i) accurate data on analysis and fore/casting of solid waste characteristics and quantities; (ii) storage, collection, handling and transport; (iii) utilization, recovery and recycling; (iv) disposal options including identification of suitable sanitary landfill sites; and (v) organization, management and implementation.

WASTE MANAGEMENT AID PROGRAMS - ARE THE PRIORITIES CORRECT?

by

David W. Jackson M.B.E.

President, European Community

Waste Management Organisation.

1.0. INTRODUCTION.

For well over 50 years a regular waste collection and street cleaning service has been an accepted municipal service in both urban and rural communities of all industrialised countries and the financial resources needed to support such a service have been provided as a matter of course. In developing countries, these conditions do not universally apply and, particularly in areas of rapid urban growth, municipal services have not kept pace with massive population increases. As international funding agencies have increased the aid to urban infrastructure and health improvement projects over the past two decades, the main initial thrust tended to address water supply and drainage problems with normally relatively small amounts of aid being allocated to solid waste management, mainly for the purchase of vehicles and equipment. It is only in comparatively recent times that it has been realised that spending money on the importation of Western technology and equipment is not the most effective way to improve environmental standards.

2.0. CASE STUDY OF A SWM PROJECT IN INDIA.

The following report on the initial appraisal of a SWM component of a proposed urban project covering nine large towns in India is indicative of the deficiencies to be found in many of the developing countries.

2.1. Background The mission commends the quality of the reports

ISWA PROCEEDINGS
VOLUME 1 ISBN 0-12-058451-4

on the solid waste management (SWM) component prepared by the project towns. In view of present deficiencies in the provision of vehicles and equipment there is no doubt that the implementation of the current proposals would result in an improvement in service delivery. However, there is also no doubt that the inherent basic weakness of the present SWM operations will greatly restrict the benefits obtainable unless a vigorous effort is made to address certain fundamental problems.

With one or two notable exceptions, acceptance of the proposals without tackling these problems will simply perpetuate the present grossly inefficient system and the improvement of service delivery will not be commensurate with the additional capital and recurrent revenue costs incurred.

The fundamental problems identified by the mission are as follows;

(i) Failure to appreciate the total cost of present poor service which accounts for 30% to 50% of total revenue expenditure of all project towns;

(ii) Inadequate senior management structure to control such expenditure;

(iii) Lack of operational database to monitor performance;

(iv) Poor primary collection system resulting in double or treble handling of wastes;

(v) Lack of integration of street sweeping and drain cleaning activities with transport operation;

(vi) No routine vehicle servicing or maintenance procedures and poor workshop facilities;

(vii) Failure to provide adequate funds for vehicle repair, resulting in high proportion of vehicles off the road.

2.2. Institutional In all the project towns, the SWM service is by far the largest employer of labour and user of transport and spends the largest proportion of the revenue budget and yet in no town was it possible to find, at senior management level, an individual officer with direct line management responsibility for all aspects of SWM operations.

Ideally each town should have a separate Conservancy or Cleansing Department with its own administrative, financial and technical support staff, headed by a technical officer with SWM experience and being directly responsible to the Administrator or Executive Officer for all aspects of SWM. At present, there is no pool of such officers within the State's resources, but training in SWM can and should be provided, under the Technical Assistance and Training component of the project for such designated officers.

2.3. **SWM Operational Data Base**. Reference has already been made to the drain on municipal revenues owing to the high cost of SWM operations. One objective of the SWM component is to raise the level of service delivery to as near to 100% of the population as is possible without substantially increasing the cost by carrying out the work more efficiently. To do this, it is necessary to have accurate data on both the cost of SWM and the performance of labour, vehicles and equipment in terms of tonnage handled and population. This is required both for project monitoring and by the responsible officer in each town as his basic management data. Site visits and discussions showed that such data is simply not available in most project towns.

2.4. **Primary Collection System**. As discussed at length with all project towns, the present system of primary collection involves double or, where non-tipping trucks, trailers or trolleys are in use, treble handling of all wastes. A video presentation was made to all concerned officers in the project towns illustrating how the principle of picking up wastes once only could be applied using simple indigenous equipment such as hand carts or pedal rickshaws working in conjunction with tractors with trailers or container handling equipment.

Unless there is a drastic change in primary collection methods, the full benefit of investment in new vehicles and equipment will not be obtained and the present health hazards, traffic restrictions and unsightly appearance of roadside heaps and open collection points will continue unabated. This is the basic root cause of the high cost and inefficient service and overcoming the problem will have maximum impact on SWM operations.

2.5. Integrated Services. Site visits indicated that in many cases the efforts of street sweepers and drain cleaners was negated by the failure to co-ordinate their work with provision of some form of transport. The result was that the neat heaps of sweepings or silt were scattered by passing traffic long before any pick-up vehicle arrived. The provision of a handcart or carts or rickshaw to serve a group of sweepers by working immediately behind them and removing the heaps to a collection point or to an area free from traffic would be more productive.

2.6. Vehicle Maintenance and Repair. Discussions and site visits highlighted the fact that with one exception, no provision was made in the project towns for such basic vehicle maintenance as routine oil and filter changing. All the indications were that the majority of towns operated on the "Fire-brigade" principle and that vehicles only received attention either following a breakdown or as a result of drivers reporting a defect or malfunction.

Many project reports indicated a very high percentage of vehicles off-the-road, awaiting major repairs. When queried, the mission was advised that it was not lack of private-sector repair facilities that was holding up the work; the main obstacle was non-availability of funds. The mission would point out that when adequate collection transport is not available, the town is still committed to paying the wages of drivers and loaders although they may not be available to carry out their normal duties through lack of transport.

3.0. PRIMARY COLLECTION IMPROVEMENTS.

In industrialised countries, primary collection systems are based on each household or commercial establishment having a storage receptacle for wastes which is emptied directly into the collection vehicle, normally on a once-weekly basis. Compare this with most developing countries - no household storage, daily or at least alternate day collection dictated by climatic conditions, wastes thrown directly into the street resulting in double or treble handling, open heaps awaiting collection and surface drains choked with garbage.

3.1. Pilot Project. To address this problem in West Bengal, a pilot project, with World Bank financial assistance, was set up in a small municipality with the co-operation of Calcutta Metropolitan Development Authority (CMDA) using the services of the All India Institute of Health and Public Hygiene (AIIHPH) Project Team under the personal leadership of Prof. K. Nath. The objective set for Prof. Nath was to demonstrate how a "handle once only" primary collection system could be provided using indigenous, low cost and easily maintainable equipment to provide a service to a much higher proportion of the population than before at no increase in cost. The end result was a primary collection mode adaptable to either door-to-door or community container collection using improved handcarts, tricycle rickshaws, small transfer stations and with tractor-trailer or tractor-container handling equipment for final transport to disposal point, elimination of roadside heaps (and subsequent reduction of choked surface drains) on a daily or alternate day frequency at a considerably lower cost per ton than previously obtained.

3.2. Replication in other Municipalities. Following the success of the initial pilot project, the opportunity afforded by a third Calcutta Urban Development Project was taken to initiate a campaign to persuade 34 municipalities, with populations ranging from 30,000 to almost 300,000, to adopt the improved primary collection method. A video film and slide presentation was prepared and shown to the chairmen of all the towns, short training courses were held for supervisors and two field project teams from AIIHPH visited each of the towns to prepare their schemes, advise on procurement of equipment and assist in physical implementation in one area or ward in each town. After three years' effort, the new system is now operational in all or part of 21 towns and a further 15 towns have schemes approved and procurement of the necessary equipment is in progress. The improvements both in service delivery and cost have been quite dramatic. The new system now provides a daily or alternate day collection, either door-to-door or from community containers in place of the previous irregular or non-existent service, daily tonnage collected has increased by some 70% and cost/ton has been reduced by 35%. In similar fashion, applying the same principles to one of the project towns mentioned in para 2.1 using an improved design of handcart and community containers served by a compression type collection vehicle has resulted in a reduction in the labour force of 125 men and 12 open trucks and the provision of a regular collection service to a much larger proportion of the population.

4.0. CONCLUSION.

The success of the initial pilot project and its subsequent large scale expansion was achieved by illustrating to both political and executive decision makers how much of their scarce resources was being wasted on a poor SWM service, by demonstrating to local community leaders that with public participation and traditional local equipment a regular collection could be given. Linking these activities to some basic SWM training and some institutional strengthening develops a confidence and pride in achievement that then provides a more receptive climate for further help and assistance in gradually introducing more sophisticated equipment and methods once the basic primary collection has been firmly established as a dependable municipal service.

MUNICIPAL SOLID-WASTE MANAGEMENT IN EGYPT

By M.M. El-Halwagi, S.R. Tewfik, M.H. Sorour
and A.G. Abulnour
Chemical Engineering Department & Pilot Plant
Laboratory
National Research Centre, Dokki, Cairo, Egypt.

A B S T R A C T

This paper endeavours to outline the historical developments current status and future perspective of the global solid waste management aspects in urban Egypt.

Over the past decade, much systematic work has been devoted to the various facets of the problem in major Egyptian cities. The approach has been based on an integrated work comprising technical, economical, managerial and social components on both the study and implementation phases. As a result a dependable data base and alternative adaptable management scheme have evolved and proven to be quite valuable assets in the various stages of properly treating the various national situations.

At present, the cleanliness level in the various Egyptian cities has considerably improved. Collection, transportation treatment and ultimate disposal are being developed. Recent assessments on the national level tend to indicate that future trends encompass the following:

1. The use of compactor trucks for transport in major cities and along main roads while other options such as tractor-trailers are recommended for smaller cities and narrow roads.
2. Sanitary landfilling will be the principal disposal method though composting with recovery of recyclable materials will be a desired option whenever funds are available to supply compost that is highly demanded for land reclamation.

1. INTRODUCTION

Egypt, like any other nation, consists of both urban and rural areas. Internal migration from rural to urban communities, coupled with modernization trends, have caused some degree of intermingling between city and village cultures and patterns. Manifestations of this phenomenon have

ISWA PROCEEDINGS
VOLUME 1 ISBN 0-12-058451-4

been evolving as gradual changes taking place in both rural and urban settings. Parts of each are getting closer in nature and in some places no sharp lines of demarcation can be readily recognized. Villages close to cities are undergoing urbanization trends; whereas poor urban areas may look like a type of a modern village. Such diffuse spectrum of conditions, with their consequent societal variations, mandate resort to a variety of solid-waste management schemes to cope with these situations.

In traditional villages, solid wastes are mostly handled on a decentralized family-scale basis. Some of the animal dung (processed into dung cakes) is used with some agricultural residues as fuel. The remainder of animal excreta (and sometimes human excreta too), is mixed with home garbage and some agricultural residues plus some silt, and is composted in heaps near to the agricultural fields. When mature and needed, the compost is used as soil conditioner and organic fertilizer. Street litter is largely taken care of by village animals. Thus, in essence there is almost complete recycling.

In contrast, urban communities with their high population densities, complex socioeconomic structures, high-rise buildings limited on-site storage space, and other known urban features necessitate adopting proper solid-waste management (SWM) systems. The prevailing systems encompass three clearly categorized types: the featuristic "Zabbaleen" system; the municipality system; and supplementary "company-type" systems. These will be described and briefly assessed, particularly from the standpoint of future developments and prospects. Solids emanating from industrial establishments and special nature entities are not included.

2. CURRENT MUNICIPAL WASTE MANAGEMENT PRACTICE

2.1. Waste Quantities and Characteristics

Data on waste composition, characteristics and generation rates, as obtained for various Egyptian Cities, are compiled in Table 1. It is apparent that, as typical for developing countries, wastes are rich in organic putrecsibles. The percapita generation rate varies between 0.325-0.73 kg/day.

Table (1)
Typical Characteristics of Egyptian Solid Wastes.

	Cairo (1)	Giza (2)	Suez Canal Region(3)	Port Said (4)	Damietta (5)
1. Composition					
Food wastes	43.7	44.3	37.1	46.2	65.6
Paper	9.2	20.4	22.6	25.0	14.9
Metals	3.0	3.8	3.1	2.3	2.6
Glass	1.9	1.4	1.8	2.0	1.1
Cloth	3.0	2.3	2.2	2.1	-
Plastic	1.9	3.5	3.0	4.5	2.1
Bones	1.3	0.2	0.3	0.4	-
Combustibles	-	-	9.6	-	-
Others	28.0	24.1	20.3	17.5	13.7
2. Distribution (%)					
Household	-	64.3	55.8	55.5	68.8
Streets	-	15.9	16.0	13.3	12.8
Commercial	-	5.8	15.7	17.0	16.6
Institutional	-	3.2	1.7	3.5	1.8
Industrial	-	5.8	10.8	4.4	Included in commercial
Others	-	5.0	-	6.3	-
3. Generation rate kg/person/day	0.325	0.69	0.47	0.73	0.72
4. Characteristics					
Moisture %	-	40	28-50	-	45
Specific gravity	0.3	.2-.25	.21-.34	0.25	0.25

2.2 Administrative Aspects

Egypt is administratively divided into 26 governorates including about 30 larger cities. Of the 21 million urban population more than ten millions reside in the three larger cities Cairo, Alexandria and Giza with a population density exceeding 50 thousand per square metres.

In most of the cities two systems are involved in the handling of solid wastes:

a) **The formal system:** in which the municipality is primarily in charge of city cleansing within the context of other utility services. This system relies basically on governmental finance in addition to a cleansing account funded by fees on households and commercial establishments.

In order to control the aggravating solid waste problem in larger cities, special organisations in charge of cleansing and beautification have been established to act as centralised facilities with flexible authorities to manage all solid waste handling activities.

b) **The informal system:** Which encompasses mostly the zabbaleen system that has been described be-

fore (1,2,7). This system offers door-to-door service mainly to high and medium standard households. The wastes are collected in shouldered baskets then transported by donkey carts to the zabbaleen settlements where maximum recycling takes place.
In smaller cities, a more or less similar system is adopted though the collected wastes are disposed of in municipal dumping sites. The major source of finance is the direct fee paid by the recipient of the service.

2.3. Collection-Transport

Various collection-transport modes are being employed in Egyptian cities covering the span from very primitive means to high technology approaches. These involve the following:
a) Door-to-door collection and transportation by donkey carts as adopted by the informal systems for households and institutions.
b) Communal collection in 3-6 cubic trailers which are tractor drawn.
c) Curbside collection in dumper trucks (4-8 m^3)
d) Curbside collection in satellite vehicles (3-4 m^3) which discharge in large rear loading compactors (10-20 m^3).
e) Communal collection in 2-3 cubic metres containers which are discharged in side or rear loading compactors.
f) Collection of street sweeping in handcarts which are directly discharged or loaded by front end loaders in various kinds of motorised vehicles.

2.4 Treatment and Disposal

Wastes collected by the zabbaleen are transported to their settlement where handsorting of recyclable materials such as paper, plastic, glass, cloth and metals takes place. The remaining food wastes are then used for raising of pigs and other small animals. The pig manure is later used as organic fertiliser. It is estimated that more than 200 tons per day of MSW are being treated by the zabbaleen system in Greater Egyptian cities.

On the other hand, about 95% of the wastes collected by the municipalities are disposed of in open dumps. However, over the last few years there has been a trend towards applying proper treatment or disposal methods. Options that are currently considered viable under Egyptian conditions are sanitary landfilling windrow composting and incineration. Table 2 compiles information on treatment or disposal establishments in major Egyptian cities. Experience with performance of these systems may be summarised as follows

a) Sanitary landfilling: Though the first properly designed sanitary landfill site has not been yet in operation, controlled landfilling with daily cover has been practiced in Cairo. Due to dry conditions prevailing in most Egyptian cities, hazards from leachate or gas generation are anticipated to be minimal.

b) Composting: The adopted windrow composting of intermediate-level technology and with recycling of recoverable materials has proved to be technically and environmentally sound. The major constraint is economical as imported plants require high capital expenditures and the sales revenues would not compensate for operating expenses.

c) Incineration: Employed incinerations are primitive and are operated on a discontinuous basis. Prospects for proper incineration with energy recovery are considered to be limited under prevailing conditions (8).

Table (2)
Major Treatment or Disposal Facilities in Egyptian Cities

Technology	City	Capacity	Status
1. Sanitary landfill	Cairo	up to 1000 tons/day	Ready for operation
2. Windrow composting	Cairo-Shoubra	10 tons/hr	Operating since 1985
	Cairo El-Salam	6 tons/hr	Operating since 1987
	Alexandria	10 tons/hr	Operating since 1985
	Giza	6 tons/hr	Operating since 1986
	Damietta	10 tons/hr	Operating since 1986
3. Incineration	Banha, Tanta El-Mahalah El-Zagazig	modular unit of 15 tons/day	Operating discontinuously since 1983

2.5. Economic Indicators

Although costs of various activities are highly variant and site specific, some general type cost indicators are given in Table 3(9).

3. KEY PROBLEM AREAS

The key problem areas facing the current solid waste management systems may be summarised as follows:

3.1. Funding Constraints

- In view of competing priorities, governmental

funds allocated for SWM are very limited.
- The community participation to the actual collection-transportation and disposal cost is only marginal.

Table (3)
Economic Indicators for Collection and Transport, Treatment and Disposal of MSW under Egyptian Conditions.

Phase	Description	Net Cost (L.E*/ton wastes)	Remarks
1.Collection and Transport	For communal collection and transport to a distance of 25-30 km.	12-15	As estimated for Giza City (2)
2.Treatment			
2.1.Composting	a) Pilot Plant (10 tons/hour) (soft loan fund in 1982)	10	As estimated on operation at full capacity (10)
	b) A conceptual 30 tons/hour plant	15	As estimated for Port Said City(4)
2.2.Incineration	For a continuous incinerator plant (15 tons/hour)	34	As estimated for Port Said City(4)
3.Disposal			
Sanitary Landfill	A 2 million cubic metres site (500 tons/day)	2.5	As estimated(10) for the Cairo site

* 1 US.$ = 1.36 Egyptian Pounds (L.E)

3. KEY PROBLEM AREAS

The key problem areas facing the current solid waste management systems may be summarised as follows:

3.1. Funding Constraints

- In view of competing priorities, governmental funds allocated for SWM are very limited.
- The community participation to the actual collection-transportation and disposal cost is only marginal.

3.2. Technological Issues

Most of available facilities have been imported through foreign grants or loans and cover a wide technological spectrum. Typical problems encountered in this area are related to maintenance, spare parts and operational experience pertinent to imported technology.

3.3. Current Use Pattern of Available Facilities

Efficiency of utilisation of collection (containers, trailers ... etc.) transportation (trucks, compactors ...) or treatment (plants) facilities is generally low.

3.4. Management and Organisational Aspects

- Both municipalities and informal systems are operating with virtually no coordination.
- Undeveloped SWM plans on both local and sectoral levels.

- Incomplete organisational structures for proper functioning at various levels.

3.5. Deficiency of Reliable Information

Despite the accumulated information obtained through the development of SWM schemes for major Egyptian cities, data regarding characteristics of current waste generators, norms of current practices, financial data are still needed. Documentation is generally inappropriate and available information lacks accuracy as well as reliability.

4. Attempts and Proposals for Improvement

4.1. Planning Consideration

Trends for improvement of current practices acknowledge the importance of sound and integrated planning for upgrading existing systems and implementing new schemes. Two key issues have been recognised in the last years:

a) Different upgrading projects for the largest cities in Egypt have been preceded by integrated management schemes encompassing other non-technical parameters such as sociocultural aspects in addition to present and expected demographic settings. Those studies have been undertaken by specialised national and international organisations.

b) For newly founded Egyptian cities in arid areas, provisions are being made for appropriate SWM facilities matching with the physical environment in addition to expected socio-economic as well as socio-cultural patterns.

4.2. Technological Aspects

Previous experience pertinent to the evaluation of the technological impact on current management practices points out to the following requirements:

a) The importance of technically matching facilities used in collection and transportation with special emphasis on the constraints imposed by prevailing infrastructures.

b) The necessity of integrating technological capabilities of the informal system with that of the formal authorities within dedicated service areas.

c) Improved current utilisation of available facilities via appropriate orientation to the suitable service area in addition to adopting appropriate codes of practice.

d) Recognising the prevailing funding constraints specifically with regards foreign currency, it deemed necessary to maximise the local technological components of the solid waste handling facilities. Current attempts include local manufacture of compactor trucks in addition to intermediate level composting technology.

4.3. Economic Issues

Current attempts to improve service level as well as service coverage require large funds which exceed the available. Major issues which are currently adopted by concerned bodies centre on the following:

a) Reducing the net cost per unit waste via improved operating schemes and maximising local inputs.

b) Increasing revenues from sales of recyclables and compost. There has been controversial issues concerning expanding the composting route. However, latest status points to increased market price for compost. Some plants have reached the point of having a long reservation list which directly reveal increased awareness to the value of this soil conditioner. It is anticipated that with local manufacture of composting plants, in addition to high demand for organic fertilisers necessary for agricultural land reclamation plans, economics of composting would highly improve.

c) Increasing the community participation for solid waste services. A major improvement in this direction is under trial in a pilot project in two urban areas in Cairo. The adopted strategy would transfer the informal system to a private formal structure with developed facilities. Households and institutions would fully compensate for the actual costs for collection and transportation. Socio economic parameters have been taken into consideration in adjusting service fees.

4.4. Supportive Research Activities

A major role is entitled to national research organisations to develop waste management facilities and schemes. Along this path the following major tasks are currently being undertaken in our National Research Centre

a) Development of Integrated Management Scheme for the city of Cairo.

b) Development of simple low cost landfill-composting technology.

c) Initiation of local manufacture of windrow composting plants with intermediate technology. A patent is being currently disclosed.
d) Formulation of national codes of practice for various SWM activities.
e) Co composting of sewage sludge and MSW in new communities.
f) Assessments of economics of incineration of selected wastes (e.g. hazardous wastes).

5. Future Outlook

Thorough analysis of the present situation would lead to the following perspectives and expectations:
a) Increased tendency for formal privatisation for household and institutional services.
b) Adoption of small covered trucks for small urban communities. Large compactor trucks would be confined to larger urban cities.
c) Sanitary landfilling would be the prevailing disposal method for almost 60% of urban wastes while the composting option would cover about 20-30%. The remainder would be treated by other methods (open dumps, landfill-composting, incineration).

REFERENCES

1. Environmental Quality International, (1981). "Solid Waste Management Practice in Cairo", Report No. 2.
2. El-Halwagi M.M. et al. (1985). "Solid Waste Management in Giza City", National Research Centre Report.
3. Norconsultant A.S. and Sherif M. El Hakim and Associates, (1982). "Solid Waste Management Study-Suez Canal Region Feasibility and Design Study", IGY/76/001-13.
4. El-Halwagi et al. (1986). "Technoeconomic Evaluation of a Composting Plant for Port Said City", NRC Report.
5. El-Halwagi et al. (1983). "Solid Waste Management for Damietta City", NRC Report.
6. El-Halwagi et al. (1987). "Solid Waste Management in Egypt; Problems, Planning and National Trends", Waste Management and Research, 75, 5(1).
7. Kodsi, J.J. et al. (1982). "Waste Management Practices in Cairo-Current Practices, Programs and Trends" in "Recycling in Developing Countries", K.J. Thome-Kozmiensky (ed.) EF-Verlag fur Umwelt-

technik.
8. Tewfik, S.R. et al. (1986). "Incineration of Municipal Solid Wastes: Prospects and Constraints Under Local Conditions", Journal of Egyptian Society of Engineers 25, 76(4).
9. El-Halwagi M.M. et al (1986). "Municipal Solid Waste Management in Egypt-Practices and Trends" in "Waste Management in Developing Countries", K.J. Thome Kozmiensky (ed.), EF-Verlag fur Energie und Umwelttechnik.
10.El-Halwagi et al. (1986). "Assessment of Shoubra-Cairo Composting Plant", NRC Report.

Biomass Gasification of Agricultural and Forestry Residues to Produce Energy for Shaft Power Generation.

L.Limbe and J. Ngeleja, Tanzania

In rural areas of Tanzania the availability of relatively small amounts of cheap energy plays a vital role in the economy of the country whose backbone is agriculture. For example thermal energy is required for cooking purposes related to human activities. Mechanical energy is required for agricultural activities and in small scale industry. Traditionally, human labour and animal traction often provide a fraction of the requirements while modern methods such as electricity and internal combustion engines require an infrastructure which is very expensive and of course the importation of the relatively expensive fossil fuels. However in areas characterized as above there is often a lot of agricultural wastes and forestry residues which if carefully deployed may provide for the needed energy resources without damage to the ecological system.

The gasification of agricultural wastes as an alternative source of energy was introduced in the rural areas of Tanzania. The first attempt was done under a research agreement between the Tanzanian and Dutch Governments and was undertaken in collaboration with the Small Industries Development Organization(SIDO) and the Twente University of Technology (THT) of the Netherlands. The project which was based in northern Tanzania in Arusha region first developed the gasification of maize(corn)-cobs to produce a gas that could substitute the use of diesel fuel in a diesel engine-grain milling machine combination in rural areas. The research programme started in 1980 and went through to 1985 with support from the International Foundation for Science of Sweden in the last two years.

ISWA PROCEEDINGS
VOLUME 1 ISBN 0-12-058451-4

In 1984 the Tanzania Wood Industry Corporation(TWICO) and the Gas Generator AB of Sweden introduced the gasification of forest residues to provide shaft power for a mobile sawmill and a landrover. The project was carried out in the same area as a measure for concentrating efforts and knowledge in gasification. This report is an attempt to address this ISWA 88 gathering of what was achieved in both projects.

2. RESEARCH OBJECTIVES:

There are approximately over 6 million hectares of arable land in Tanzania. Of the produce grown in this arable land by far the major part is left out as residue in form of waste. Some of the agricultural crops grown in Tanzania are **coffee, maize, sorghum, millets, coconut, cotton etc.**

A survey for different agricultural wastes was done in the country and the following were identified as technically potential for gasification; sorghum - straw, coffee-husks groundnuts shells, coconut-husks and maize cobs. Among these the maizecobs were found to be not very useful for **other purposes and have virtually no v alue as fertilizer or** soil improver as they consist purely of cellulose and some silicate rich ashes. The cobs do not fit for animal feeds, on the other hand the cobs have enough energy to act as feedstock in a gasifier. The heating value for maizecobs is 10 to 20 MJ/kg.

Maize production in the country is about 1 ton/ha/year and the crop produces two types of waste namely the empty cob and the stem. The stem which is normally left on the land **either ploughed down into the ground as manure and/or is** eaten by cattle. The cobs which are estimated to be 0.5+/ ha/year are left as waste.

During the seventies due to the rising costs of oil products on the World market, there was a critical shortage of energy for oil in the country. Tanzania was forced to spend more than 60% of her foreign earnings towards the importation of oil. After a survey of available options for decentralised power production with local available resources, it was found that the forestry residue/agricultural waste gas generator be used in connection with internal combustion engines as a measure towards alleviation of the energy oil crisis especially in remote rural areas of the country.

The main aim of the project was to develop and implement waste gasification in the country, by way of

a) introduction
b) demonstration and
c) local manufacture of small scale gasification units fuelled by maize cobs residues for village applications.

3. METHODOLOGY:

The project was carried out in two phases.

Phase I involved other technological development of gasification equipment which was constructed from local raw materials (b) demonstration of the prototypes in rural areas in sites which were identified for field tests.

c) training of local mechanics on operation and maintenance of gasification plants.

Phase II involved (a) implementation of five gasification plants in selected village (b) evaluation of these plants with respect to the social acceptability, economic benefits and technical feasibility (c) preparation for a wider scale dissemination of the gasification technology and possible use of other feed-stocks.

The Socioeconomic monitoring of the project was done by the Bureau of Research Assessment and Land Use Planning (BRALUP) of the University of Dar es Salaam, the country's highest institution of learning. Village surveys were conducted by students of the same University.

3:1 Technical Feasibility:

Tests at the test site in Arusha showed possible diesel savings of about 80%. However under real village use, diesel savings of about 50 per cent or even lower were recorded. The reason behind these differences were considered as follows:-

a) the test site figures were based on a continuous process while in a milling business atmosphere, operations are intermittent accompanied by long periods of idling. Experience shows that dual fuel gas operation diesel savings are maximum at full engine load and minimal during idling.

b) The test site experiments were based on engine speeds of around 1500 RPM which was considered as maximum thermal efficiency of gas use, while in villages in order to raise the engine output and reduce milling time in support of demands of the customers the engines ran at 2200 RPM

3:1:1 Filter

A glass wool fibre bag filter was used as an effective device to protect the engine against poor quality of gas which was occasionally generated.

3:1:2 Engine Performance

It was found possible to maintain power output of the diesel engine by using a gasifier as much as 10-20% loss of rated power could be experienced. But in real village grain milling this power seemed adequate. The recommended servicing of the diesel engine was every 400 hours. Tar and dust led to an increased number of servicing of diesel engine in an attempt to avoid engine-break down.

4. General observations

Maize cobs if adequately stored have proved to be a very good fuel. Estimates of proportion of cobs required for gasification for a given quantity of grain to be milled range from 10% to 20% of the cobs. Mechanical grain milling despite being a village's basic need is intermittent in Tanzanian villages and therefore does not seem to be a feasible application in gasification technology. The economics of gasifier application in continuous processes seems more promising as will be described below in a saw-milling business.

5. Gasification of forest residues. Enormous quantity of wood residues in form of slabs and sawdust exist in Tanzania. The yearly estimate of the available quantity is over 100,000 litres of fossil fuel equivalency. In this process the residues are converted into charcoal which is in turn used as feedstock in the gasifier. Application was made to fuel a landrover.

5:1 Gasifier powered landrover. The model of the landrover which was used for testing is series III 1982. The following are the technical specifications of the vehicle:-

1) SI engine with four cylinders
2) Volume of cylinder 2286 c.c.

3) Piston displacement - 88.9mm

4) Diameter of cylinder 90.49mm

5) Compression ratio 7.0:1/8:0:1

6) Maximum torque 17.5 Kgm/2500 R.P.M.

7) Firing order 1:3:4:2

Tests on soft wood charcoal have been carried out for a distance of about 6000KM.

On average 400gm of pine or cypress charcoal could cover a distance of 1KM while on the other hand 370gm of Eucalyptus charcoal covered the same distance under heavy load variations. Long distance driving in plain countryside recorded about 7 kilometres to a kg of charcoal. There is a reported loss of power of about 30% when the landrover is in gas operation.

5:1:2 The gas powered saw mill

The equipment used to generate gas here is a drawndraft gasifier meaning that the flow of the gas is in the same direction as the flow of the feedstock. The calorific value of the gas obtained is about 5,200 KJ/NCM. The gas after mixing with air in the required ratio is introduced into the manifold of an engine which drives a 900mm diameter circular saw-blade. The power output is about 32-35KW at 2200 to 2400RPM. Tests revealed that there were no major problems with the engine.

Occasionally damp charcoal caused condensates to the spark plugs making ignition very difficult and sometimes impossible. There was a daily routine cleaning of the filters which seemed cumbersome.

6. Conclusions and Recommendations:

a) More than 90% of the population of Tanzania live in rural areas where they are engaged in subsistence agriculture. It is recommended that the agricultural wastes and forest residues available in these environments be given due attention by the government and donor agents and help to prevent what could be the biggest environmental mismanagement.

b) The fact that Tanzania like any other country in the third world has considerable biomass resources in form of forest residues, wood processing wastes. Surplus softwood, agricultural residues, animal waste etc. when carefully used can offer a near to medium term economic conversion to energy and other products which can bring about a better quality of life to the rural community.

7. References

1) The Potential of Biomass based gas fuel in the substitution of fossil fuel in the forestry industrial sector of Tanzania.
(the case of a self energized saw-mill)
James Ngeleja 1987

2) Gasification of Maize cobs to a fuel usable in internal combustion engine-Proceedings of materials and Energy from refuse-Antwerp.
Lawrence Limbe - 1986.

COMPARISON OF SOLID WASTE MANAGEMENT IN TOURISTIC AREAS OF DEVELOPED AND DEVELOPING COUNTRIES

Kriton Curi
Head, Pollution Control Research Group
Faculty of Engineering
Boğaziçi University
Bebek, İstanbul, Turkey

1. INTRODUCTION

Natural beauty and pollution free environment are usually the main reasons for making certain areas attractive to the tourists. This attraction however, may have detrimental effects if the necessary infrastructure for handling the needs of this extra population is not available. Most of the time tourism generated due to the natural beauty has a seasonal character. In other terms during a certain period of the year the population of the touristic areas becomes several times more than it usually is. This has as a result the need of facilities which will be used only for a couple of months per year and remain idle for the remaining period. Cases like that although may be handled easily in economically developed countries, it consists an important problem for the economically developing ones which have limited resources. The situation becomes even

ISWA PROCEEDINGS
VOLUME 1 ISBN 0-12-058451-4

more critical when it is considered that the economically developing countries while giving their struggle for development and for increasing their income of foreign currency very often neglect or underestimate parameters like the "bearing capacity" of a given touristic district, and they try to attract more and more tourists, thinking the short term benefits and forgeting that in the long run this may have destructive effects. Kocasoy (1988) for example has proven that the uncontrolled and not properly planned increase of tourists may have negative effects on the seawater quality of recreational resorts.

Among the many problems which are encountered in touristic areas in relation to the maintenance of the attractive environmental beauty the management of solid wastes has a particular importance. In the present paper this subject will be examined, making a comparison whenever possible between the situation existing in developed and developing countries emphasizing the conditions prevailing in the touristic areas of Turkey.

2. SOLID WASTE GENERATION

There is a striking difference in the quantity as well as composition of solid wastes between high and low seasons in the touristic resorts of developing countries. For example, in the Princess Islands located in the Sea of Marmara near İstanbul, although during the winter the solid wastes generated per day is in the order of 12 tons, this value becomes 250 during the summers. In a similar way in Çeşme, another touristic resort along the Aegean Coast of

Turkey the per day production of solid wastes changes from 11 tons to 32, in the Greek Islands Santorini (Fira) from 10 m^3/day to 50 m^3/day, in Mykonos from 20 m^3/day to 80 m^3/day etc (Grammatikopoulou, 1988). This difference in quantity is not only due to the increase in population but at the same time due to the higher per capita production of solid wastes observed during the touristic season. For example, in Princess Islands, the solid waste production rate which is around 0.76 kg per capita per day during the winter in summer reaches the value of 1.33 kg/cday. This value is higher than the average value for İstanbul which was 1.15 kg/cday in 1987. In Bodrum this value changes from 0.92 to 2.36 and in Çeşme from 0.71 to 1.88. The corresponding value for Monaco is 2.58 (SMA, 1988). This difference cannot be attributed to seasonal variations, because in the city of İzmir, for example, this value is constant around 1.07 throughout the whole year while in İstanbul is a little higher during the winter. The difference observed between the two seasons can be contributed to the difference in habits and standards of life of the local population and those coming for touristic reasons. This is verified also from the fact that in Princess Islands where the summer population consists mainly of people from İstanbul who have almost similar habits with the local population the difference between summer and winter values is small.

The composition of the solid wastes as can be seen in Table 1, shows also significant differences between summer and winter (Curi, 1986). As can be seen in this table there is a significant increase in the amount of paper disposed during the touristic season. In a similar way an increase in the quantity of plastics is noticed during the same period.

Table 1 - Composition of Solid Wastes in Touristic Areas of Turkey

	Princess Islands		Çeşme		Bodrum	
	Summer	Winter	Summer	Winter	Summer	Winter
Food Remains	48.60	53.87	45.62	58.07	44.14	62.18
Paper and Cardboard	16.08	6.18	29.17	9.13	34.77	8.12
Textile	2.44	3.25	4.66	5.31	3.97	3.33
Plastics	19.13	5.16	8.73	7.85	8.24	6.37
Glass	4.93	2.18	3.12	1.77	4.04	4.31
Metals	3.61	3.03	2.81	1.56	2.35	4.63
Fines	1.14	21.31	1.49	13.05	0.40	6.80
Miscellaneous	4.07	5.02	4.40	3.26	2.09	4.26

The composition of the solid wastes of İstanbul and of some other cities/countries is given for the purpose of comparison in Table 2. As can be seen the composition of the wastes of the touristic resorts of Turkey during the summer months resemble those of the developed countries, while the same observation is not valid for the seasons when tourists are not plenty. Information obtained from scientists from Greece Yugoslavia and Egypt verify that similar results are obtained in the touristic areas of these countries.

Table 2 - Composition of Solid Waste in Different Cities

	İstanbul	France[1]	Rio de Janeiro[2]	Monaco[3]	Jakarta[4]	Egypt[5]
Food Remains	44.62	15-35	36.75	35-15	93.88	60
Paper and Cardboard	11.98	20-35	38.92	38.40	8.24	13
Textile	4.28	1- 6	3.07	1.90	3.16	2.5
Plastics	11.87	3- 6	6.83	1.25	5.52	1.5
Glass	3.42	5-10	3.79	6.47	1.78	2.5
Metals	2.28	5- 8	3.84	2.47	2.08	3.0
Fines	14.12	10-20	-	3.38	-	-
Miscellaneous	7.43	-	6.80	1.53	5.34	17.5

[1] *Le Ministre Délégué Chargé de I'Environnement, 1987*

[2] *Paraquassu de Sa, 1988*

[3] *SMA, 1988*

[3] *Bebassari, 1986*

[5] *El-Halwagi, 1986*

3. COLLECTION OF SOLID WASTES

Most local authorities, regardless of the degree of development, try to collect the solid wastes in an effective way, because this activity in public mind is an important criterion which can effect their re-election. Still there are several important differences in relation to solid waste collection in developed and developing countries. For example, although in the developed countries most of the time the refuse is collected in standard receptacles which are loaded mechanically to the collecting vehicles, in developing countries the storage of refuse is done at any type of container or plastic bag and they are loaded maunally by door to door collection. In the Princess Islands for example, 83 percent of the domestic solid wastes are collected by door to door collection, while the remaining part through a voluntary delivery to containers which have a capacity usually of 1100 L and are located at some central points.

The type and the number of refuse collecting trucks is another important question for developing countries. Usually to own a large vehicle with a compactor is considered as a sign of prestige for local authorities and because of that this type of a vehicle is usually purchased regardless if it is appropriate for the local conditions. Usually the representatives of some companies manufacturing these type of trucks convince the municipalities to buy some vehicles which can not even function properly. As a result of that, a rather high number of refuse collection trucks (as high as 60 percent of the total number available) is staying in garages waiting for a spare part or being out of order entirely.

Due to the lack of proper coordination between different small municipalities located very near to each other, it is very common to have duplications in instrumentation and other facilities in an illogical way.

4. RECYCLING

Little attention is paid to recycling in the touristic resorts of the developing countries. Separation at source and separate collection is something which is practiced very rarely. The only recycling which may be encountered is mainly realised by scavangers which are active in the solid waste disposal sites (Curi, 1982). Recycling by collection from the refuse receptacles - something which is widely encountered in large cities of developing countries, is not common in the touristic communities, mainly due to the efforts of local authorities. From time to time, charity organizations in order to raise funds organize campaigns of collecting separately some scrap material - usually paper - which have commercial value. Actions like that however, take place very rarely.

5. DISPOSAL

There is a sharp difference between the methods of disposal used in the touristic resorts of developed and developing countries. In the first ones incineration in plants with heat or without heat recovery is very common. Beyond that composting and sanitary landfilling is used for the same purpose. In Monaco, for example the 70 tons of solid wastes produced per day are completely incinerated. In France from 1980 onwards there was a sharp increase in the number of incineration plants with heat recovery, while a slowing down in the commissioning of small capacity incinerators was observed (Le Ministre Délégué Chargé del'environnment, 1987). In contrast to this in most touristic resorts of developing countries of the Mediterranean like Turkey, Greece and Yugoslavia, the solid wastes are still disposed in open dumps without any control. There are few exceptions of some composting or incineration plants most of which do not function properly. According to information obtained from the authorities of these three countries more than 95 percent of their solid wastes is disposed presently at open dumps, without taking into considerations the effects on the environment. Recently, however in all these three countries activities have started for replacing the present system with a more hygienic one. In the Princess Islands of Turkey, for example, attempts for adapting the sanitary landfill technique or for transferring the refuse to the main land have started. Attempts for construction of sanitary landfills are taking place also in Bandırma, İzmir, Çeşme, Bodrum and many other touristic resorts of Turkey. A similar action is taking place recently in the Greek Islands Mykonos, Santorini, Paros, Tinos and Syros (Grammatikopoulou, 1988) while a study has started recently in Rhodos for finding the most appropriate technique for solid waste disposal (Papahristou, 1988).

The situation is not too much different in other developing countries rich in touristic resorts. In Egypt, for example, 95 percent of wastes collected by municipalities are disposed in open dumps (El-Halwagi, et al, 1986); in Thailand most of the municipalities employ open dumping and occasional burning as their disposal method. In Bangkok out of the 5.3 million cubic meters of solid wastes produced per year only 0.8 millions are composted while the remaining is disposed in open dumps (Pairoj-Boriboon, 1986). In Brazil out of the 42000 tons of garbage collected in 1984, 5.2 percent was disposed by dumping to rivers or mangroves, 45.4 percent in open dumps, 23.8 percent in controlled landfills, 21 percent in sanitary landfills, 3.8 percent in composting plants and 0.8 percent in incinerators (Veit, 1986). In Rio de Janeiro, out of 4600 tons of solid wastes produced per day during 1988, 4400 are disposed in partially controlled dumps while the remaining 200 tons/day are disposed by composting (Paraquassu de Sa, 1988).

The present way of solid waste disposal in the touristic resorts of developing countries, if it is not improved may damage the environment, and as a result of that these places may lose their popularity.

6. PROBLEMS OF DEVELOPING COUNTRIES IN RELATION TO SOLID WASTE MANAGEMENT IN TOURISTIC RESORTS

Solid waste management in touristic resorts is an important problem for developing countries. The most important reasons contributing to this problem are summarized below:

a) Local authorities usually desire to be efficient in collecting the solid wastes, but they are reluctant in allocating funds for their proper disposal. They prefer instead of that to spend them for activities which may contribute in a more direct way to their reelection.

b) The lack of standards in most developing countries in relation to machinery and instruments used for solid waste management make easier a wrong decision.

c) The lack of a national solid waste management policy in developing countries is detrimental to the environment. The acceptance of these countries to import or to produce new goods without studying their impact on the environment causes many problems. The acceptance in the last two years for example, to use one way containers (disposable plastic bottles or tin cans) in Turkey, Greece and Lebanon resulted to a new type of pollution for the beaches as well as the bottom of the sea of the touristic resorts of these countries.

d) The admiration of local authorities for "foreign experts" and the insistence of some international organizations to ignore "local experts" ends up many times to failures like it happened with the incineration plant of the Greek Islands Zakynthos, or the composting plant in Mersin, Turkey.

7. CONSLUSION

The touristic resorts of any **country do not be**long only to the people of that country but to all humanity, which takes advantage of its natural beauty or historic interests. Because of that a close cooperation of everybody is required in order to minimize the detrimental to the environment effects of the improper management of solid wastes.

REFERENCES

Bebassari, S. et.al,(1986) "Waste Investigation and Refuse Composition in Central Jakarta" in Thome-Kozmiensk, K.J. (Edit.) Waste Management in Developing Countries, EF-Verlag für Energie und Umvelttechnik GmbH, Berlin.

Curi, K. (1986) "Solid Waste Recycling in the City of Istanbul and Possible Applications in Other Developing Countries" in Thome-Kozmiensky, K.J. (Edit.) Waste Management in Developing Countries, EF-Verlag für Energie und Umvelttechnik GmbH, Berlin.

Curi, K., Kocasoy, G., (1982) "Waste Recovery with Sorting Technologies in the City of Istanbul", Thome-Kozmiensky, K.J. (Edit.) Recycling in Developing Countries, E. Preitag Verlag für Umvelttechnik, Berlin.

El-Halwagi et.al (1986) Municipal Solid Waste Management in Egypt-Practices and Trends" in Thome-Kozmiensky, K.J. (Edit.) Waste Management in Developing Countries, EF-Verlag für Energie und Umwelttechnik GmbH, Berlin.

Grammatikopoulou, N. (1988) Local Authority of Island Syros, Personal Communication.

Kocasoy, G. (1988) Bodrum'da Turizmin Deniz ve Sahil Kirliliğine Etkileri, Ph. D. Dissertation, İstanbul University.

Le Ministre Délégué Chargé de I'Environnement, (1987) "Solid Waste Management - French Policy, Neuilly-sur - Seine.

Pairoj-Boriboon, S. (1986) "State of the Art of Waste Management in Thailand" in Thome-Kozmiensky, K.J. (Edit.) Waste Management in Developing Countries, EF-Verlag für Energie und Umwelttechnik GmbH, Berlin.

Papahristou, E. (1988) Aristotelion University of Thessaloniki, Personal Communication.

Paraguassu de Sa, F., (1988) Comlurb, Rio de Janeiro Personal Communication.

SMA (Societe Monegasque d'Assainissement) (1988), Monaco-Personal Communication.

Veit, M.A., (1986) "Solid Refuse Management in Brazil" in Thome-Kozmiensky, K.J. (Edit.) Waste Management in Developing Countries, EF-Verlag für Energie und Umwelttechnik GmbH, Berlin.

RECYCLING AND RECOVERY OF MATERIALS FROM WASTES-PHILIPPINE EXPERIENCE[1]

Dr. Juanita A. Manalo[2]

Increasing volumes of refuse and the scarcity of disposal sites are persistent problems of good environmental management everywhere. Solid wastes originate from all sorts of human activities resulting from man's continuous utilization of material resources to satisfy his needs.

In the Philippines, growing population and migration to urban cities have complicated the waste management problem. Although residents of Manila throw away less than one third of what New Yorkers discard the amount generated by more than 10 million Metro Manila residents also constitutes big disposal problem. While conservation of raw materials and energy has been the main objective of the reclamation of municipal or household wastes in the industrialized countries where better living standards have generated wasteful societies, recycling for livelihood has been the main reason for scavengers and junk shops in the Philippines. It has offered Filipinos the opportunity to trim the waste disposal needs. Wastes considered useless or of very little value by people who discard them are of significant value to another. Waste available for recycling include all consumer discard. Analysis of Metro Manila's solid wastes shows it is made up of 31.8% food and kitchen wastes, 9.8% paper, 7.7% yard waste, 6% plastic, 4.9% metal, 4.7% cardboard, 2.7% glass, 1.3% textile, rubber 1.1%

---

[1] Paper presented at 5th International Solid Waste Conference and Exhibit, September 11-16, 1988, Copenhagen, Denmark.

[2] Director, Science and Math Centre, the Philippine Women's University, Manila, Philippines.

ISWA PROCEEDINGS
VOLUME 1 ISBN 0-12-058451-4

and other combustibles 6.0%. Complementary activities like street sweeping and water ways sanitation add to the volume of waste collected.

SOLID WASTE COLLECTION AND DISPOSAL

Metro Manila generates 3,400 tons of garbage daily of which 1,000 tons are dumped in creeks, canals, small rivers, burned or recycled. It has a system of solid waste collection and transport but is not quite efficient as only 70% of the generated waste is collected and disposed of in one way or another.

The wastes are dumped in nine open dumpsites within the Metropolis; the biggest being the controversial Smokey Mountain in Balut, Tondo. It is around 15 hectares where some 2,500 families live on scavenging. It has been the welfare target of both local and international aid agencies.

A pilot recycling project was initiated in Manila called "PERA SA BASURA" for the purpose of lessening the stigma attached to the house to house scavenging. Workers were trained and provided with uniforms and carts to collect the garbage. The project has long faded out but scavenging from the low income group still exists. Fund raising schemes of school and charitable institutions always include sale of recyclable items.

Scavenging is the main source of income for some 12,000 to 15,000 persons at the nine dumpsites. About 75% of the residents inside and around the dumping area are wholly dependent on material recycling as a major livelihood. Dumpsite owners do not collect any form of garbage fee but get their income from the share obtained from their authorized middlemen who purchase recyclable materials from the dumpsite scavengers and refuse collection crew.

In addition to those at the dumpsite there are itinerant scavengers with pushcarts who roam the street retrieving discarded materials with recyclable value from the household refuse bins, commercial containers, and illegal dumps.

TABLE I

SUMMARY OF MATERIALS RECOVERED BY SCAVENGERS AT DUMPSITE THEIR SELLING PRICES

Recyclable materials	Selling Price P/Kg
Paper and cardboard	1.50
Glass Bottles	0.80 - 2.50
Hard Plastics	0.05 - 3.00
Plastic Films	0.10 - 1.20
Iron	0.10 - '10.00
Bronze	8.50 - 10.00
Aluminum	3.00 - 4.50
Broken Glass	0.05 - 7.00
Rubber Scrap	o.o5 - 7.00
Tin Cans	0.05 - 0.10
Bones	0.15 - 0.40

TABLE 2

SUMMARY OF SCAVENGERS EARNING/DAY

Earnings	% of Scavengers Earning Amount
P 5 - 10	40%
11 - 15	22%
16 - 20	17%
21 - 25	5%
26 - 35	10%
35 - above	3%

Based on 200 scavengers (1980 Figures)

Another group of ambulant pushcart buyers are out of school youths provided with about three hundred (P300.00) peso daily capital by junk dealers. They go around towns and cities to buy old newspapers, cardboards, sacks, plastics, bottles, scrap metal and all kinds of junk directly from house owners. Each article is bought at a standard price and resold to the junk dealers with a small amount of profit representing their incomes. If lucky, one can earn as much as P40.00 - P50.00 a day from a haul. Favorite sites are affluent housekeepers, who wanting to get rid of their junk, merely give them away for free or at very low prices.

All these recyclable materials from whatever scheme finally find their way to Chinese junk shops where they are resold at higher prices, 2-10 times more than what they cost at source, to those who need them. When cleaned, they command higher prices. Catsup bottles are sold to oil, vinegar or soy sauce local entrepreneurs. Mayonnaise bottles are sold to small food processors who cannot afford to buy new ones. Nescafe and other bottles are recycled as drinking glasses or containers of other products. Even big companies buy recycled bottles in bulk at half or one third the price of new ones for their products.

One private garbage collector who collected the garbage from a millionare's residential subdivision made a lot of money from the sale of recyclable materials to scavengers. He used the non-recyclable refuse to fill his empty lot which accumulated through the years and eventually raised its level to accommodate an office building which he rented out later.

San Miguel, a beer and softdrink company buys all kinds of glass bottles, even broken ones which are reused as raw materials for their glass factory. Their own products require deposits for the containers which are forfeited if lost or broken. All softdrink dealers demand the same from consumers to insure return of bottles.

Old newspapers, magazines, or other kinds of paper are sold by the kilo and recycled as additional pulp for the paper industry. They are also used to wrap other commodities or made into small packaging bags for seeds, grains, or eggs when bought in small quantities in the local markets.

Old and rusty iron of all kinds are bought and recycled. These are resold to iron factories where they are re-melted and manufactured into iron grills, furniture, utensils or materials for housebuilding. Even metal from crashed airplanes are made into scissors and kitchen utensils. Steel covering of drainage and railroad tracks are stolen and also sold for recycling.

Thermoplastics are classified and grouped together, melted and remolded. Others are cleaned and resold for use.

OTHER RECYCLING ACTIVITIES

Packaging materials like cement bags are resold to paper factories for pulping. Sugar or rice bags are reused as such or converted into smaller marketing bags. The old textile bags of animal feeds were recycled into dresses, pillow cases, or hand towels depending upon the buyer's needs. New synthetic materials now used are recycled just like sugar and rice bags.

Textile trimmings are made into rag dolls, pot holders, floormats, or hand mops which are in great demand to jeepney drivers and housewives.

Another big recycling business is the resale of old things. Clothes are exchanged for other articles or sold by piece or by the kilo for direct use or recycled into other forms of outfits. Old shoes, umbrellas are repaired and cleaned, so are the electric fans, radios, televisions or refrigerators. All are resold to the less fortunate ones at prices within the affordable limits of the poor consumers.

A new livelihood source at Smokey Mountain started by the Department of Labor and Employment in coordination with the Institute for the Protection of Children is recycling fluorescent lamps. They are repaired and sold at lower prices.

Tires which normally are converted into slippers, flower pots, and doormats have found another use for environmental purpose. They are now being made into artificial reefs to provide a safe haven for marine life. Coral reefs which have been damaged due to illegal dynamite fishing method are now being artificially substituted with rubber tires and in areas where they were set up, fish catch has gone up.

FARM WASTES

In the provinces, wastes from farm and agriculture such as poultry, or crop farms have found better use. Besides being used for mulch, animal feed, and fuel, small scale farm entrepreneurs now use rice straw to produce high quality paper. Foliage from dried cassava which are discarded are now being sundried to remove the toxic cyanogen content, ground and sifted to form cassava meal which maybe used to substitute for copra-meal. Banana stalks which are just thrown away from plantations are now utilized as raw materials for fibers when combined with polyester. They are used in making table cloths and placemats but maybe used for garments because of its sheen and gloss, resembling silk.

BIOGAS PRODUCTION

Farm wastes like straw trimmings, peelings, and manure are recycled for energy generation as in biogas production.

Biogas production is already an established means of fuel production at the Maya Farms in Angono, Rizal where all the energy needs are supplied by biogas from pig manure and all other wastes produced by the Integrated Farm System.

ORGANIC FERTILIZERS

Peelings, fallen leaves, or other organic household or industrial wastes are first converted into compost, later mixed with tobacco dust, burned rice hulls, and dried chicken manure, covered with plastic and allowed to ferment for a few days before use.

ANIMAL FEED

Big eateries sell or give away food wastes as animal feed. Most eatery owners grow pigs fed with their own food wastes and slaughter them for use in their own establishments.

Another source of animal feed are fruit peelings, trimmings, etc. which are chopped and steamed and fermented. Commercial urea and a mixed culture of *Saccharomyces sp* and *Aspergillus niger* organisms are added, mixed thoroughly and covered with cheesecloth to ferment. Resulting product is dried, milled, and used as supplementary feed to chicken or pigs.

INTEGRATED FOOD PROCESSING SYSTEM

Another economical scheme designed to maximize utilization of the raw materials is the Integrated Food Processing System (IFPS). Main features of the system are 100% utilization of raw materials and diversification into multi-product line thus reducing problems of food wastage and waste disposal.

When applied to fully ripe fruits, major products produced are jams, jellies, bottled preserves in syrup, dehydrated glazed fruits, or juice. Outputs of by product utilization are wine or vinegar, animal feeds or fertilizers from the seeds, peelings, and trimmings.

Citrus microcarpa Bunye is utilized for its juice as nectar when mixed with sugar but its fruit wastes of peel, pulp and seeds after extraction are utilized in the production of pectin, essential oil, "kalamansi" jelly, and feed meal.

From coconut, a number of products are developed. Shells are converted into charcoal, coconut water into vinegar or as substrate for a useful organism used for dessert called "Nata". The solid grated residue left after oil extraction can be made into candies or viands. When not used as such, may still be used for hog feed or as raw material for fertilizers.

For chicken, recycling is illustrated in the use of feathers for stuffing pillows and in the preparation of decorative materials like fans. Heads and legs are cleaned and barbecued while the skin is deep fried into crispy flakes and eaten as delicacies by beer drinkers.

CONCLUSIONS AND RECOMMENDATIONS

This recycling business benefits a lot of people. The original user, the repair man, and the middleman all profit from the business, while the end user can afford the item at reduced price.

Because the Philippines is poor, deprived of capital and has a high unemployment rate, this lowly form of business has become acceptable. It maybe degrading, unsanitary, and unhealthy, but with no other recourse to livelihood to some it is the only hope for survival. The recent announcement to level Smokey Mountain and convert it into a public market and industrial site has evolved a protest mass from the scavengers against their relocation.

FUTURE OUTLOOK

By the year 2,000, an 80% increase of 5,010 tons per day (TPD) in total refuse generated in Metro Manila has been projected. Recommended method for sewage disposal is the Integrated Recovery System which involves sanitary land fill, resource recovery, composting, and possibly incineration if affordable. Of the total waste to be generated, about 950 tonnes per day are theoretically expected to be recovered for recycling purposes. Other benefits to be derived are compost fertilizers and energy from incineration. Another recommended solution to the garbage collection at the present dumpsite is to convert the garbage into "green charcoal", a compacted fuel which can even be exported to other countries.

All these recommendations require capital investment but until such time that the country has fully recovered economically, scavenging and recycling will remain big business for the poor but enterprising Filipinos.

REFERENCES

Ang, Jose (1984). Unpublished Term Paper in a course on Solid Waste Management.

Anzaldo, Felicidad E. and Don Juan Barbara. (1987). Management and Utilization of Citrus Wastes. Paper presented at the FACS Asian Conference, Kuala Lumpur.

Banana Stalk For Fiber. Technical Process Review Vol. 3, No.7. (1987).

Borjal, Arturo. (July 8, 1987) Smokey Mountain. The Philippine Star.

Cassava Leaf Petiole as Feed Substitute. (1987) NIST Journal Vol.2, No.4.

Florescan, Roman. (1987) Old Tires Put to Good Use. The Philippine Star.

New Livelihood Source at "Smokey Mountain". (1987) Philippine Daily Inquirer.

Paper from Rice Straw (1987). Forest Products Research and Development Institute, NSTA. Technical Process Review. Vol. 3.

Pollack, Cynthia (1987). Mining Urban Wastes. The Potential for Recycling Worldwatch paper '76.

A New Solution to Solid Waste Problems in Developing Country Cities

David C Wilson, *Technical Director*, **Environmental Resources Limited**
106 Gloucester Place, London W1H 3DB (United Kingdom)

Jean-Marie Bourgeois, *Managing Director*, **IBH** (Belgium)

Khun Ta-noo Vicharangsan, *Director*, **Energy Industry Development Office**, Ministry of Industry (Thailand)

Introduction

Urban Solid Waste (USW) is a problem in most cities. In developing countries, the present disposal method is generally open dumping, with associated water pollution and public health problems. Upgrading open dumps into properly managed, environmentally acceptable, sanitary landfills is difficult, due to the lack of suitable sites (particularly where the water table is near the ground surface), a shortage of cover material and the presence of scavengers.

The traditional technology-based solutions suffer from high capital costs and the need for skilled operating and maintenance personnel. The USW is often too wet to give a net surplus of heat from incineration, while the soil conditioner produced by composting plants is often difficult to market due to contamination with plastics and glass.

This paper presents the results of a study, conceived by the Ministry of Industry of the Royal Thai Government and the Commission of the European Communities (CEC), to find a new approach, an innovative solution to the problem of USW in developing country cities, using Bangkok as a case study.

The theme of the study was to promote private sector involvement in the production of energy and other useful products from urban solid waste (USW) in Bangkok, with possible application in other ASEAN cities.

This study was seen as being practical in nature. The central objective was to develop specific pilot project proposals to demonstrate the feasibility of business schemes involving private sector participation and based on USW as a raw material. The aim was to produce an action plan capable of immediate implementation, rather than just another study report.

A parallel objective was to provide practical training and experience for

ISWA PROCEEDINGS
VOLUME 1 ISBN 0-12-058451-4

representatives from Bangkok and other ASEAN cities. This they could use to promote private sector schemes for the collection and treatment of USW in their own countries.

The Study

The study was undertaken by a core-team comprising staff from two specialist European consulting firms, Environmental Resources Limited (ERL) and Bureau d'Ingenierie Bourgeois et Harris sa (IBH), and representatives nominated by the Governments of 4 ASEAN countries (Malaysia, the Philippines, Singapore and Thailand), with local support from ACT Consultants Company Limited.

The project commenced with a desk study to identify and evaluate relevant USW schemes and an initial fact finding visit to Bangkok. The ASEAN representatives then spent one month on a team preparation programme in Europe, after which the core-team returned to Bangkok for 3 months. Detailed information gathering was undertaken in three main areas.

(a) The present system in Bangkok for USW collection, treatment and disposal as operated by the Bangkok Metropolitan Administration (BMA), including the quantities (currently around 4600 tonnes per day) and composition of the USW.

(b) The present materials recycling system (*Figure 1*), on which the team's proposals were intended to build.

(c) An industrial survey, which was aimed at establishing the existence of present or potential markets for material or energy products recovered from USW, and of non-hazardous industrial waste streams which could be co-processed with USW, e.g. to increase the calorific value of a fuel product.

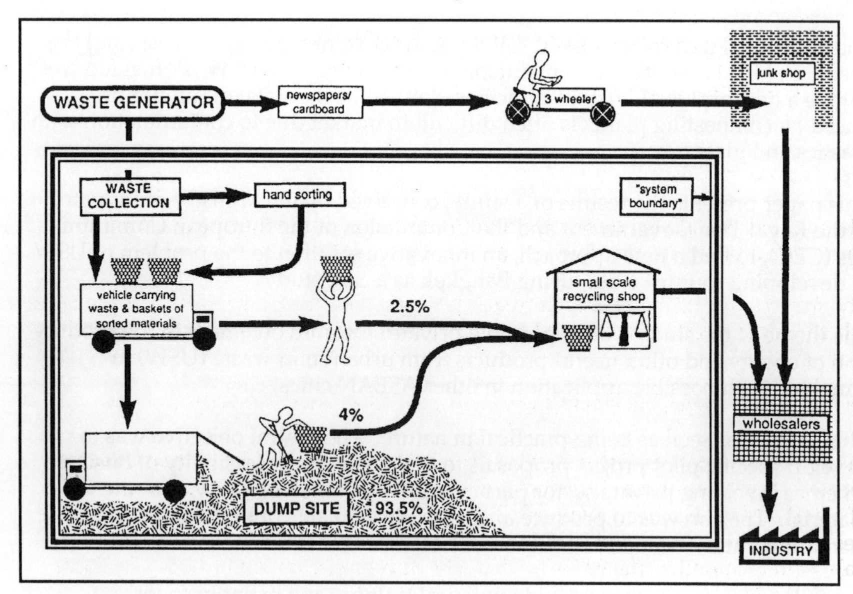

Figure 1 The present materials recycling system in Bangkok.

Alternative Business Schemes

Definition of Schemes

The study team developed four alternative business schemes, all of which start with a combination of a transfer station with a basic manual and mechanical sorting plant to separate three streams:
- Paper, plastics, glass and metal for recycling;
- Organic matter; and
- Reject materials.

The organic matter can then be made into compost, or digested in the absence of air to produce both biogas fuel and compost (anaerobic as opposed to aerobic composting); the reject fraction can simply be buried in a landfill or the combustible portion separated and processed into refuse-derived fuel (RDF).

The four business schemes are as follows:
Scheme A1 - Sorting with aerobic composting
Scheme A2 - Sorting with anaerobic composting
Scheme A3 - Sorting with aerobic composting plus RDF production
Scheme A4 - Sorting with anaerobic composting plus RDF production

Comparison of the Alternative Business Schemes with Existing Options

Table 1 compares the performance of the four alternative business schemes developed here with three types of treatment plants currently being operated or considered by the BMA, namely:
- Existing primary compost plants;
- New compost plants; and
- Transfer stations serving distant sanitary landfill sites.

The major performance criteria in *Table 1* are discussed below.

Net Treatment Cost

The aim has been to examine the medium term position, comparing future commercial plants in, say, five years time.

The first row in *Table 1* is the net treatment cost. This is defined as the tipping fee which would need to be charged for each ton of incoming USW to cover costs, that is,

Net treatment cost = Operating cost + Capital charges (including land) - Revenue from sale of products

In order to compare transfer stations on an equitable basis with compost plants located remote from the city, an estimated savings in collection costs to the BMA of 100 Baht/ton has been allowed. Thus, the cost for new transfer/landfill facilities for the BMA is entered in *Table 1* as 200-400 Baht/ton, rather than the 300-500 Baht/ton included in the BMA's 1987-1991 development plan (1 US$ = 25 Baht).

Three of the alternative business schemes developed here, A1, A2 and A3, have a projected net treatment cost at or below the bottom of the range of BMA alternatives, while the fourth, A4, compares favourably with existing or proposed new compost plants. This suggests that a tipping fee could be negotiated which would be attractive both to the BMA (through cost reductions compared to similar quality treatment plants) and the private sector participants (through worthwhile commercial returns). In the long term, private sector provision of USW services would free scarce public sector capital resources for other much needed social projects.

Environmental Impact

Two measures of environmental impact are included in *Table 1*.

- The volume of waste remaining for landfill disposal is important, as it is difficult to find suitable sites for sanitary landfilling within an economical transport distance from Bangkok. The two alternative sorting and composting options, A1 and A2, leave 33% by volume of the original USW for landfill disposal, while the two options incorporating RDF production, A3 and A4, reduce this to 10-15%. This compares very favourably with the alternatives available to the BMA (>40%) and even with incineration (about 10%).

- A less tangible environmental benefit of the transfer/sorting schemes is that they all facilitate the phasing out of open dumps and their replacement by sanitary landfills. Residues remaining for landfill are unattractive both to human and other scavengers, making management of the site easier; indeed, it may be possible to use the residues as an intermediate cover material. Both the quantity and the strength of leachate produced from the residues will also be much reduced, making the siting of new sanitary landfills easier.

Materials and Energy Recovery

An important criterion for comparing alternative schemes for the treatment of USW is their contribution towards valorizing the materials and energy content of the waste. *Table 1* records both the extent of material recovery and the net energy efficiency of the alternatives. *Figure 2* illustrates the overall mass balances for alternative business schemes A1 and A4, as compared to the present open-dumping system and to the existing compost plants at Nong Khaem.

The present open dumping system results in the recovery of about 4% of the input USW as materials separated by scavengers. It is estimated that all of the alternative business schemes developed here could increase materials recovery to about 16% by weight. Little materials recovery results from the existing compost plants or from a simple transfer station, while the new compost plant proposed by BMA would include a short manual separation belt, resulting in a recovery of perhaps 5%. It should be noted that all of these figures are in addition to the estimated 2.5% separated by the collection crews prior to delivery to the treatment plant.

There is an established infrastructure for materials recycling in Thailand. The present demand exceeds domestic supply, while the quality of, for example,

Table 1 Comparative evaluation of alternative schemes

Criteria	A1 Sorting + aerobic composting	A2 Sorting + anaerobic composting	A3 Sorting + aerobic composting + RDF	A4 Sorting + anaerobic composting + RDF	Existing compost plants	New compost plants	Transfer/ landfill
Net treatment cost[1] (Baht/ton)	0	240	180	440	300-550	650	200-400
Environmental impacts							
• Volume of waste left for landfill (%)	33	33	14	11	70[2]	40-50[2]	100
• Eases the phasing-out of open dumps	++	++	++	++	–	+	–
Materials recovery (%)[3]	16	16	16	16	0	5[2]	0
Net energy efficiency (%)							
• Energy products only	–6	8	13	38	–9[2]	–15[2]	–8[2]
• Including energy savings from recycling materials	63	79	83	110	–9[2]	15[2]	–8[2]
Employment of scavengers	++	++	++	++	–	+	–

[1] Assumes a saving in collection costs of 100 Baht/ton for all transfer stations
[2] Estimates by the Consultants
[3] Excludes materials presorted by collection crews

Figure 2
Comparative mass balances for:
PRESENT SYSTEMS
(1) The existing open dumping system
(2) The existing compost plants at Nong Khaem (a small proportion of the rejects to landfill are first passed through an incinerator)
PROPOSED SYSTEMS
(3) Scheme **A1** with sorting and aerobic composting
(4) Scheme **A4** with sorting, anaerobic composting and RDF production.

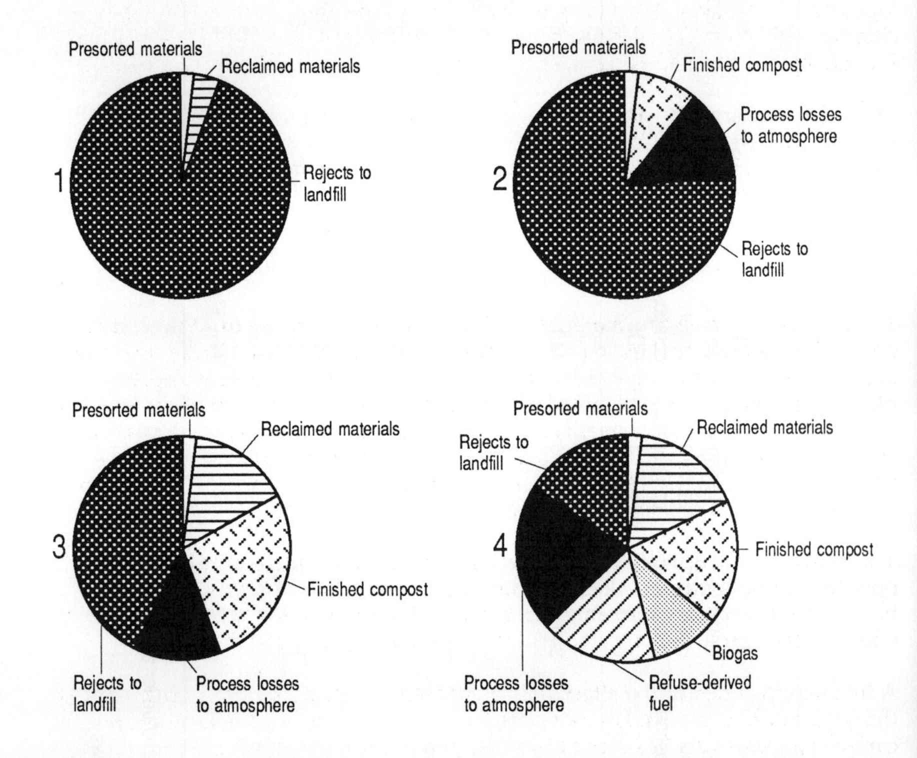

imported waste paper is poor. Based on the study team's market studies, it was concluded that the materials produced from the proposed sorting plants will be saleable, although some market development work may be necessary to develop new outlets for hitherto unavailable grades of recycled material (eg, some mixed plastics).

As shown in *Figure* 2, the alternative business schemes result in a total conversion of USW to useful products in the range of 40-60%. In addition to reclaimed products (paper, plastics, etc) there are several transformed products (compost, biogas, RDF). The compost should be relatively free from the finely ground glass and plastic fragments which limit the market opportunities of the present BMA product (no pulverisation of the USW is envisaged, while manual separation will be used to remove contraries from the organic fraction). The industry survey showed considerable interest in energy from waste, although extensive market development work would be required for either biogas or RDF.

Two measures of the net energy efficiency of the USW treatment process are included in *Table 1*. The first represents the percentage of the energy content in the input USW which is incorporated in energy products (biogas, RDF), after allowance is made for energy utilisation in the process itself. The second allows also for the indirect energy savings through recycling material products. As the denominator in each case is the heat content of the USW, this second figure may exceed 100%.

Considering only energy products, all of the alternative BMA treatment plants, together with Scheme A1, sorting with aerobic composting, have negative efficiencies. Schemes A2 and A3 have positive efficiencies of around 10% while that for Scheme A4, incorporating anaerobic composting plus RDF, approaches 40%.

The indirect energy savings from materials recycling are, by comparison, relatively high. This results in a small positive efficiency for a new compost plant, estimated at 15%, while the efficiencies of the four alternative business schemes developed here range from more than 60% for Scheme A1 to 110% for Scheme A4.

Benefits to the Thai Economy

The benefits of recovering material products and energy values from waste are only partially measured by including the revenue from their sale in the financial analysis. As Thailand imports most of the raw materials it needs for paper, plastics and metal production and for energy use, the development of indigenous raw materials from USW benefits the economy by reducing imports. It has been estimated that, if 30% of Bangkok's USW was processed through the business schemes developed here, that there would be a reduction in imports of between 550-650 million Baht (about 25 million US$) per annum.

The final row in *Table 1* indicates the relative effect of the alternative treatment options on employment. All of the transfer/sorting schemes have a positive benefit in creating employment, while most of the alternatives have a negative effect, simply displacing scavengers from the existing open dumps.

A further advantage of the alternative business schemes developed here is that, at the stage of full commercialisation, it is anticipated that most, if not all, of the capital equipment could be manufactured and maintained locally.

The Next Step

In the short-term, a pilot project is required to convince investors that the predicted quantity and quality of recovered materials can be realised, and therefore that there will be a worthwhile commercial return. To achieve this most effectively, one would wish to involve the private sector at an early stage, and thus to keep capital costs to a minimum. The proposed pilot plant in Bangkok will focus on proving the technical and commercial feasibility of a mechanical/manual separation and treatment plant, rather than either refuse collection and/or transfer.

Part I comprises a sorting plant as in option A1, with an average daily throughput of about 120 tons per day. The aim would be to involve private sector participation in this semi-commercial part of the project from the outset.

Part II is a smaller scale, pre-commercial, pilot plant aimed at establishing the technical feasibility, market potential and long-term commercial viability of biogas and RDF production from the output streams from the Part I sorting plant.

It is proposed that the pilot plant would be owned and operated by a joint venture public/private sector company, comprising the Thai Authorities, local private companies with an interest in the scheme (eg, contractors, waste producers, users of recycled products) and one or more overseas promoters (equipment manufacturers). Finance would be raised from local commercial institutions, from development agencies and from the Governments of the overseas promoters. The Thai authorities would contribute, for example, by providing land, a long-term contract for delivery of USW, and guarantees for the commercial loans.

The company equity would be held initially mainly by the public sector and the overseas promoters. As the uncertainties regarding the quantity and quality of recovered products from the Part I plant are removed over, say, the first two years of operation, the aim would be to increase the participation of the local private sector, both by increasing its equity stake and by taking over Government guarantees on the commercial loans. The private sector will thus be well placed to initiate the next generation of replication projects on a more or less fully commercial basis.

Conclusions

The study reported here has resulted in an innovative solution to the urban solid waste problem in developing country cities which could find widespread application. It displays the advantages of, for example, incineration with electricity generation:
- Low volume remaining for landfill;
- High energy efficiency.

It also overcomes the disadvantages in the following manner:
- Capital costs are relatively low;
- Sophisticated technology is kept to a minimum;
- Established materials recovery systems are built on rather than undermined;
- There are no implementation problems due to high moisture content of USW.

Additional advantages include:
- Creation of employment for scavengers from the open dumps;
- Easing the phasing out of open dumps; and
- Stimulating new Thai industries by making available additional raw materials and energy products, and by local manufacture of plant and equipment for future commercial plants.

The next step is to implement a pilot project in Bangkok to demonstrate the concepts in practice. It is hoped that a detailed feasibility study will commence in Autumn 1988, with the formation of a joint venture public/private sector company to build and operate the pilot plant, and that construction will start in late 1989.

Reference

(1) Wilson D C (1981) *Waste Management – Planning, Evaluation, Technologies*, Oxford University Press.

Session 10

Collection and Public Cleansing

Session 10

Collection and Public Considerations

COLLECTION AND TRANSPORTATION OF WASTE IN THE U.K. MEETING THE CHALLENGE OF CHANGE

BY

DENNIS TAYLOR

F.Inst.W.M., F.B.I.M., M.I.W.H.M., M.I.M.T., M.R.S.H.

DIRECTOR OF CLEANSING, CITY OF MANCHESTER

PRESIDENT, BRITISH INSTITUTE OF WASTES MANAGEMENT

1. Introduction

The collection and transportation of domestic and commercial waste in the United Kingdom has traditionally been a public sector activity, undertaken by local authorities.

In recent years, this monopoly position has gradually been broken down as more and more private contractors engaged principally in commercial waste collection. From next year (1989) the position will change even more dramatically, following the introduction of Government legislation which will require all local authorities to expose their refuse collection services to competitive tendering.

The decision to impose Central Government 'control' over locally administered and delivered services has created considerable controversy at both political and officer level and there is a very real chance that, however justified the protestations may be, service

ISWA PROCEEDINGS
VOLUME 1 ISBN 0-12-058451-4

provision may suffer, if too much time is spent debating the merits and de-merits of the legislation and not enough spent on preparing an organisational structure to meet public needs.

In recent years the Audit Commission, in association with Her Majesty's Government, have been undertaking 'Value for Money' exercises, aimed at highlighting ways in which the delivery of refuse collection services, to the public, can be improved by increased 'economy, efficiency and effectiveness'. At the same time the Government have imposed sanctions and financial penalties on high spending Councils (however well meaning their motives), aimed at ensuring that there is a radical re-appraisal of service provision with a view to improving the effective use of available resources. The 'Political' debate, although a necessary part of any democratic process, centring as it does on the conflicting ideologies of 'agents of centralism' or 'locally determined, democratic choice', cannot, in itself, resolve the issue of the needs and demands of service provision. Ultimately, if the public are not to suffer, the talking must give way to action.

2. Refuse Analysis

Analysis of the composition of domestic refuse has been undertaken in the U.K. for over 40 years and the changing trends have been used in the development of refuse storage facilities, collection vehicles design and ultimate disposal methods. Higginson's (1) pioneering work in the late 1960's dramatically illustrated the changing nature of waste and proved invaluable source material, which influenced collection vehicle design and the evolution of large waste receptacles to meet the expanding waste generation. Gulley's (2) analysis of waste classifications, in 1983, plots the fluctuating percentage, by weight and density, over 45 years and his results show a significant stabilisation over the last 10 years. Various socio-economic arguments have been advanced for this apparent standstill but whatever the cause it is significant that during the same period (1974-1984), there was relatively insignificant change in refuse collection vehicle design and refuse storage facilities. Bickerstaffe's (3) examination of packaging and domestic waste whilst confirming statistical trends,throws up the fact that almost one third of all domestic waste is packaging. Accepting that reduced packaging will not produce a directly proportional saving on collection and disposal it, nevertheless, gives greater credence to the lobby for multi-trip packaging aimed at reducing unit costs of products to the householder and reducing waste generation.

3. Refuse Storage

Throughout the 1970's as the volumetric increase in domestic refuse began to slow down, so refuse storage facilities stabilised. Gone were the 2¼ cu. ft. galvanised steel bin to be progressively replaced by 3¼ cu. ft. bins with a transition to the lighter, cleaner and quieter plastic variant.

The introduction of paper and plastic sacks, on economic grounds led to reduced crew sizes or increased productivity, eliminating the return trip with the empty bin. Few would argue that savings made from increased productivity in sack collection services were often severely eroded by increased littering, arising from interference by animals and itinerants at the storage point; poor presentation, bursting etc. at the collection point and lost time due to employee injury arising from unsafe preparation.

As Councils continued to strive for further economies in refuse collection costs there evolved a growing practice in the early 1980's to move away from traditional back door collection to kerbside collection of refuse. Incorporated into the one-trip sack service this has resulted in major cost cutting and significant labour reduction.

Nothing can dissuade me from the view that kerbside collection of refuse, in any form of presentation, is not a public service (you are asking the public to carry out, often at great inconvenience, a large part of the service you claim to be providing). Kerbside collection of sacks and the assorted boxes etc., which house domestic waste, is, I believe, the single largest contributor, to the ever increasing litter problem which is consuming society today.

In the last few years the 'continental wheeled bin system' has manifested itself throughout the U.K. There is little doubt that, in terms of health and safety, environmental protection and improved service provision they are having a major impact on our communities. It is unfortunate that the arrival of the 'innovatory' concept comes at a time when the financial aspects of our service are 'under the microscope'. A multitude of salesmen and users tell us that:- large capacity bins reduce the need for extra collections; wheeled bins enable operatives to work until retirement with reduced stress factors; hinged lids reduce litter spillage; the rigid containers reduces risk of injury and illness to operatives e.g. tetanus, hepatitis, AIDS etc. and all this may be true. However, the major selling point of kerbside collection as a cost saving exercise often ignores the capital cost of bin provision and vehicle adaptation. It has been argued that capital expenditure has been balanced by public participation in the collection activity and employee reductions in order to justify implementation. It is pleasing to see that, amidst many authoritative works written commending the economics of introducing the scheme Mansfield (4) concludes that transition to 'wheeled bins' will 'in isolation' marginally increase collection costs, but this can be justified, initially on environmental and economic grounds. Thereafter,as the service is refined, long term savings will accrue.

The introduction of wheeled bins into the U.K. domestic and commercial refuse collection industry is the single most important development in refuse storage in the last 40 years. Every vehicle manufacturer is responding to the transition and the range of options open to the customer is a significant step to tailoring the service to meet the needs of the client. I would be interested to hear delegates views on the merits of providing this environmentally acceptable service option on a collect and return basis rather than relying on the support of the public.

5. Transportation

It is in the field of refuse transportation that the importance of refuse analysis has a very significant impact. Whewell (5) has chronicled transport development in a major British City from 1552 to the present day and developments have kept pace with weight and volume variants. Many U.K. towns and cities did not introduce compression vehicles until the late 1960's and during the next 15 years there was considerable developments largely centred around packing mechanisms. It is significant that as refuse composition stabilised in the mid 1970's so too did the refuse collection vehicle manufacturing industry. Gardner (6) claims that looking at today's vehicles in profile it could be difficult to identify one machine manufacturer from another. Hollyhead (7) (like Gardner - a manufacturer), goes to great pains to emphasise that today's vehicles are 'designed with the end user in mind' and highlights the main customer needs as follows:-

Operational
- Reliability
- Payload
- Application
- Acceptability to Crew
- Material to be collected
- Low fuel costs

Crew
- Fast and easy loading
- Ease of operation
- Cab accessibility and comfort

Engineering
- Ease of Maintenance
- Low repair costs
- Good service back up
- Standardisation

Financial
- Fuel costs
- Maintenance and repair costs
- Vehicle life expectancy

Whilst we may, as customers, have bias towards one or more of these considerations, these factors still hold good 6 years on and the U.K. manufacturing industry continues to provide the highest quality products. We are undoubtedly spoilt for choice in terms of the number of manufacturers and the wide variation in models available to the customer. Perhaps a reduction in the number of companies in the market place with limited model options, might result in greater opportunities for hi-tech research aimed at improving vehicle development. I cannot imagine this situation arising by choice, but the rapid reduction in the number of R.C.V.'s sold in the U.K. may well force the manufacturer's hand. As councils prepare for competitive tendering with the added uncertainty of not knowing who will deliver the service two years hence, few authorities are purchasing in large quantities. Vehicle life is being extended; working days are lengthening. Double shifting is being introduced and the manufacturing industry is seeing a diminishing market place.

In my paper to the British Institute Conference in 1984 (8) I examined trends in vehicle design at that time. Although the market was down by 33% from 1978 (1250 vehicles to 825), 91% were of a fixed hopper variety. The improved safety elements is self evident and today the moving hopper is virtually extinct. The trend then was for the 'normal' vehicle to be a single

axle, 16 tonne G.V.W. machine. Variation included narrow bodied machines for restricted access; bulk loading machines for 'paladin' collections with side loading and front end loading machines fitting into minority areas of the market place. Today with 'economy, efficiency and effectiveness' keywords in our vocabulary there has been a significant shift towards
(a) twin rear axle vehicles to cut down on trips to the disposal point or increase loading productivity.
(b) multi purpose vehicle - to collect paladins/continental bins/hand loading (sack or dustbins) and in some case rear end loading skips.
(c) the attachment of lifting gear to handle the continental wheeled bin system.

The future may not hold significant changes in store, in terms of product development, but European and Domestic legislation will place our service under close scrutiny. Already the Health & Safety Executive Document PM52 (9) has laid down strict guidelines on safety in the use of refuse collection vehicles which are being incorporated into new vehicles and existing fleets are being modified.

Overloading problems may well be slightly alleviated by the introduction of 17 tonne G.V.W. vehicles. This particular subject is analysed in great detail by Stoke (10) in a recent paper to the I.W.M.

Starting with the 1977 E.E.C. Directive on noise and subsequent updating legislation, considerable advances have been made in suppressing noise levels both 'in cab' and 'drive past'. A great deal more work is, however, needed in this area. No doubt with competitive tendering in mind, specialist vehicle hire companies are springing up in the U.K. which may lead to councils and contractors cutting their spare pools of vehicles .

6. Disposal

Although the subject of my brief does not include waste disposal, it is imperative that collection agencies, be they public or private, liaise regularly with the Disposal Authority. As disposal sites become scarcer and transfer loading facilities increase to facilitate economic collection round radii, there is a need for both the collection and disposal agencies to acquaint each other with every step of their service development. Vehicle size, total fleet, number of trips to tip, timing of arrival, quantities of waste anticipated, nature of waste, extension to service, special waste collected etc. are vitally important elements a waste disposal authority needs to be aware of, if they are to continue to provide the best possible economic service to the collection authorities they serve. Organisations like the British Institute do much to bring these bodies together in common purpose.

7. Management

It is a sad indictment of our industry that in the last 20 years municipal waste management has not kept pace in performance, skills and vision, with the best in private industry. Much is said regarding the intervention of politicians, work study engineers and undue influence of trade unions. None of these 'red herrings' can hide the fact that in organisational skills, human and industrial relations and maximising resources, municipal management have been woefully deficient. The misuse of public money has not, until recently, taken onboard the air of seriousness with which the private sector view unwise husbandry.

Whether competitive tendering legislation is intended to provide a viable alternative to traditional municipal Service delivery or simply 'a shot across the bows' aimed at making councils more efficient one cannot help feeling that the surge forwards injecting entrepreneurial skills and 'best practices of the commercial world' into local government management may very well be 'too little - too late!'.

8. Competitive Tendering

In my opening remarks, I made reference to the 'political war' that is being waged over the Government's competitive tendering legislation.

Speaking as a serving local government officer and not as President of my National Institute, the very great sadness seems to be that there can be no winners at the end of the day. The tragedy of the situation is that the pitfalls are numerous, their resolution will be both protracted, and of necessity, largely a compromise and there will be casualties on all sides during the exercise. I have attempted to highlight some of the more obvious difficulties:-

i) The political overtones of Central Government <u>interfering</u> in local government service provision threatens to become an issue of principle which may mask the real issue of service provision.

ii) The legislation has been drafted to include clauses designed as a defence against what the Government consider may be 'restrictive practices' aimed at keeping work 'in house'.

iii) The time scale for implementation has not been thought through e.g.

a) Some councils have insufficient time to prepare proper specifications and contract briefs.

b) There are insufficient private contractors with appropriate experience and expertise to tender for the work.

c) The splitting up of services which in many areas are inexorably linked e.g. refuse collection/street collection/vehicle maintenance could have serious ramifications if each tender was won by a different contractor. Each could be dependent on the other to operate efficiently. In simple terms one's destiny will be in others' hands.

iv) There are insufficient safeguards to councils and the public they serve if a contractor fails to perform.

v) The effect that the legislation, its timing and its phasing, will have on the supply industry could be devastating. Erratic peaks and troughs of demand could lead to the demise of suppliers unable to survive the lean times.

As President of my Institute I am concerned that our role in the immediate future will be to arbitrate on contractual differences, many of which will result more from ideological differences rather than service failure, when there is so much we ought to be doing in providing a first class professional service to our industry.

9. Future

The British Institute is, as I am sure delegates know,

drawn from both the public and private sector of the service industry and manufacturers who supply both parties. It is generously endowed with consultants who have many years' experience in all these fields.

Our Institute stands aside from politics, both local and national. It carries no particular brief for either the public sector or the private sector though it cares for both. The I.W.M. is the single most authoritative voice on waste management in the U.K. A voice that is sought and heard in all corners of the world. Our aim is to help our members improve the quality of life for anyone and everyone whom they are privileged to serve.

****

REFERENCES

1. Higginson A.E. - 'Analysis of Domestic Refuse' I.W.M. monograph (1987)

2. Gulley B.W. - 'Progress in the Forecasting and Analysis of Household Waste' D.O.E./I.W.M. Seminar (1983)

3. Bickerstaffe Mrs J - 'The Impact of Packaging on Domestic Waste' D.O.E./I.W.M. (1983)

4. Mansfield R — 'Wheeled Bins - The Bolton Way' I.W.M. N.E. Centre Paper (1987)

5. Whewell R.G. — 'The Development of Transport for Cleansing Services in Manchester' I.W.M. N.W. Centre Paper (1985)

6. Gardner D — 'Refuse Collection Vehicles A Manufacturers Viewpoint' I.W.M. Scottish Centre Paper (1982)

7. Hollyhead G — 'Factors Affecting the Design of Refuse Vehicles' I.W.M. Paper (1982)

8. Taylor D — 'Refuse Collection Vehicles and Ancillary Equipment - The Customers View Point' I.W.M. Conference Paper (1984)

9. Health and Safety Executive — PM52 'Safety in the use of Refuse Collection Vehicles' H.M.S.O. Publications (1987)

10. Stoke J — 'The Changing face of Transport for the Refuse Collection Service' I.W.M. Spring Meeting (1988)

6 1/2 YEARS OF EXPERIENCE IN COMBINED COLLECTION OF SOURCE SEPARATED MATERIALS

Ole Vennicke Christiansen, Enviroplan A/S, Denmark
Kjeld Christiansen, Renovadan system transport a/s, Denmark

Experience in the 1970's showed that the costs of collection of newspapers and bottles from households, largely exceeded the value of the collected materials. Cheaper systems where the households themselves should bring the materials to containers, however, would in certain cases be economical, but the efficiency measured as the amount of material collected compared with the total potential amounts at the households, would normally be very low.

In 1981 Enviroplan A/S presented the theory, that a system collecting paper and glass combined with the normal collection of household refuse from single family houses would minimize the collection costs. Further, to make it easy for the household to use the system, and thereby improve the efficiency, a special container to be placed at the household had to be developed.

The system developed to try the theory consisted of two elements:

- A special designed sack holder for the traditional paper sack for the collection of the household refuse and two cassettes, one for the collection of newspapers and magazines and another for the collection of used bottles and other glass packaging.

ISWA PROCEEDINGS
VOLUME 1 ISBN 0-12-058451-4

- A special designed collection vehicle with a traditional compression unit for the household refuse and replacement containers, one for paper and one for glass.

The paper sack for the household refuse will be either a 110 l sack or a 125 l sack holding 12 - 14 kg of refuse, the two cassettes have a capacity of 30 l corresponding to approximately 6 - 8 kg of glass or paper.

Using the special sack holder, the household would not need to store the newspapers and bottles inside, but put it in the cassettes, when desired without having to care about which day in the week the materials are collected.

The vehicle is a standard chassis equipped with a 11.4 m^3 compression unit with a capacity of approx. 400 sacks or 4.0 - 5.0 t of refuse.

The containers for paper and glass are of steel with a capacity of 2.4 m^3. The glass container is equipped with inside slanting drop plates which reduces the drop height of the bottles and keep them from breaking. This is to ensure as many unbroken bottles as possible, as a Danish tax system enforces a tax on any new bottles produced, which makes it economical feasible to sort reusable bottles and rinse them. The containers have a capacity of 450 - 500 kg both for paper and glass.

The materials are loaded onto the container by an elevation system placed on both sides of the vehicle, which makes it possible for two persons to carry out the collection, each operating his side of the vehicle. When the container is full, it will be transported to an area in the district, where it is reloaded onto a 30 m^3 open container. The containers on the vehicle are emptied from the bottom. When onloading the material, the container is lifted out of the vehicle by two hydraulic arms. The materials are then emptied into the open container.

The collection of paper and glass are carried out every other week, one week paper and the other glass. The sack containing household refuse is emptied every week. The renovation workers will bring an empty sack and an empty cassette for the sack holder, and exchange these and load the filled ones onto the vehicle.

The system was established in February 1982, as a full scale project in an area with 2.000 single family houses in the municipality of Farum, north of Copenhagen. Following a test period of 8 1/2 months, the system was extended to cover 3.500 households and turned into a commercial system operated by Renovadan system transport a/s, as contractor. Not all households have the special sack holder, some have two buckets, but the municipality intends to replace these buckets with the special sack holder shortly.

The original vehicle was designed with a side loader compression unit and the changeable containers removed when full. Today, a second generation vehicle has been developed with a backloader compression unit as this reduces the lifting height for the renovation personnel. The containers with bottom opening presented lower reloading costs. This improvement was carried out by the contractor, who is currently planning a third generation vehicle with special emphasis on reducing the falling height of the glass and thereby reducing the number of broken bottles.

The sack holder has the draw back of containing the normal refuse sack which means that you have to scrap the old sack holder which is not necessarily worn out, if you are introducing the system in an area with an existing refuse collection system.

This has led to the design of a cheaper unit, called OK unit. This unit only contains the two cassettes and can easily be attached, with bolts, to the existing sack holder.

Results and Economy

During the test period of 8 1/2 months, 3.2 kg of newspaper and magazines were collected every fortnight, from each household corresponding to more than 90% of the potential amounts from the households. The figure for glass was 2.8 kg every fortnight, corresponding to more than 90 % of the potential amounts. The system has now been in operation for 6 1/2 years and the efficiency of collection of paper and glass has stabilized at approx. 80% respectively 85%.

The composition of the collected paper is very good; 75% newspaper, 9% magazines, 15% other paper and cardboard fibres and only 1% of other refuse.

The paper is sold to papermills producing tissues and egg cartons.

During the test period, 39% of the glass was broken in the vehicle and during emptying of the removable containers. After the redesign of the containers to bottom opening, this figure was reduced to approx. 30% and Renovadan transport system a/s hope with the third generation vehicle to reduce this figure further to 20 - 25% broken glass.

The glass is sold to a glass sorting plant, where unbroken, reusable bottles are sorted and the broken glass and non-recycle bottles are passed on to the glass works, where it is used as raw material for the production of new bottles.

The additional cost for the vehicle is approx. DKK 200.000 (DKK 100 corresponds to US$ 16) compared with a normal compression vehicle with the same capacity, and additional man-hours for the combined collection are, measured based on time and motion studies, approx. 10%.

The additional collection costs are calculated to be DKK 65/household/year, including the additional labor costs and depreciation for the vehicle, but excluding the cost for the unit at the households.

The costs for the units at the households are DKK 800 for the special designed sack holder and DKK 350 for the OK-unit alternatively used.

Total costs including depreciation of the OK-unit are DKK 106/household/yr. Sale of the collected materials provides an average income of DKK 250/t and approx. DKK 100/t is saved on incineration fees. This reduces the costs to (net) DKK 50/household/yr collecting 150 kg/household/year of reusable materials. This is an increase of the total refuse collection costs for the household of 5 - 10% utilizing 20% of the household waste.

The system has proved to be the most effective reuse collection system in Denmark, collecting more than 80% of the potential amounts, because it is very easy to use for the household and right now, it is introduced into several municipalities in Denmark.

New Systems

The experience with the system in Farum has led to the design of other systems to match other types of areas and other types of waste and materials.

In the municipality of Albertslund, west of Copenhagen, the sack holder idea is copied and a system designed to collect paper, glass, cardboard and a mixed fraction of PVC, metal cans and batteries combined with household refuse.

The area serviced by the system is a new strip building and the collection is performed by the renovation personnel in the area using small collection vehicles. The collected materials, as well as the refuse is transported to a centrally located container area, where it is reloaded.

In the municipality of Moen, one of the southern islands of Denmark, the cost saving idea with the combined collection vehicle has been copied, and a system designed, collecting paper and glass combined with garden waste and bulky waste from households. The collection is carried out with a special designed vehicle. The vehicle is equipped with a crane and a big container for the garden waste and bulky waste in front and a double small container for paper and glass, with a special elevation system placed in the rear.

When unloading the paper and glass, the crane is used for lifting the double container down from the vehicle. With the small container unmounted, the garden waste and bulky waste can be tipped off the big container.

The system will minimize the collection costs, but it has just been initiated, and so far no results are available.

The Working Conditions at Danish Sorting Plants

P. Malmros, M.Sc., Technological Institute, Department of Biotechnology, Box 141, DK-2630 Taastrup, and
C. Petersen, M.Sc., gendan ltd., Holbergsgade 26, DK-1057 Copenhagen K.

This paper gives an account of the Danish surveys carried out during the last two years of the working conditions at sorting plants for waste and recyclable materials. Space does not permit a detailed review of the surveys. However, at the conference supplementary material to this paper will be made available.

In Denmark discussion of the working conditions at both manual and mechanical sorting plants has run high since the autumn of 1986, when cases of respiratory ailments were recorded in certain of the employees at the mechanical sorting plant 4S at Kåstrup (which we will call 4S in the following).

Against this background a study was begun in 1987, with the support of the Danish National of Environmental Protection Agency, in order to survey working conditions at the Danish sorting plants (most of which are based on manual sorting). The survey includes a study of the literature, interviews with employees at Danish sorting plants and personal records from a large number of Danish and foreign plants. The study showed in general that a number of features of the working environment are unsatisfactory.

Results of the interview survey

The interviews among other things accomplished 3 sorting plants where industrial waste is being sorted out manually. The results from these plants are presented here.

The interviews showed that the employees at manual sorting plants were exposed to both physical and chemical effects and that their health has suffered a number of ill-effects as a result of work at the plants.

ISWA PROCEEDINGS
VOLUME 1 ISBN 0-12-058451-4

One type of ill-effect was due to the materials sorted:

- a lot of dust often develops when the material is tipped on to the belts and in the actual process of manual sorting;

- there is often food waste among the so-called "sortable" materials, which contributes to the proliferation of microorganisms and odour nuisance;

- sharp and pointed objects as well as oil and chemical waste are occasionally discovered in the materials, and constitute an accident risk.

Another type of ill-effect is due to the physical arrangements at the workplace and the actual organization of the work:

- working positions at the belts are inappropriate and the work is monotonous; this results in ergonomic strain;

- draughts and cold exist in several places; the ventilation is inadequate; the thermal indoor environment is often unsatisfactory;

- normally, noise only occurs in work with bale presses etc.

On average, each of the employees at manual sorting plants specified two health nuisances as a result of the work. One third of the ill-effects concerned muscles and joints (back, neck, legs, wrists). One quarter concerned respiratory ailments (among other things). The other symptoms concerned, among other things smarting eyes, fatigue and occasional nausea.

Respiratory ailments and microbiological readings at Danish sorting plants.

Parallel with the implementation of the above-mentioned survey, work hygiene readings were taken at a mechanical sorting plant for mixed household and commercial/industrial waste - the 4S plant - in Denmark. The readings are described in the following. The reason for this was that among the employees at least five had suffered occupationally induced respiratory ailments.

The work hygiene readings demonstrated endotoxins in excess of recommended normal values for air content. It is known from other sectors that the presence of endotoxins can result in symptoms like those observed at the 4S plant.

The readings at the plant were taken in the reception area, the automatic sorting hall and the area for manual sorting of cardboard, paper and plastic from mixed sortable commercial/industrial waste. High endotoxin values were measured for all functions.

Other possible health hazards in the working environment (dust, noise, light) do not exceed Danish limits.

There have been no reports of recorded respiratory ailments in other Danish sorting plants where manual sorting is done of socalled sortable commercial/ industrial waste is carried out on belts.

Microbial contaminants in the working atmosphere of waste sorting plants

In this part of the paper, the preliminary considerations on what would be the right agent of parameters to control is reviewed.

The plant 4S is a new built mechanical sorting plant which receives households refuse and industrial waste. The plant is equipped with a manual sorting line which has been stopped due to the Danish labor inspection authorities, but was the sorting line working during the measurements.

The observed cases of illness, which originally called attention to the plant, is mainly asthma, but there have also been flue-like symptoms with fever (possibly allergic alveolitis) and eye- as well as skin irritations. To summarize, half of the staff have had symptoms from the respiratory system.

Inspection at the site revealed that many parts of the plant contained accumulated wet refuse, and that this refuse also was mixed with the material intended for reuse.

It was also clear that many of the processes performed in the plant would cause severe aerosolisation of the material lying in the plant. Thus were for example all the conveyor belts open and compressed air was used to clean up the facilities at the end of the working day.

These findings and presions findings in other environments with microbial contaminations suggested that at least one of the responsible agentia was of microbial origin and would be airborne.

In table 1 are preliminary results from investigations on etablishing the human dose response relation upon inhalation of endotoxins presented.

	Effect-level relatively certain	ug endotoxin/m3 air uncertain
Symptom:		
Fever	0,5 - 1,0	
Acute bronchoconstriction	0,1 - 0,2	
Chestpain	0,3 - 0,5	
Chronic bronchitis		0,05
Hyperreactive airways		?

Ragnar Rylander, Göteborgs Universitet pers.comm.

In order to compare these findings with measurements of airborne endotoxins table 2 shows some results from measurements of airborne endotoxins in areas with problems at least partly due to airborne endotoxin or gram-negative rods.

Table 2. Airborne endotoxins.

	ug/m3
Dried sewage sludge	1.3
Sewage water pumping	1.2 - 3.5
Composting	0.001 - 0.4
Handling moldy hay	0.3 - 6.7
Refuse handling	- 1.0 *

* results from this study.

The presence of large amounts of wet refuse implied high microbial decomposition activity, and a very high fraction of these microbes would be expected to be gram-negative rods and fungi, and the bacteria are the source of endotoxins.

Endotoxins are not new as a problem in the working atmosphere, as they were identified as at least part of the problem arising from working with dried sewage water sludge in 1976 (1). But in the light of the problems experienced at the installation in Skive, and because of the expected increase in re-use activities, the toxins will be shortly presented in the following.

The symptoms following small doses of endotoxin are: fever, chills, headache, muscular stiffness, muscular pain, loathing of food, nausea, vomiting and diarrhea.

It is furthermore known that inhalation of dust containing endotoxins can give symptoms from the respiratory system, such as allergic alveolitis, asthma and possibly chronic bronchitis.

As mentioned above, there is good reason for taking an interest in endotoxins. But other microbial parameters are also present in the working atmosphere. Fungi are known to be a problem in composting plants (4), and there is also reason to take an interest in the thermophillic actinomycetes, as this group is known to cause respiratory problems.

These two parameters will be investigated in more detail in the research project to follow this pilot study.

Table 3.

Place	Total count cfu/m3 1)	Endotoxin ug/m3 1)	Total count cfu/m3 2)	Endotoxin ug/m3 2)
Reception	>20.000	0.48	7.600	u.d.
Manual sorting working	>20.000	0.48	1.000	0.110
Manual sorting not working	-	-	2.600	u.d.
Primary magnet	-	-	8.400	u.d.
Second magnet	>10.000	0.55	-	-
Pill press	-	0.99	7.100	u.d.
Office	3.000	0.026	1.400	u.d.
Unpolluted exterior	1.500	0.0003	-	-
Truck	-	-	-	0.003
Outside	-	-	450	-

1) Results from September 1987 - before rebuilding of the plant.
2) Results from March 1988 - after rebuilding of the plant.

For all these measurements an Andersen Sampler was used.

Results and discussion

The results obtained are shown in table 3. Table 3 represents the status of the plant before and after encapsulation of the conveyor belts and the establishment of a central vacuum cleaning device.

The figures show great variation, and the following interpretations are made from the generel picture more than from single, specific results.

The results from September 1987 verified, that the observed problems could be due to microbial contaminants. The medical examinations carried out to confirm this are still taking place during the preparation of this manuscript.

As a result, it was decided to rebuild the plant with special emphasis on minimizing the release and small amounts of dust, and to increase the level of cleaning in the plant.

The results from March 1988 gives a status for the plant.

On the basis of this it was decided to further improve the plant with a new ventilation-system, to establish routine microbial control programs and to perform investigations in order to establish how fast the new systems are able to eliminate microbial contaminants from the working atmosphere after a mechanical breakdown.

The new research programme, and the preliminary results hereof will be presented at the ISWA 88 Conference.

The above mentioned work on microbial contaminants was performed at the Technological Institute, Taastrup, Copenhagen and the financial support was granted by the Municipality of Skive.

Further work in Denmark

Before we carry out any further expansion of the sorting plants, we need to make a survey of the health hazards involved in the sorting of specific materials and mixtures of material, and of the standards that should be set for the materials so that they will not constitute a particular health hazard.

There is therefore a wish in Denmark to carry out the following two study programmes.

1. A study programme to:

    - establish the relationship between microbiological pollution and the specific type and composition of waste;
    - carry out occupational medical tests on employees at both manual and automatic sorting plants.

2. The implementation of projects aimed at ensuring purer materials for the sorting plants.

At the ISWA Conference the authors will present programmes for these studies and the results that have been achieved.

References

1. Petersen, C., Arbejdsmiljø ved sorteringsanlæg. Miljøstyrelsen 1988.

2. Sigsgaard, T., L. and Malmros, P. 1988: "Endotoksiner, En kort udredning specielt med henblik på affaldsbehandling". Miljøstyrelsen 1988.

3. Rylander, R. et al., 1976: "Sewage Workers Syndrome". The Lancet 28: 478-479.

4. Clark, C.S., et al. 1981: "Evaluation of the health risks associated with the treatment and disposal of municipal waste water and sludge". E.P.A. - 600S1-81-030, May 1981.

THE MANAGEMENT OF STREET AND PARK CLEANSING: PROBLEMS AND NEW TRENDS

Eng. A. MAGAGNI - Managing Director A.M.N.I.U.P. - Padova

Dr. A. PERONI - Managing Director A.M.I.U. - Modena

1. GENERAL CONSIDERATIONS

1.1 The street as "receiver" of some services

The planning and the management of cleansing services concerning town streets is today characterised by the introduction of a modern principle, which considers the street as "receiver" of some services or interventions.These operations can be divided in interventions:

- on the street
- above the street
- under the street

The operations on the street mean the cleansing of streets and pavements, with diurnal and nocturnal sweeping, the maintenance and/o periodical and radical cleansing, the management of the necessary waste-paper baskets, the particular cleansing for animal wastes (dogs, pigeons, ecc.), the maintenance of street cloak, the restoration of road signals for vehicules circulation, the treatments concerning disinfection and grass cutting, the maintenance of flower-beds and traffic islands, the watering of flower-beds and trees.

The operations "above the street" mean the maintenance of public lighting, the maintenance and installation of the vertical signals and traffic-lights, the management of advertising spaces, the control of air pollution due to vehicles and heating plants, the disinfection and actions against flies, mosquitos and insects in general, the compulsory maintenance of building front face, the co-ordination of networks installation (gas, electric power, telephone, and so on).

The operations "under the street" mean the maintenance of collection of rain water sewages, the cleansing of trapdoors, the cleansing of street wells, the co-ordination of

ISWA PROCEEDINGS
VOLUME 1 ISBN 0-12-058451-4

pipes, cables installations in order to have contemporary interventions.

So the street become the "receiver" beyond the traditional sweeping service, of some services which define the life quality in the environment of the street.

In the street men and means live and work. This mouvement represents the life of the street and of town and this justify the proposal of the co-ordination of all the interventions on the street, in order to improve the population mobility in town.

From the mouvement of men and means two important problems come out:

- the cleansing of street
- the vehicles parking

The traditional task of street sweeping service has become in modern times a problem of "hygienical cleansing" and of "propriety cleansing".

On of the most important problem which conditions the street cleansing is the vehicles parking on the street, which very often interfere with the cleansing service.

It should be necessary to regulate (with limits and or periodical prohibition) the parking of vehicles on the street. It should be useful assure not only the street cleansing but also the greatest number of other services which interest the street, in order to make people understand that it is worth to endure the limitations to parking if the conterbalance is the accomplishment of all the other necessities of the street.

This involves the citizen in a new, rational and complete consideration of services which make the street become receiver and a producer of services.

1.2 The propriety and the hygienical action

The propriety cleansing concerns the removal of all things which dirty the aesthetical aspect of the street. For instance a piece of paper thrown away concerns the aesthetics of street.

The hygienical cleansing concerns the maintenance of good hygienical conditions in order to avoid the growth of insects, microbes, virus that can represent a risk for public health. These are two different operations which complete each other:

the organization of street cleansing divided in "propriety " and "hygienical" is important also for the correct information of public opinion which lead to a considerable cooperation.

The aesthetical aspect of a town is still more important that the hygienical one, when the street sweeping service organisation is being planned.

Nevertheless it is always more important the hygienical aspect which must be guaranteed to operators and citizens.

The street sweepings, in most cases, are formed by mineral elements (rubble from flaking of paving or of buildings); by vegetal wastes (paper, leaves, and so on); by wastes produced by animals (dogs, cats and so on) and by birds (problem of pigeons proliferation) and in industrialized towns, by the falling on soil of dusts and ashes (this facts determines a new aspect of the problem).

A particular importance must be given to liquid wastes, either for the high content of organic substances, or for the presence of pathogenous micro-organisms, above all in liquids of human and animal origin.

The average quantity of "street" wastes for citizen/year is not easy to determine. The values can be variable from place to place and they depend on many factors such as the paving of public areas, the species and quantity of existing animals and above all the hygienical and civic education level of the local people.

There can be daily and seasonal changes in these values in relation to the life habits of the local people.

So it is difficult to establish exact values, but within

certain limits, it is possible to affirm that the average quantity of street wastes for citizen/year is about 50/150 kg. with a specific weight between 700 and 1.000 kg/mc.

In the hygienical action it must be considered the necessity of removal of sedimentary motes, infact they are dangerous to men and to his health because they arise for natural ventilation, for vehicles transit, and by the sweeping and removal operations.

They can be dangerous to men and to animals for two reasons:

- epidermic consequences in a short time
- cancer in a medium-long time

The infections danger is connected to excrements, to pathogenous virus, to salmonella, to tetanus and so on. If the public areas are soiled by human excrements (which is not an exceptional case) there is also the possibility of spreading of many virus.

Beyond the microparasitic, it is easy the spreading of worms, in the excrements of all animals included dogs, which are responsible in most cases of dirtying streets.

The more evident remark, above all during the winter months, of dust falling that we can note walking on the streets, confirm the importance of this problem.

These fallings are in direct relation with the air pollution intensity and are formed above all by ashes and unburnt solid among which it is possible to find tarred substances which contain anthracene and benzpyrene derivatives.

These pollution kinds are different from place to place from season to season and in the same place there are differences among urban, industrial and rural areas.

In the urban areas, while during summer there are more mineral dusts, during winter there is not only an increase of dust containing carbon remains that provoke pulmonary diseases, but also in sedimentary dust there is the presence of aromatic hydrocarbon polycyclic. The heating systems and internal combustion engines are responsible for the forming of

these potential cancerous products.
These dangerous substances, according to some stable data, are to be found in dusts, in food, in waters and in soil. Wind and rain cause a contamination of all earthly elements. In industrial areas the problem is more complex and it is in relation with the kind of activities established and to working methods.
Together with dust there are often toxic substances of various nature, in relation to the development of technology. So it is certain that as far as the technical aspect of street sweeping problem is concerned, the hygienical aspect is more important than that of propriety.

2. The productive elements and their evolution

2.1 The service and the equipment

In the european towns there is not an homogeneous behaviour concerning the holding of service, as well as the obligation of its accomplishment.
The results of an inquiry carried out in 15 european towns confirm that there are different systems and situations which must be examined to find the common features, given that the problems have essentially the same characteristics. In some towns the pavement sweeping must be done by the owners of the house facing the street , in some other towns the municipal society provides for pavements which are wider than a certain limit. In other towns the sweeping service is accomplished only in the centre of the city.
The mechanization level of service is also very different from place to place. There is the simply manual service and the partially mechanized one, the use of water is more or less remarkable either as auxiliary means or as the principal instrument of the service.
It is enough to say for instance the roads conditions, the infrastructures state, the vehicles increase, the traffic, the parkings and the streets development of urban centres. The street sweeping will not have radical changement like

those that have influenced the urban wastes collection systems, that use since ten years the computerized containers mouvement.

It will be necessary to operate with:

1) the vehicles and the traffic
2) the mechanization of productive systems with more use of water and operating during night
3) the use of suitable equipments to allow a correct service with parked vehicles.

Among the european towns the above mentioned principles are particularly applied in Barcellona where there is a high mechanization level, not only for the quantity of employed means, but also for their specific use according to the characteristics of the areas in which the town has been divided.

In Italy the 86% of the 8200 Municipalities has a population lower than 10.000 inhabitants. The street sweeping service, where it exists, is managed by the Municipality. The mechanization level is not very high and the sweeping of parked areas is a true problem.

Besides the knowledge of accomplishment methods it is necessary to know the methods which control the quality of service, either for the served area or for the times and for the route vehicles. Intervention and calculation methods, that contribute to increase the quality of service, have been worked out.

The mechanization has always interested all types of street sweeping, including the manual one. The use of mechanic means can be found for the cleaning "from wall to wall" and maintenance cleaning. The vehicles usually employed are:

- power-sweepers vehicles of medium (2 mc) and high (6 mc) capacity, either sucking, mechanic or mixed, with 2-3 mt of advancing lenght, with 2 or 3 lateral brushes;
- cleaning-sweepers vehicles of little (< 1 mc) capacity for particular roads and pavements;

- pavement-washing vehicles with manual loading of materials;
- street-sprinkler vehicles with different operation pressure, with sprinkling bars;
- sucking-leaves vehicles of high capacity with equipment mounted on vehicles for various uses.

Which can be the further development in the management of service ??

As for the other public services it is necessary to affect on the various elements to improve the results of each operative intervention.

It is useful to point out briefly the following:

a) <u>External elements to the service system</u>

- to provide the citizens with the traditional instruments, as the waste-paper basket
- to promote campaigns in order to press the active participation of people
- to put into action a good coercive system
- to know the development of politic and admini-strative strategies concerning the choice for state of the roads, parkings and so on. The indications must be valid at least for an average period of 6/7 years, to optimize economically the productive structures to manage;
- to agree with the competent municipal office the infrastructures to realize for new installations and the possible intervention on the existing in order to praise the service mechanization.

b) <u>Internal elements to service system</u>

- to employ vehicles and equipments less noisy (possibly < 75 db) to make easier the nocturnal service of sweeping from wall to wall;
- to use with increasing frequency the surfaces washing, swept or not, with bars or other systems, at high pression;
- to improve and to use continually vehicles and equipments

that can operate with parked vehicles. There are already working some prototypes which have given interesting results. They work with high pressure water or with high pressure air.

The pushing action of material towards the middle of the road is due by means of a bar which operates under the cars or by a jet which provides to remove the wastes from pavement to the middle of the road;

- to improve the handling and the control of the traditional sweepers vehicles wich must pour off the wastes in the located containers during the routes;
- to increase remarkably the operation speed of sweepers vehicles when the operation is done in particular streets in which there are no parked cars;
- to rationalize the employment of a set of vehicles and equipments for particular places in the urban routes;
- to mechanize the maintenance service. There are already some new equipments in the market.

  It is a little equipment, driven by a little electric motor which can be placed on a trolley, electrically driven.

  The equipment weight is moderate. The sucking system allow the accomplishment of service with more hygienical-sanitary safety, either for people or for the operators;
- to employ feeding systems better than diesel and petrol.

For vehicles and equipments of little potentiality, little motor-trucks, little power sweepers vehicles, and so on, it is better to use electric motors. For a medium and high capacity equipments, the use of methane allows interesting results at technical and economic levels.

Moreover with the use of methane something is done to control air pollution. The use of alternative feeding systems for public vehicles should be taken into consideration more seriously.

2.2 Computerizing of service

The new data processing technologies are quickly finding new applications, even in the urban cleansing services field, as it is widely presented in another paper.

For the sweeping services it is necessary a knowledge of the factors connected to soil conditions to permit an elaboration according to the equipment exigences.

This is the reason why the methodological analysis of the territorial census is very important.

The improvement (referring to elements: times, quantities, consumption, and so on) must be done also for streets cleansing service particularly for power sweepers vehicles routes, and for streets washing vehicles. A lot of variable elements change very often the roads conditions such as : direction changement of one-way , and more generatly the circulation rules, transit interdiction transit because of public works and so on.

So it becomes necessary to modify the service according with the changements of external conditions.

The use of computer can be a valid help for the management of data for statistical aims about streets.

The street must not be considered as a single unit, infact usually the different interventions are carried out without considering the street in a global sense.

So the street is considered in its minimum unit, that is the "part" limited by cross-roads. These "parts" limit the referring unit, that is block of houses and are identified to be elaborated, by the nearest civic numbers to the cross-road.

In the initial preparation of census cards the streets are divided in different blocks of houses, either because the street is interrupted by cross-road, or because the right or left side of road are different block of houses borders. In every "part" of the street all necessary data for the accomplishment of urban cleansing services are taken such as:

- lenght and width of the street

- lenght, width and height of pavements
- lenght and width of pedestrian crossing and of cycle tracks
- lenght and width of arcades
- number and type of trap-doors
- number and types of trees
- number and placement of waste-paper baskets
- dimensions of sewers and of drains

Every "part" of street is then codified in different ways for the various services.

This codification demands as next step the reassembling of zones for the different services.

It is possible to foresee that the computer will give two different types of data:

- "descriptive" type with street listing, and connecting data to the streets
- "graphic" type traced with suitable graphic conventions

So there is an exchange of information:

- for each street, it is possible to have the belongings to a certain zone corresponding to different services and to have all data collected
- for each zone, it is possible to have all the streets or "parts" of streets included in the same zone and the collected data necessary to the service of that particular zone.

2.2.1 Some management examples

a) Streets sweeping service

- Display and printing of all territorial data for each zone, their dimensions or quantity with adjournement possibility;
- each information is connected to the corresponding zone and for each zone it is possible to look for one or more data, for instance (streets kilometres, numbers of waste-papers baskets or number of waste-paper baskets for each street, reassembled for each block of houses, type of

frequency, type of method applied and so on).

b) Leaves collection service

- total number of trees in each street, types of plants in each zone
- indicative programming of leaves fall according to types plants and to streets and consequent intervention
- disinfection programming according to type plants and to streets and storage of data
- possibily of intervention of operative team to examine the situation.

Each element is sufficient to be used for interventions including for each type of service, during operation, maintenance, traditional phases: street sweeping, snow service, disinfection leaves collection, sewers cleaning and so on.

3. Some significant indications: parameters and average index

Each parameter taken for the calculation of work charges, of productivity and of efficiency index, depends on the particular considered zone and on local standardiżed habits.

Nevertheless it is possible to show for each index a representative space, at least for the towns with efficient system.

The index that follow depends on the above mentioned conditions and refers either to streets sweeping service or to auxiliary services.

a) manual sweeping, including pavement, five hours of real work, nr. 1 operator with a little motor-truck:

- linear metres of swept street

in center	mt. 2500/3500
in the suburbs	mt. 5800/7000

b) mechanized sweeping, five hours of real work, nr. 1 power sweeper vehicle with nr. 1 operator:

- linear metres of swept street

in center	mt. 16000
in the suburbs	mt. 22000

c) power sweeper vehicle, nr. 1 operator for 1 hour work (advancing lenght: mm 1200):

- treated surface mq. 1600/h mt. 1300/h

d) sewers cleaning with mechanical means

nr. 1 driver with a sucker loader, trap-doors cleaned for each turn: 45 - 60

e) hydrodinamics sewers cleaning (diameter mm. 1300):

- 3 units (2+1) with nr. 1 sucker loader and nr. 4 hydrodinamics from mt. 150 to mt. 300
- 2 units (1+1) with nr. 1 combined from mt. 170 to mt. 200

f) pavement-washing vehicle (advancing with cm 90) nr. 2 operators with 1 washing unit

- washed surface mq/h 250

g) nocturnal washing (with two jets five hours real work)

- nr. 1 driver, nr. 2 street-sprinklers, nr. 1 power sweeper unit with nr. 1 tank truck capacity mc. 10 mt. 4500 treated each turn

water filling for each work turn: nr. 3

h) snow service

- salt-cellar vehicle, capacity mc. 5 km. 19 covered for each charge, spread salt: 20 - 25 gr/mq
- snow-plough blades: operation speed mt/h 3400, street width mt 2.50-3, cleared surface mq/h 26600

i) forest trees treatment

- nr. 1 operator with nr. 1 tractor and nr. 1 atomizer: 80/h with 2000 lt of liquid

SPECIFIC SUMMER AND WINTER TASKS OF ROAD CLEANING IN BUDAPEST

Janos Banhalmi

Budapest, the capital of Hungary is a metropolis tan of 2 million inhabitants in Central-Europe. The city is divided by Danube, the second largest river of Europe, into two geographical areas: the plain Pest and the hilly Buda having 400 to 500 m high mounts.

The capital is the seat of the governmental and state administration agencies and, at the same time, it is the core of the scientific and cultural life. In congruency with the distribution of inhabitants, about 20 to 25 per cent of the industrial production capacities are located in Budapest.

Including the hundred thousands of people who go to their work from various settlements and the Hungarian and foreign visitors, there are about 2.5 million people working, going out, using transportation means and, unfortunately scatter rubbish as well in Budapest.

Even though the number of passenger cars has been rapidly increasing recently, for instance, between 1975 and 1986 it was more than doubled, mass transportation has still a decisive importance in the daily transportation of the inhabitants and employees. The underground network interconnecting the inner districts in East-West and North-South directions is the backbone of the mass transportation network. In the city of 525

ISWA PROCEEDINGS
VOLUME 1 ISBN 0-12-058451-4

sq km large basic area the public tram and bus lines reach the underground network. Since Budapest is the node point of the national main roads and highways, its network of roads is loaded with a significant amount of target and transit traffic, until the highway surrounding the city is built. For the key data of Budapest's public road traffic, see table No.I.

The aesthetical landscape of high culture assumes clean public areas and the dust-free roads improve the **hygiene** of settlements as well as the health condition of the inhabitants.

Table No.I.

Some typical data of the road infrastructure in Budapest

Denomination	Value
Roads of solid cover:	
length	3092 km
surface	22027 thousand m^2
Length of the underground network	30.6 km
Length of the electrical network	169.1 km
Length of the bus network	716.7 km
Number of passenger cars	389.5 thousand
Bridges over the Danube for public transport	5
Total length	3170 rm

In the precipitative winter conditions, it is essential to provide the reliability of the public transport for running the mass transportation, plants and institutions.
These demands specify the tasks of the Public Area Maintenance Company of Budapest ("FKFV"), cleaning and running the public roads in winter and summer.

First, I would like to say a few words about the summer cleaning tasks of the road cover of 17 million sq meter area and of other public areas and then I would give an introduction to the winter tasks including the snow disposal and anti-slip operations.

The summer road cleaning is decisively based on mechanization. It is well known in the industry that this job can never be fully mechanized at any place. To hire people for manual road cleaning is met with increasing difficulties in Hungary as well and it is limited to special areas, such as cleaning of under-passages and vehicle parking lots.
The basic technology of summer road cleaning is the suction-sweeping and the road washing at night. There are two types of suction sweeping equipment; the mobile and flexible medium category machines with a garbage bin of 2.5 to 4.0 cu m and the large size, high capacity vehicles fitted out with a body of 9-12 cu m.

Both types meet the environmental protection and **hygienic requirements and they suck the garbage** dust-free by spraying. The medium category special machines are used to clean the surface and deeper parts of internal roads of 2 to 4 tracks, the high power machines are used for sweeping of main roads running into the national road network, public over-passages and bridges.

The winter road cleaning tasks are similar to those in other large cities of Middle Europe, with some unique features originating in the geographical characteristics of Budapest. Because there is a difference of altitude between the plain Pest and the hilly Buda, sometimes highly different climatic situations develop in the two parts of the capital.
For instance when the air temperature is about zero C^{o}, it is raining in Pest and heavily snowing in Buda and the bridges connecting Buda and Pest become heavily slippery due to the vapour freeze. Neither the extreme weather conditions, are excluded in Hungary of continental climate. For instance the winter of 1986/1987 was extremely cold and precipitative, while the winter of this year was extremely mild and practically free of snow.

The main weather characteristics of the afore mentioned two winter periods were as follows:

	1986/87	1987/88
number of snow falls	23 times	10 times
height of snow fallen	101 cm	22 cm
number of slippery days	70 days	28 days

In the average winter periods, about 10 to 15 days may be snowy and 60 days slippery.

Following the weather characteristics, we developed our winter machine fleet configuration including various combinations of anti-slip (spraying) and ploughing machines. The basic machine used in the Buda hills is the UNIMOG type all-wheel drive truck equipped with snow-plough and an automatic spraying equipment. The small volume, highly mobile UNIMOG is especially suitable for snow disposal over the steep and narrow mountain roads but it can be well utilized in the narrow streets of the downtown in Pest.

For cleaning the round boulvars and lanes, the overpassages, bridge tracks, highways and city stages of motorways, special machines of super size and power are developed with 4 m wide ploughs and 6 x 6 all-wheel driven Soviet chasses.

The anti slip technology is based on using the solid and liquid chemicals and the solid scragging agents. As far as we know, the liquid anti-slip agent dissolved in Magnesium Chloride ($MgCl_2$) is used in the largest quantities in Bp all over Europe. Perhaps it is not irrelevant that I give a little more introduction to the solvent anti-slip technology.

Over a period of 10 years, the company has made many financial and intellectual efforts to develop the technical and technological facilities of the solvent anti-slip technology and the complex technological system.

The bases of solvent anti-slip technology are the container plants installed in three reasonable points of the capital. The open air installed carbon steel containers have a storage capacity of 9500 cubic meters. Spraying is regular in the course of summer road cleaning, with a 7 cu m tank mounted on a SKODA watering cart.

The technological advantage of the solvent anti-slip technique over the solid industrial salt is the stronger melting effect, more steady **distribu**tion of the material. The practical experiences of several years show that the magnesium chloride solution's corrosion is less over the iron metals than the sodium chloride corrosion and the flora is less deteriorated in the areas treated with the solution than in the roads treated with salt or sand mixture. The company regularly tests the existing and potentially utilizable chemical agents. For instance, based on our internal researches, we introduced the regular use of urea in order to **minimize** the corrosion damages of the essential bridges over the Danube. We investigate the potentials to introduce the wet salt technology in Hungary which is suitable to reduce the amount of industrial salt used.

However, comparing everything to the international practices, the use of sodium chloride is essential in the winter anti-slip operations, we do not find a potential to substitute the overall mass of salt with an other agent because of the **economic and** technological considerations.

For typical use of anti-slip agents in Budapest, see table II.

Table II.

Anti-Slip Agents used in Budapest

Denomination	Qty	Remarks
Industrial salt (sodium chloride)	16-17000 tons/season	In the form of salt and sand mixture
Sand	28-35000 tons/season	
Magnesium Chloride solution	10000 tons/season	
30 % MgCl	120-150 tons per season	

In order to develop the salt spraying technology, the company increased recently the number of up-to-date automatic spraying equipment and built the first enclosed salt container project of 4500 cu meters.

Finally, I would like to outline the key organizational principles of the winter and summer tasks and illustrate the magnitude and facility requirements of the tasks. The basic principle is the areal one. The capital is divided into 4 sectors and in each sector is a technical plant installed from where the special machines are sent to operate according to a schedule. In winter, for the faster and more intensive action, the areal principle is completed with work done in stages. In the first step, the mass transportation is taken care of or the other important roads (leading to hospitals for instance) are cleaned from snow and anti-slip treated and afterwards the main roads of housing estates are cleaned. The operative action plan, the winter work order, the central duty service, the system of short term weather forecasting, the round-o' clock shifting and the use of up-to-date (VHF) telecommunications promote to the coordinated work operations in winter.

...cont.

Table III.

Denomination	Value
Total public areas cleaned	17.6 million m^2
Length of roads cleaned in winter	2.090 km
Number of medium category sweepers	37 pcs
High power sweepers	8 pcs
Special sweepers (pavement cleaners)	6 pcs
Washing, watering, MgCl spraying machines	50 pcs
Medium power spraying/ploughing machines	105 pcs
High power ploughs	22 pcs
Auxiliary machines (loaders, small and large power machines)	38 pcs

THE RECYCLING CENTER : AN ORIGINAL SOLUTION TO BULKY WASTES

by Christian METTELET - General Manager
LES TRANSFORMEURS
French Agency for Waste Recovery and Disposal

"A permanent center where waste are brought voluntarily and on purpose" "a special waste recycling and recovery center", a "dumping service" or "**déchetterie**" ...("déchet" is waste in French), the term has developed but the concept itself and the so designed product are evolving and asserting themselves. Since the first "décheterries" appeared by the end of the seventies, there are enough of them today to be able to appraise the impact of this new type of collecting bulky objects.

SO WHAT IS A "DECHETTERIE" ?

This dumping center or "déchetterie" could be defined as a properly layed-out, **equipped,** and guarded area, well-known **by the public and intended to receive waste whose origins** and quantities are set by the administrator; its basic principle is simple : citizens find there the possibility, at any time, to get rid of their wastes and particularly of bulky objects which the traditional household collection system will not take into account, and this by means of containers left on purpose at their disposal and adapted to the nature of any waste (paper, old iron, oil, glass ...)

Together with this basic notion of "reverse self-service equipment" one will keep in mind the notion of recovery of materials, which this selective dumping of wastes done by the citizen himself allows.

HOW DID THE FIRST DECHETTERIES APPEAR ?

The first ones were built **at** the end of the 70s, through the impetus given by the Waste Law from July, 15th 1975, which started a general policy of fight against uncontrolled dumping. Important means were developed by different regions with the technical and financial help of **the French Agency for Waste Recovery and Disposal-Les Transformeurs** (regional programs for the suppression of

SWA PROCEEDINGS
VOLUME 1 ISBN 0-12-058451-4

uncontrolled dumps and rehabilitation of the areas). Local collectivities and cities volunteered and put forward different means for prevention, among which the first project of recycling centers.

THE ADVANTAGES OF THE "DECHETTERIE"

The **success** of the "déchetteries" and their quick development can be explained by all the advantages this new product offers :

. it provides an easy and free access for the private citizen needing to get rid of his bulky objects and being a nuisance such as : rubble, mowing waste, iron, oil ...

. it is a key element in the protection of environment and fighting against uncontrolled dumping. For example in Bordeaux City (South-West France) 8 "déchetteries" have been installed since 1980. In 1980, 12 000 m3 wastes were collected, today it has decreased to 2 000 m3. A noticeable difference knowing that one m3 collected from dumps costs 85 francs whereas one m3 freely brought to the recycling center or "déchetterie" costs only 10 francs, and of course to this should be added a far better cleanliness of the area because of the disappearing of dumps.

. recovery of products such as paper, cardboard, iron, glass, oil .... is possible by means of selective sorting done by the user himself.

. the cost of it can be considered as rather low regarding the service offered : the investment represents 50. 000 to 100. 000 francs, running costs represent on an average 10 francs per inhabitant and per year.

. a noticeable saving induced on collection and treatment of household wastes due to a reduction of quantities entering the traditional recovery system.

FITTING-OUT AND RUNNING OF A "DECHETTERIE"

<u>FITTING-OUT</u> : A large range of possibilities are offered to towns as far as fitting-out is concerned. More frequently, one will find a **heightened** platform with a special path. The platform gives access to containers placed below. A lay-out which makes loading and unloading of wastes easier and guarantees the best security for users in separating

areas accessible by "customers" from areas and drives taken for the removing of wastes and containers.

Input and output devices should be planned too : keeper's house, fences, gates, an information point intended for users on the center itself, landscaped gardens, ...

FOR WHICH KIND OF WASTES ?

A "déchetterie" should admit any kind of refuses produced by a household . This means that any kind of wastes can be disposed of properly.

As an indication, following types can be distinguished :
. unrecoverable wastes : ordinary stuff, rubble, earth, unsorted refuses, and
. recoverable wastes, mainly : iron, paper and cardboard, glass ... and, according to the local context : tyres, plastics, wood, trees and branches, or fuel (in case the city is equipped with a special recycling center with recovery of energy).

Special household wastes can be admitted too : unused medecines, batteries, solvents ... The "déchetterie" is then a neutralization place for those toxic wastes which can be sent to special centers for treatment.
Of course, this means that required conditions of security are respected (a well-kept, locked and ventilated room).

Craftsmen (cottage industry as a whole) and shopkeepers (local trade industry) should be admitted to the "déchetterie" as they produce kinds of refuses similar to household wastes.

FREQUENTING

The impact range of a "déchetterie" depends on settlement, means of access and on the information campaign launched before. As a rule, one can retain an average distance from 5 to 10 kms or 10 to 15 minutes in towns. Although this will change according to the place and the period of the year, a "déchetterie" will admit between 8 to 10 visitors per week for 1000 inhabitants concerned. If it is opened on Saturdays and Sundays, frequenting during those 2 days will represent 40 to 60% of total frequenting.
A few examples :
. ANGERS (Western France) 190 000 inhabitants :
500 visitors/week
. AVRILLE (Western France) 13 000 inhabitants : 200 to 400

visitors/week
. BORDEAUX (South West France) 600 000 inhabitants :
6 900 visitors a week (for 7 centers).

Besides, it has been noticed that frequenting rises with time. Visitors come more often and for smaller quantities. They consider the "déchetterie" as a new service being at their disposal and no longer as a compulsory and restricting trip.

WHICH QUANTITIES ARE BROUGHT ?

From "déchetteries" now in operation, one can say that 2/3 of refuses brought are composed of ordinary stuff (rubble, mowing and gardening waste, ...) and 1/3 is composed of recoverable waste (iron, paper, cardboard, ...) the repartition of which -in kilo- can be calculated as following :
. iron : 2 to 4 kg/inhabitant/year
. paper and cardboard : 2 kg/inhabitant/year
. glass : 1 kg/inhabitant/year in case a selective collection is carried out, and 10 kg/inhabitant/year if not
. oil : 0.1 liter/inhabitant/year

Those figures will change according to the neighbourhood, the period of the year, and the impact on population.

GUARDING, CARE-TAKING, OPENING HOURS

"Déchetteries" must be kept or guarded -depends on their size.
Care-taking will be required as a guarantee of the reliability of the "déchetterie" for the local authorities as well as for the owner and "customers" ... and neighbourhood too.

THE LEGAL ASPECTS

In order to clarify the legal status of the "déchetterie" and to strengthen the guarantees against nuisance, "déchetteries" have been classified by Law as **Classified Equipments for Environmental Protection.** A project makes provision for : a simple declaration for a "déchetterie" whose size is inferior to 2.500 m2 and a special authorization for larger ones (> 2.500 m2). The project investigation linked with the authorization could be simplified, thus restricting the geological, hydrological and hydrogeological aspects.

PROSPECTS AND DEVELOPMENT

A QUANTITATIVE APPRAISAL

The rise of the number of "déchetteries" has been very quick : 40 "déchetteries" by the end of 1985 and nearly 80 by the end of 1986. Today, 180 "déchetteries" are in service or being installed and the development of this equipment will increase. Local authorities are interested in this solution and the inventory of short term projects rises to 50.
Our purpose is to reach 1000 "déchetteries" in France within a few years. Rather realistic, in regards to the certain success of the "déchetterie". This would mean 60% of the population having a "déchetterie" at their disposal, that is to say 30 to 35 million inhabitants concerned, which enables us to say that at this level the average population concerned by a "déchetterie" would be from 30.000 to 35.000 inhabitants. And to prove it, let us compare this figure to the percentage of the French population served by a household waste collection system : 98 %, with those having a treatment equipment : 87 %.
Besides, if the first centers have been installed in urban areas, today, such equipments are developing in rural and urban areas. Big towns tend to be equipped with a larger network : 8 "déchetterie" in BORDEAUX (600.000 inhabitants), 5 in LA ROCHELLE (100 000 inhabitants- (Western France). One should notice, however that such a service could hardly be possible with a population level below 3.000 inhabitants.

EVOLUTIONS

. Local authorities are no longer the only ones concerned : private promoters are now in a position to offer local authorities a new service, against payment.
. Experiments have been made in order to reinforce recovery and recycling of more and more materials : recovery of garden waste by composting in Angers, recovery of used car batteries in Taverny, Nantes and Amiens.
. A new global approach of the setting up of "déchetteries" on the scale of a region or a département (special programmes in Jura- Eastern France- and Finistère- Western France-)

A NEW IMAGE TO ESTABLISH

So that the "déchetterie" goes on developing, it is necessary for local authorities and citizens and for users also, to develop an attractive and enhanced image.
In this aim, **LES TRANSFORMEURS, French Agency for Waste Recovery and Disposal** has worked out :
. a special book defining the signal system and the norms to be respected "Signalétique Déchetterie : livre de normes" and to be proposed to local authorities for a harmonious, effective, reliable and uniform signalization on the center itself and in the neighbourhood.
. a special architectural package, worked out with architects, and which can be adapted to any case, integrating both functional and economic aspects with design.
. an advertising package for information to make public opinion sensitive to the problem with posters, messages, ... to be used by the local authorities.

The "déchetterie" has been created because it answers the need of local authorities and citizens faced with the problem of uncontrolled dumps and bulky wastes. The concept asserts itself, gets more precise and evolves in order to answer as precisely as possible the need of the users. It cannot be denied that today, this dumping service : the "déchetterie" **has found a right place among the techniques** which local authorities have at their disposal to fight against waste.

Session 11

Resource Recovery and Recycling

RECYCLING AND OTHER ALTERNATIVES FOR WASTE VOLUME REDUCTION

Masaru TANAKA, Ph.D.
The Institute of Public Health
Tokyo, Japan

1. Introduction

A serious problem for the management of wastes is how to secure waste treatment and disposal facilities. It is particularly difficult to procure a landfill site near an area where a large amount of waste are generated. This difficulty is attributable to these two factors:

1) It is physically and economically difficult to secure land for disposal facilities.
2) It is also hard to obtain consensus among residents about environmental protection measures and the disposal method.

Even if site can be secured, it is not a permanent solution to the problem of treatment/disposal of waste because landfill space will be consumed within a short period.

The most effective solution to this problem is volume reduction of waste for landfill. Among the measures for that purpose are recycling, and recovery of valuable materials from the waste. But the effects of recycling of waste and its influence are not fully recognized. Successful recovery of such valuable materials depends largely on various factors, therefore, many unsuccessful instances have been reported in the past.

ISWA PROCEEDINGS
VOLUME 1 ISBN 0-12-058451-4

To reduce the landfilling of waste, it is essential to select the optimum system for reduction after comparing recycling and other alternatives.

This paper will discuss these subjects:

1) Present situation and problems for recycling and material recovery in Japan
2) Strategies of waste volume reduction for landfill
3) Future activities to solve the problem of waste management

2. Present Situation and Problems for Recycling and Material Recovery in Japan

(1) Recycling in Japan

Recycling activities in Japan differ widely according to the kind of waste -- municipal waste or industrial waste. In the case of municipal waste, such as beer bottles and dry cells (mercury and silver dry cells), existence of a recovery route at the retail distribution stage is highly effective for promotion of recycling. Collection contractors and secondary material dealers also play an important role in recycling. One example of the contractor is tissue paper barterer who collects waste paper and waste textiles. Disposal of industrial waste is regulated by the Waste Disposal and Public Cleansing Law. Such waste is disposed of by the business operators discharging the waste or by contractors with a permit of the prefectural government, and at that stage, the waste is recycled in some cases.

Generally, waste paper sorted out and discharged by house-holds are collected by waste paper collecting dealers.

Thus it is not collected by the local government in many cases. The total quantity of old paper, empty cans, empty bottles, etc. recovered by collecting dealers in a year is estimated at approx. 10 million tons; thus the recycled household waste accounts for at least 20% of 40 million tons of municipal waste handled in a year.

On the other hand, only 3% or less of the waste collected by the local government, which includes waste re-used as compost and as feed, is recycled.

Among the waste from households, aluminum scrap, iron scrap, waste glass, waste paper, waste textiles, etc. are recycled to a substantial extent. Use of the heat obtained by power generation by the thermal energy from the combustion process or in some other way is one form of recycling in a board sense.

1) Aluminum scrap

The aluminum scrap recovered is used as a raw material for a secondary aluminum alloy for parts of an automobile and others. Aluminum products recovered from households include aluminum pans, but no statistical data on them are available. Approx. 40% of empty all-aluminum cans are recovered. Most of them are presumably recovered by retailers through the reverse-distribution recovery route. Aluminum is used as the material for end lids of two-piece cans with an iron body, but recovery of this aluminum scrap has not yet been started.

2) Iron scrap

Iron scrap generated in Japan can be divided into processing scrap and old scrap. In addition to them, some quantity of iron scrap is imported.

The total quantity of processing scrap and old scrap recovered throughout Japan accounts for 65% of the total quantity of iron scrap.

The iron scrap is delivered ultimately to electric furnace producers and others via collecting dealers and intermediaries.

Approx. 40% of the empty cans are recovered, and this high recovery rate contributes much toward prevention of littering with empty cans. These empty cans recovered include various commercial cans.

3) Waste glass

Flat glass and all other glass products except glass bottles are hardly recovered, and no statistical data on them are available. Waste glass are recycled into returnable bottles and cullet primarily.

Ninety to ninety-five percent of returnable bottles for cola, soda pop, juice and others and beer bottles are returned by retailers through a reverse-distribution recovery route; thus their recovery rates are somewhere very close to a limit.

In Japan, 2,150,000 t of glass bottles are produced in a year (1986), and 550,000 t of cullet in market are collected, in a year. Bottler cullet with a total weight of 150,000 t is collected in a year. Thus a total of 700,000 t are collected by citizen groups and by resource collecting dealers. This total figure and the quantity of factory cullet, which is recycled inside factories, add up to 1,200,000 t -- the total quantity of cullet recycled per year.

4) Waste paper

The total annual paper (include paper board) production in Japan is approx. 20 million tons, and 50% of waste paper is recovered. Their utilization rate is also 50%. Thus 10 million tons recovered are used as the material for paper manufacturing.

Waste paper is collected from households, small-scale shopping centers, factories, publishing/newspaper companies, department stores and others. Waste paper is collected by tissue paper barterers and buyers, passed on to intermediaries and delivered to paper manufacturer by direct contractor. The price of waste paper once dropped to such a low level that tissue paper barter business almost died out. Thus the price of waste paper changes quite wildly.

5) Waste textiles

Waste textiles are mostly recovered by waste paper collecting dealers and are reutilized as second-hand clothes and as rags (for wiping a window and for wiping off machine oil, etc.). Also waste textiles are frayed out into a cotton-like state, and then the threads obtained are woven back into textiles. Textiles so produced of recycled wool fetch a high price on the market and did much to promote the waste-textile business.

Waste textiles are used as the material for small textile articles, mops and roofings and as the packing for a doll or the like, and they are exported also. Waste textiles weighing approx. 1 million tons are discharged from households, and approx. 200,000 tons are presumably put on the recovery route.

Approx. 300,000 to 400,000 tons are obtained from the production process of the textile industry, and almost all items fit for reutilizing are recycled.

(2) Collection by local governments

Waste collection activities prior to the local government's refuse collection are carried out mainly on an economically justifiable basis. Hence there is a limit to recovery of useful materials from waste subsequent to the point of collection by the local government.

In Japan, municipal waste is usually divided into combustible refuse and incombustible refuse.

The former is incinerated, and the latter is disposed directly to landfill. Recycling for thermal energy recovery and volume reduction of refuse can be accomplished by incineration for combustible refuse.

To reduce refuse further, it is essential to recover valuable materials from those portions of incombustible refuse and bulky refuse which are used directly for landfill. Refuse is often collected without source separation (mixed refuse), and the entire quantity is incinerated. A substantial volume reduction has been accomplished in this way, but recovery of valuable materials from incombustible refuse (empty cans, glass, etc.) in the incineration residue will be needed to reduce the refuse further.

In Japan, the waste collected by local government authorities are disposed by taking into overall consideration the merit of valuable material recovery, the merit derived from the securing of source of alternative energy and the merit derived from preservation of landfill space resources by reducing the material for landfill.

3. Strategies of Waste Volume Reduction for Landfill

(1) Alternative for Waste Volume Reduction

Measures to preclude the possibility of generation of refuse include measures associated with production and manufacturing activities, measures related to citizens' consumption and also placement on the resource recovery route for waste and disused articles. There is a limit to the local government authorities, acts of intervention on production activities and consumption.

From the standpoint of local governments, incineration of the entire quantity of directly combustible refuse is the best waste volume reduction measure at present. The national government extends a financial assistance in construction of incineration facilities. Refuse not directly combustible can be used as material for landfill after shredding, or combustible materials sorted out after recovery of valuable materials may be incinerated.

Therefore, the waste volume reduction actions in which local government authorities take part include a measure involving use of the resource recovery route for waste articles and disused articles to control generation of refuse and activities for volume reduction at the collection, transportation, intermediate treatment and final disposal stages.

All these actions are following

1) Various measures to control generation of refuse
   a) Supplier's voluntary recovery of waste hard to dispose properly.
   b) Trade-in
   c) Recovery of empty containers by dealers

d) Used article exchange program

e) Collection by citizen's group

2) Source separation - recovery of valuable materials

3) Building up facilities for treatment of incombustible refuse, bulky refuse and others

4) Utilization of incineration residues

(2) Study of measures to reduce waste volume further

Among the common factors detrimental to proper management of waste in the 12 major cities in Japan are increase in quantity and diversification in quality of refuse, shortage of treatment/disposal sites because of the need of highly profitable use of land, increases in the public cleansing cost, sophisticated environmental protection needs, hindrance to build-up of treatment facilities due to the foregoing factors, etc. The measures taken in the 12 cities are divided into three types:

(A) Top priority is given to the strengthening of incineration facilities. (Mixed refuse collection)

(B) Importance is attached to the reduction of loads on incineration facilities and prevention of pollution (prevention of secondary pollution due to incineration), and source separation refuse is exercise as a rule. (Refuse is sorted into two types; combustible and non-combustible)

(C) Inadequacy of treatment/disposal abilities is compensated by material recovery from sorted out refuse with the cooperation of residents (Refuse is sorted into different types; combustible, non-combustible and others for recycling)

In the cities taking the measure described in (A), for example, Yokohama, Kitakyushu, Kyoto and Kawasaki, 94 to 99% of the refuse is incinerated, and the reduction rate (by weight) is 72 to 81% (including bulky waste that is shredded and sent to landfill). Virtually the entire quantity of combustible refuse is reduced. To reduce the refuse further, it will be essential to recover bottles, cans and other incombustible materials which are relatively easy to recycle and to develop a method of re-using the residue after incineration.

In the cities taking the measure described in (B), e.g., Kobe, Fukuoka and Nagoya, the incineration rate (by weight) is approx. 72 to 85%, and the reduction rate is 61 to 69% (including bulky waste that is shredded and sent to landfill). These rates are lowered to an extent corresponding to the plastics sorted out and to combustible materials contained in the incombustible refuse collected. In addition to the tasks mentioned for the type (A) of cities, the incineration of plastics and combustible materials in the refuse collected will be essential in order to reduce the volume of refuse further.

The city of Hiroshima takes the measure described in (C). In Hiroshima, the incineration rate is 71%, and the reduction rate (by weight) is approx. 68% (including bulky waste that is shredded and sent to landfill). The disposal activities in this city are comparable to those in the type (B) of cities. Thus source separation of valuable materials from refuse by residents for recycling is making a significant contribution.

In this connection, it should be noted that recovery prior to the municipal authorities' collection is promoted in this city. Source separation of valuable material for recycling, however, depends heavily on residents' cooperation, the market prices of recycled articles, etc. Therefore, stabilization of the recovery activities will be essential. To reduce the refuse further, it will be imperative to incinerate shredded bulky refuse and combustible materials in incombustible refuse collected, etc.

In large cities where good incineration facilities are provided, substantial waste volume reduction has already been accomplished by incineration. Further reduction by recovery of valuable materials will be a major task in future. In this connection, attention should be directed to sort-out of bottles, cans, etc., their collection and recycling in Nagoya, Sendai, Kawasaki and others (carried out as a model practice), extraction of recycled articles handled by dealers at the bulky refuse station, re-use of incineration residue in Yokohama, etc.

4. Future Activities to Solve the Problems of Waste Management

Here is a brief remark about selection of recycling or other measures for waste volume reduction.

When a local government is studying its future waste management strategy, it is important to set up specific purposes of waste management. In the past, "improvement of the public health level" and "protection of living environment" were generally considered to be the purposes of waste management. In view of the need to ensure consistent stable living for all people on the earth, which is a new requirement in this age of internationalization, "conservation of natural resources" should be added to the purposes of waste management. To attain this new purpose, the local government must solve the problem under many restrictive conditions.

Some of those restrictive conditions include:

1) Restriction of the landfill space resource
2) Environmental protection
3) Considerations concerning the effects on health
4) Degree of demand for conservation of natural resources
5) Restrictions on the equipment, materials, labor and cost needed for treatment and disposal.

The strategic points essential for efficient attainment of the purposes under the restrictive conditions are as follows:

1) Control policy for waste generation at its source.
2) Volume reduction of the waste to be collected.
3) Proper treatment and disposal of the waste to be disposed ultimately, in a way helpful toward environment conservation.

Among the measures to carry these out are recycling, control at the source by material recovery, waste volume reduction to be treated or disposed of and efficient treatment/disposal by better use of the equipment, materials and cost. To establish a waste management method suitable for environment conservation and health protection purposes, it is imperative to select a management system based on risk assessment and suited to the environment or the community. A waste management method combined with recycling or a waste volume reduction strategy is selected with all the factors taken into consideration. The personnel in charge in the local government should study these measures and the factors to be considered.

It is essential to push forward study of the measures to

cope with diversification in quality and increase in quantity of the waste in the future, conserve the natural resources which is exhausted in these days and improve our living environment by use of restricted cost, equipment and materials, which is the initial object of waste management.

References

1. Report on Promotion Planning for Volume Reduction of Non-Combustible Waste, Water Supply and Sanitation Department, Ministry of Health and Welfare, Japan, 1986.

2. Recycling '87, Clean Japan Center, 1987.

RECYCLING OF REUSABLE WASTE MATERIAL IN THE GDR

Dipl.-Ing. Volker Matthes

Institute of Municipal Affairs, Dresden

In the German Democratic Republic great importance has always been attached both to the prevention of waste and the utilization of reusable waste material. Characteristic of this attitude is the "Bill for the intensive utilization of reusable waste material" which was enacted on 11 December 1980. This bill specifies that all waste and residue from production and manufacture, as well as that arising from communal and individual consumption is, fundamentally, to be considered as reusable waste material. This means that all waste and residue is to be regarded in principle from the point of view of its utilization and potential for recycling into the raw material cycle of the economy. This factor is also the basis for safeguarding the material growth of production; for, in essence, expansion of production can still be achieved whether or not utilization of primary resources is constant. Concurrent with the practice of waste material recycling is the fact that prevention of waste in this way plays an important part in the protection of the natural environment.

The bill referred to above prescribes the following responsibilities for producers or creators of reusable waste material (hereinafter RWM):

- Responsibility for the collection of RWM in the

ISWA PROCEEDINGS
VOLUME 1 ISBN 0-12-058451-4

absence of other undertakings with this direct responsibility.

- Responsibility for utilizing such RWM in its own factory or undertaking, or, if this is not possible, responsibility for passing it to another firm or undertaking for utilization. This also applies for products no longer serviceable: in this respect, for instance, the VEB Reifenkombinat Fürstenwalde, as the largest tyre manufacturer in the GDR, is responsible for retreading and remoulding of worn-out tyres.
- Responsibility for scientific-technical research into the development of effective waste collection methods and treatment processes.
- Responsibility for the provision of the necessary collection, treatment, and utilization capacities.

There are special factories for the collection and treatment of scrap metal arising from manufacture elsewhere, or discarded when unserviceable (VEB Kombinat Metallaufbereitung), and for the collection and treatment of RWM discarded by households (VEB Kombinat Sekundärrohstofferfassung).

Useless waste must be disposed of in such a way that it does not become a pollutant. Proof of non-reusability of waste material must be furnished by the GDR equivalent of a condemnation board: this board issues a "negative certificate" authorizing disposal.

Special depositories are in operation for the storage of more than 20 kinds of RWM which can not at present be treated, but for which treatment will be possible in the course of the next few years.

In pursuance of this policy the following results are achieved through utilization of RWM in the processing industry:

Crude steel	70%	
Copper	69%	
Lead	100%	
Aluminium	69%	
Waste paper	49%	
Cullet	39%	
Glass containers	70%	(standard assortment of bottles & jars)
Building materials	12%	

The collection and processing of RWM amounted to 30 million tons in 1985: it is expected to amount to 35-36 million tons in 1990. The proportion of RWM of the total resources important for the national economy in 1985 was 12%: this is expected to amount to 14-15% in 1990. The achievement of these results can be attributed in great measure to the GDR policy of central planning of the national economy: industrial production is so planned that it includes the utilization of a fixed proportion of RWM.

A considerable proportion of RWM accrues through State purchase of RWM accumulated in private households. The means of collection of such RWM is by an extensive network of reception points set up by the VEB Kombinat Sekundärrohstofferfassung; collection by commercial and private collectors; as well as in social service installations and organizations - hospitals, schools, canteens, and similar institutions.

There is at present a total of 15,600 collection points in the GDR; 1 for every 1,079 head of the population.
In 1986 the following average quantities of RWM were recovered per head of the population from households:

Scrap metal	25.0 kg	
Waste paper	17.1 kg	
Waste textiles	4.1 kg	
Bottles [1]	45.9)	ca. 25 kg
Jars	30.1)	
Thermoplastic waste 0.42 kg (collected in special net containers outside retail stores and other establishments)		
Total ca.	70 kg	

[1] excluding bottles for which a deposit is charged and refunded on bottle return.

Further examples of the type of RWM collected from households, shops, and other commercial establishments, include unserviceable chemical-fixing baths, old car and other heavy-duty batteries, bedding feathers and hair from hairdressing saloons.

The work of over 11,000 employees of the VEB Kombinat Sekundärrohstofferfassung, that of those employed in the approximately 12,000 reception points run by commercial traders, and that of private collectors, as well as the possibility which exists for collection of some types of RWM in schools and in the immediate vicinity of housing estates, contributes to the fact that arising of what would otherwise be rubbish are prevented to the extent of 70 Kg per head of the population per year.

Further, in the GDR, parallel to the municipal refuse collection service there is also a scrap-food collection service. Special bins for scraps are located alongside rubbish bins: these scraps are processed for pig feeding, and the total of such feed amounts to 55 Kg per head of the population per year. Overall, therefore, a total of 125 Kg of RWM per head of the population per year is processed - the sum of the actual refuse from households and commerce being reduced to 175 Kg per head of the population per year. This means that about 40% of potential refuse is processed at the outset.

Other examples of both waste prevention and RWM recycling are:

- The comprehensive system of returnable beer and non-alcoholic beverage and milk bottles.
- The principle of "new for old" in respect of car and other heavy-duty batteries and cells for quartz clocks and watches, as well as worn tyres for retreading or remoulding.

In comparison with other countries which employ different strategies, the GDR is able to recycle household RWM in but modest volume. Limitations in this respect are:

- The fact that, in general, large conurbations are non-existent in the GDR; and, therefore, the accumulation of specific refuse is limited.

- Household refuse has a lower calorific value (ca. 5,050 kJ/kg).
- It also contains a relatively small proportion of organic substance (33 mass%).
- The fact that about 75% of household refuse arises in dwellings with individual solid-fuel heating, so that only the refuse arising during nonheating periods is suitable for processing.

3-4% of the total of 3 million tons of solid refuse arising in housing estates is processed by incineration and composting.

The agricultural demand for organic substance is great. For this reason priority is given to recycling household refuse to obtain fresh compost with the simultaneous retrieval of scrap iron. In the next few years it is intended to build further plant on the lines of that designed by the Institute of Municipal Affairs, and also to establish decentralized compost clamps to increase RWM recycling: this will contribute to meeting the demand for soil-improving organic substance, and compost quality will be assured by set standards.

Recovery of Materials from Waste from Demolition and Construction Activities

J. Bjørn Jakobsen, Morten Elle, COWIconsult, Copenhagen, Denmark.

1. INTRODUCTION

In 1984 the generation of solid waste from construction and demolition sites in Denmark was estimated to 1.25 million tons/year. This corresponds to 250 kg/inhb/year. In addition approximately 0.75 million tons were generated at civil works corresponding to 150 kg/inhb/year.

Approximately 80% of the construction and demolition waste were bricks of tile and concrete. 15% was wood/timber and the remaining 5% was miscellaneous (cardboard, metal etc.).

The mentioned amount of waste in Denmark is expected to increase in future, especially owing to increased activity in concrete repair works at multistorey buildings and in demolition of factory sites, wharfs and old harbour facilities.

There are a number of advantages in recovering materials from demolition, as well as construction works; among others, in evidence, the reduction of landfill area and consumption of gravel resources. In spite of these advantages there are, however, some obstacles in the extended recovery of materials from demolition and construction waste. These obstacles vary from country to country.

Typical obstacles are:

- Virgin resources of gravel are inexpensive and easily accessible.
- Disposal of waste on landfill is inexpensive and without any limitations.
- The demolition and construction works are not designed for materials recovery, giving mixed waste materials not suitable for any utilization.
- The need for technical guidelines or standards for utilization of recovered building materials.

Recovery of materials will not take place without overcoming one or several of the above-mentioned obstacles.

ISWA PROCEEDINGS
VOLUME 1 ISBN 0-12-058451-4

For some years COWIconsult, Consulting Engineers and Planners, has conducted a number of investigations within the programme of recycling of construction and demolition waste. These have included:

a) Surveying the quantity and composition of waste/recoverable materials from demolition-, building construction and restoration activities.
b) Testing source-separation on the building site of paper, cardboard, plastfoils, wood and bricks of tile.
c) Testing on-site utilization of crushed and sorted tile and concrete bricks originating from demolition work.

2. INVESTIGATIONS ON QUANTITY AND COMPOSITION

Building activities can typically be divided into:
- demolition works
- building constructions
- renovation works/repair works

The generation and composition of waste from the mentioned activities differ very much from each other. In order to establish a set of specific waste production figures, surveys have been conducted on selected building works of the three categories.

The surveys comprised:
a) two demolition works (multistorey dwelling of a total of 5,300 m²)
b) two building constructions (multistorey dwellings of a total of 17,500 m²)
c) two repair works (multistorey dwellings of a total of 3,700 m²)

All generated waste from the works have been examined by sorting and weighing.

The following results on composition have been achieved through the investigations [1]:

The specific waste production from the mentioned building activities was for:

Demolition Works 900 kg/m²
Building Constructions 16 kg/m²
Repair Works 13 kg/m²

Table 1. Percentage distribution of waste materials.

Building Activity / Fraction	Demolition	Building construction	Repair works
Tile and concrete	85%	59%	59%
Other non-combustibles	1%	22%	20%
Wood/timber	13%	13%	20%
Other combustibles	1%	6%	1%

3. SOURCE SEPARATION AT THE CONSTRUCTION SITE

Investigations on source separation were carried out at one site for building construction and at another site for repair works.

At the site for building construction a total of 241 tons of solid waste was generated. An amount of 8.1 tons (≈ 3.4% of total) of paper/cardboard/plastfoil was sorted out at the site by the handicrafters. This amount corresponded to 80% of the potential amount of the said materials. The quality of the materials was excellent fit for use in paper/plastic regenerating factories.

At the site for repair works a total of 41 tons was generated. The amount of sorted paper/cardboard/plastfoil was 0.6 ton (≈ 1.5% of total) corresponding to approximately 80% of the potential amount. The sorted amount was very small and of poor purity (polluted by other waste materials).

At the same sites wood was sorted out for combustion in incineration plant in stead of a disposal at landfill with the non-combustible materials.

The source separation was reasonably successful indicating future possibilities for source separation at building sites in order to achieve more pure fractions of recoverable materials, combustibles as well as tile and concrete bricks for crushing.

4. ON-SITE UTILIZATION OF CRUSHED DEMOLITION DEBRIS

4.1 OBJECTIVES OF THE PROJECT

Through local recovery of demolition debris, a number of advantages can be achieved. These are:

- Reduced external transport of waste to landfill sites (i.e. reduction of traffic).
- Low risk of mixed and non-controlled waste materials, resulting in upgraded materials of higher quality for later utilization.
- Reduced transport of gravel to sites for new construction.

There will, however, be a possibility of negative impact to the surroundings from noise, dust and vibrations from the crushing and sorting of the waste.

A project on local recovery of demolition debris has been conducted in Copenhagen. The overall objectives of the demonstration project were defined as:

- To design and survey a demolition project in which bricks of tile and concrete are to be upgraded to gravel.
- To upgrade (crush and sort) the sorted bricks from the demolition.
- To test the produced gravel.
- To utilize the gravel for constructing a parking area.
- To test the quality of the upgraded materials within the mentioned utilization.

The demonstration project was designed and supervised by COWIconsult and Demex, both consulting engineering companies. Financing was made by the Danish Agency of Environmental Protection and by the Danish Council of Technology.

4.2 DEMOLITION PROJECT

An old factory building was selected as the object for the demonstration demolition. Specifications for the factory building are as follows:

- Building area approx. 3,400 m²
- Storey area approx. 7,800 m²
- Waste quantity approx. 8,600 tons
  - . Tile approx. 3,500 tons
  - . Mixed demolition waste approx. 5,000 tons

. Wood	approx.	70 tons
. Steel	approx.	470 tons

The layout plan for the factory and surrounding areas is shown below.

The time period for the demolition work was 6 weeks. Steel as well as good quality timber were sorted out separately. Tile walls, concrete storey floors and roof constructions were demolished as individually as possible.

Approximately 3,400 tons of tile and concrete waste were sorted out for later crushing. The residuals (5,000 tons) were disposed of on a landfill.

4.3 CRUSHING OF DEMOLITION WASTE

The crushing of the sorted tile and concrete was carried out by a mobile crushing plant, "Combi Screen" produced by "Vedbysønder Maskinfabrik". The "Combi Screen" has a Hazemag impact crusher with a capacity of 120-180 tons/h. Dust from the crushing operation was cleaned through a cyclon precipitator. The crushing plant had a two-man crew. The waste was fed to the plant by means of a hydraulic shovel periodically assisted by a dozer.

The total time of operation for the crushing plant was 152 hours. Running time for the crusher was 70 hours giving an output of 56 tons of demolition waste per hour. A total of 3,900 tons of demolition waste was crushed, comprising:

. 250 tons of concrete (non-reinforced),
. 300 tons of mixed concrete and tile from foundation of the buildings,
. 3,350 tons of tile.

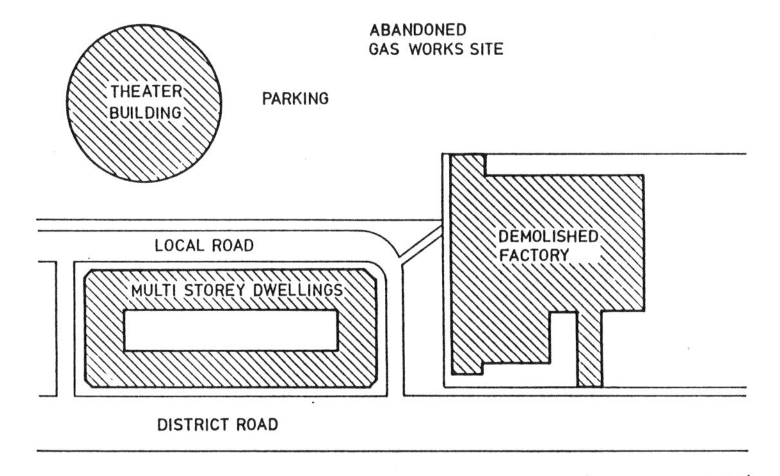

Parameter	Analysis no.	Tile 1	Tile 2	Mixed tile/concrete 3	Concrete 4
Uniformity		52	47	42	7.5
Screened >16 mm(s)	(%)	20	18	21	34
Sand equivalent (SE)	(%)	69	49	67	70
Capillarity (h_c)	(cm)	35	50	30	20
ASTM $_{d.max}$	(t/m³)	1.68	1.75	1.79	1.79
Water content (W_{opt})	(%)	19.0	17.0	15.7	11.2
Modified proctor					
. $_{d.max}$	(t/m³)	1.68	1.69	1.74	1.79
. W_{opt}	(%)	17	13	14.5	13.5

Noise level, in the surrounding area, was measured during the crushing. Valid information was registered about noise levels from the crushing plant for use in other similar projects. The noise level at a distance of 60 metres from the crushing plant ranged between 71 and 76 dB(A). The A-weighted noise level was calculated to L_{WA} = 117 dB.

4.4 QUALITY OF THE CRUSHED MATERIALS

The quality of the crushed materials was examined by the Danish National Road Laboratory. Samples were taken from three types of crushed materials: tile, mixed tile and concrete as well as concrete. The results from the laboratory analysis are shown in the table below.

Sieve analysis for the crushed tile was comparable to that of high quality gravel. The sieve analysis for crushed concrete showed a quality that had a too high content of medium and coarse gravel.

4.5 CONSTRUCTION OF A PARKING AREA FROM CRUSHED TILE AND CONCRETE

The crushed materials from the demolished factory form the basis for the construction of a parking area on space next to the factory site.

The purpose of this part of the project is to assess the crushed materials as sub-bases and unbound layers within low priority utilization. The parking area will not be paved with asphalt.

A total of approximately 1,000 m^2, corresponding to 42 parking slots, is established as shown on the sketch below:

The parking area is divided into five separate sections. Different pavements are used in each field using crushed tile, crushed mixture, crushed concrete, respectively.

During the execution of the parking area selected parametres are measured. For each layer the intention of these measurements is to establish:

- the deformation characteristics,
- the obtainable density,
- the compaction requirements for standardized density.

The results from the measurement of density are given in the table below.

The density of the tile base after 12 passes was found to be far below the 100% Standard Proctor requirement, which was the ultimate goal. The tile base was therefore compacted with 12 more passes. The density thereafter reached an average of 95.6% (nuclear densimeter) corresponding to 98.3% standard proctor.

As seen from the table, the compaction effort is higher for tile than for concrete. However, tile has properties which makes it as suitable as concrete for the actual project.

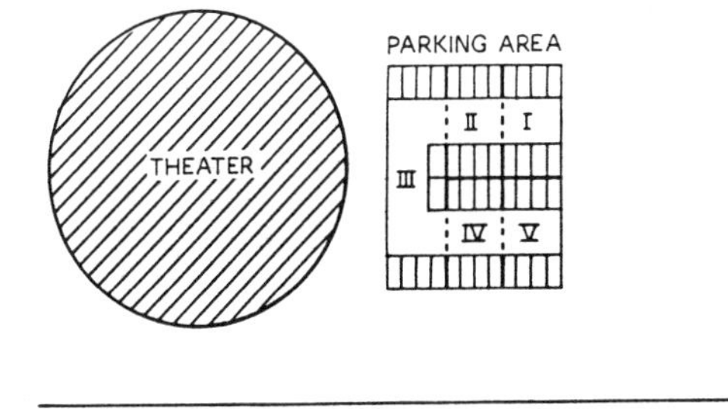

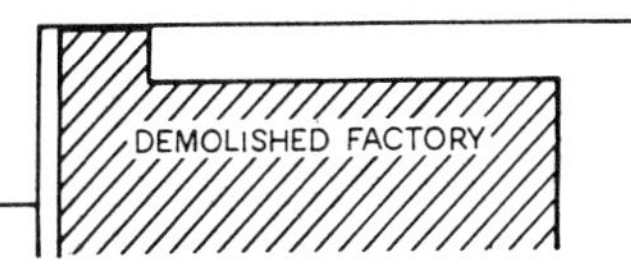

NOTE: I - V: TYPES OF PAVEMENT

	Density (Standard Proctor)	
	Nuclear Densimeter % Subbase/Unbound	Sandfilling % Subbase/Unbound
Concrete	96.7/98.7	99.0/103.4
Mixture	93.2/	95.0/-
Tile	94.2/92.5	94.3/91.8

5. CONCLUSION AND LONG-TERM VISIONS

Through the mentioned projects recovery of materials from construction and demolition sites is proved possible and realistic. Paper, cardboard and plastfoil can be sorted out at source (site) without any difficulties resulting in good quality materials for the recycling industry. Further blocks of tile and concrete can be taken out during the demolition operations for later crushing. The crushing can take place locally or regionally. A good quality gravel substitute can be produced on a crushing plant.

Owing to the rense of the regenerated demolition debris approximately 3,500 m³ landfill volume and 2,100 m³ gravel were saved. Furthermore, a total number of 340 trucks heavily loaded with the debris and gravel was avoided on the local and regional road system.

The future recovery of waste from construction and demolition debris will very much depend on the overcoming of the previously mentioned obstacles. The conditions for high grade recovery vary very much from country to country.

The present conditions in Denmark are not of a kind which enforces high grade recovery. The price of virgin gravel resources is comparatively low. Further, the technical guidelines and standards for utilization of recovered building materials are not prepared.

It is, however, a high priority area for development indicating a fast increase in the degree of materials recovery. Within a 5-10 years period it is anticipated that 40-60% of the construction and demolition waste in Denmark will be recovered each year.

References:

[1] COWIconsult, October 1985, Recycling of waste from building activities.

POSSIBILITIES FOR EFFECTIVE COLLECTION, SORTING AND RECYCLING OF SPENT BATTERIES

Mikael Backman & Thomas Lindhqvist, TEM/University of Lund, Sweden
Kim Christiansen & Stig Hirsbak, Technological Institute, Denmark

Background

The dumping of mercury and cadmium is one of the most serious environmental problems of our time. These metals are highly toxic and have a highly **cumulative** effect in living organisms. Their effect on health and the environment is richly documented and is extremely frightening. The problem area is therefore one **on** which the attention of environmental protection bodies is focused. This is true for the majority of industrialized countries.

The continually increasing use of batteries containing heavy metals constitutes a major source of the dumping of the above mentioned metals. A consequence of this is that a lot of countries run some type of battery collection programme, which might even include subsequent sorting and (attempts at) recovery of the constituent metals. All these have a

ISWA PROCEEDINGS
VOLUME 1 ISBN 0-12-058451-4

number of things in common: firstly, collection systems which are not very effective, secondly, costly manual sorting systems with obvious work environment risks, and finally, costly recovery systems with disturbingly poor results.

Introduction

To investigate the possibilities of finding more effective solutions to the battery treatment problem, three research institutes have developed a cooperative research project, which will closely examine potential improvements to the existing systems. The Technological Institute and TEM are jointly exploring the possibilities of more effective collection and sorting of spent batteries, while TNO in Apeldoorn, The Netherlands, is responsible for developing a recycling process based on hydrometallurgical principles.

In this paper we will present the work being performed in Sweden and Denmark, i.e. a study of deposit based collection on the Danish island of Bornholm and the development of automatic sorting processes at the TEM-Centre in Sjöbo, Sweden. This research is supported by the Commission of the European Communities, the Nordic Council of Ministers, the Danish National Agency for Environmental Protection, as well as by several Swedish organizations, including: the Foundation REFORSK, the Swedish National Board for Technical Development, the Stockholm City Waste Recycling Company and the Swedish National Environmental Protection Board.

The Bornholm Collection Study

The island of Bornholm is situated to the east of the Danish mainland. It has 50 000 inhabitants, corresponding to approximately 1% of the total population of Denmark. Bornholm was chosen because of the well defined geographical borders, due to the isolated location of the island. It was intended to perform a complete experiment with a battery deposit system, including the transfer of a deposit sum amounting to DKK 5.-. This plan necessitated the total agreement and voluntary cooperation of all shops selling batteries. The general local readiness to participate in a complete deposit experiment was broken by the two main grocer's chains, which after consultations with their central organizations refused to introduce a deposit on a voluntary basis. The collection programme was then transfered to a voluntary collection, but the main emphasis of the study, to identify and to develop the collection aids and the organizational and administrative routines necessary in a deposit-based battery collection system, was retained. This could be accomplished in a voluntary collection by simultating the particulars of a deposit system, e.g. collection of batteries in shops selling batteries only and the separate treatment of the batteries from the different shops until counted and recorded.

As the collection started the vast majority of the shops selling batteries joined the programme, i.e. a total of 138 shops, among them 88% of the camera shops, 86% of the radio stores and 77% of the grocer's shops. Some preliminary results of the study will be given in this paper. A most important observation is the improved condition of the collected batteries as compared to batteries collected in outdoor boxes and the absence of disturbing and dangerous items among the collected batteries. The consumers are encouraged to sort the batteries into four fractions, i.e. alkalines, rechargeables, button cells and others (notably Leclanché cells). There are corresponding separate compartments in the collection box, as shown below along with the information poster. This consumer sorting has largely failed, e.g. 75% of the alkalines has been misplaced. In a survey only 30% of the consumers considered themselves able to distinguish the different types of batteries. A survey of the shop owners shows that 64% of them favour a labelling of the batteries (4% against), 92% consider a deposit system as more effective, while 52% think it would be a good idea to introduce a deposit on batteries. Among the consumers, 87% want spent batteries to be collected (3% against), 86% favour a labelling (5% against) and 49% a deposit (35% against).

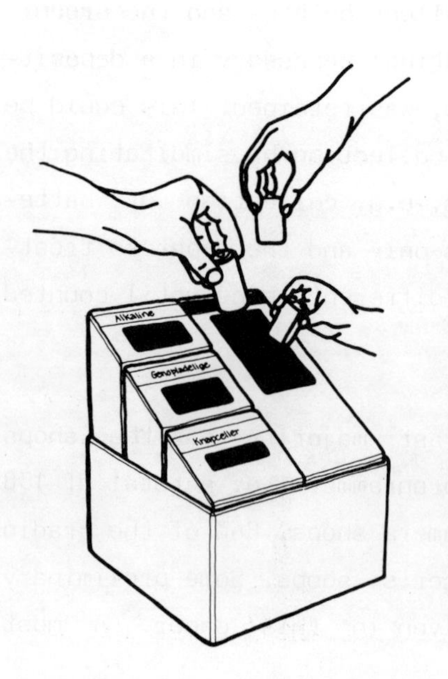

Automatic Sorting

For many recovery processes for spent batteries, it is viewed as essential or at least advantageous to have the batteries sorted into different fractions according to which chemical system the batteries are composed of. Manual sorting of the collected batteries is connected with high costs and low efficiency. Preliminary measurements of the mercury vaporization from collected spent batteries has focused upon the importance of considering the work environment, on a broad level, when handling spent batteries. Further investigations are needed before final conclusions are made.

A satisfactory sorting technique requires, according to ongoing research, the use of several sorting parameters. An initial sorting according to size facilitates the subsequent process steps. The weight distributions of the common battery sizes show significant differences between the Leclanché cells on the one side and the alkalines and the NiCd-cells on the other side. To distinguish alkalines and NiCd-cells a couple of different methods have been studied with encouraging results. These methods include x-ray-fluorescence, ult-

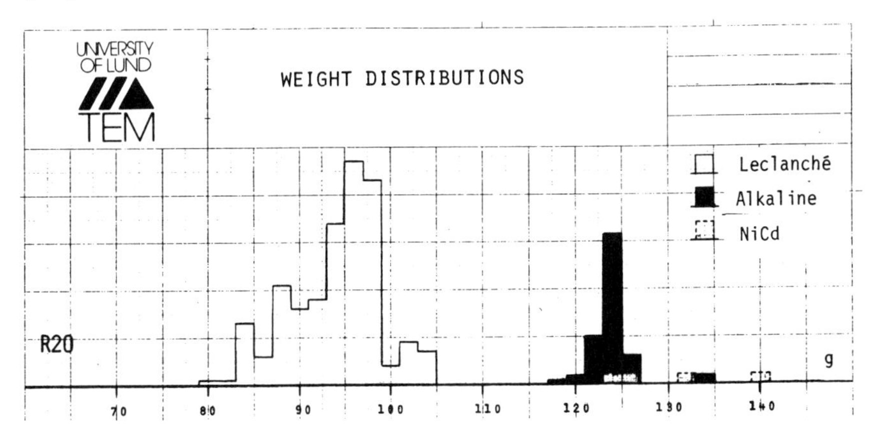

ra sound propagation, dynamic heat capacity characteristics, as well as resistive and inductive properties of the batteries. These preliminary studies have led to a sorting model which is able to treat the overwhelming majority of the batteries successfully, as illustrated by the figure below. This model is now being further developed and implemented at the TEM-Centre.

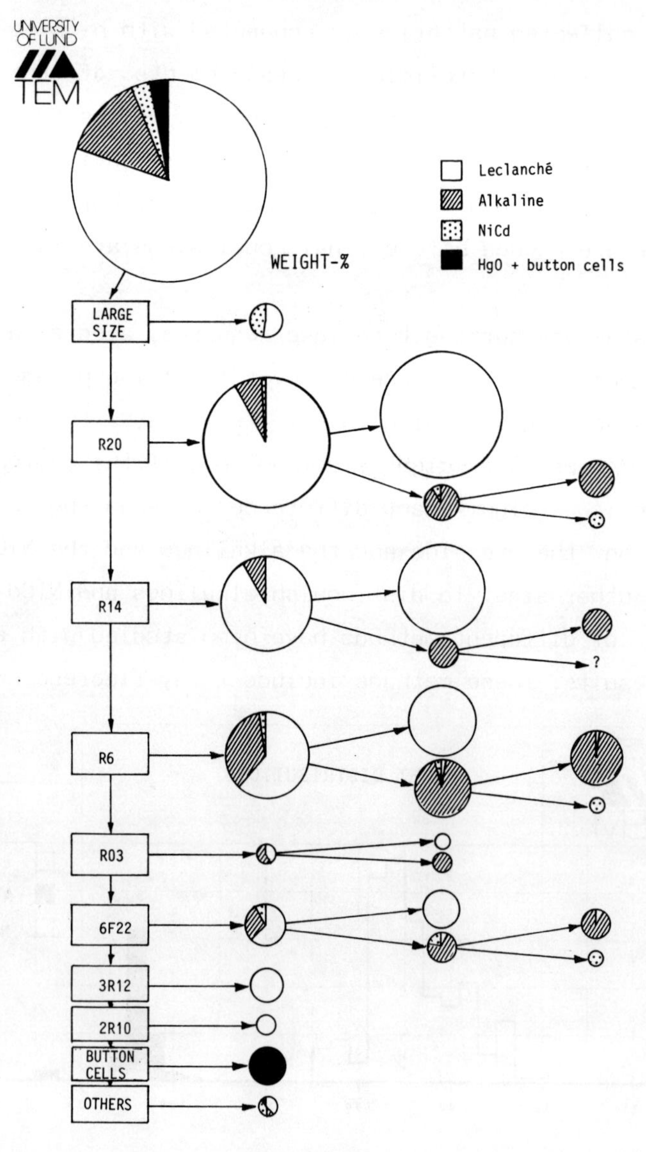

Session 12

Waste Reduction and Clean Technology

PROMOTION AND DEVELOPMENT OF CLEAN TECHNOLOGIES IN INDUSTRY

Jean-Claude NOEL*

* Manager of Industrial Department
Artois-Picardie Water Agency
764 Boulevard Lahure - 59500 DOUAI - FRANCE

SOME EXPLANATIONS

The manager of a firm has to consider all the technical and economic elements that will guide him towards the best investment choices, in order to face the pollution problem generated by his plant. Then, he has to put into question again the purification of this pollution so called "ineluctable" (the pollution is a loss of matters) and to think over the nature and the means by which polluting matters would be reduced or won't be created. The expression "Clean Technologies" includes the whole technological means aiming at reducing or eliminating the pollution coming from industrial equipments at the very level of production.

Clean technologies gather various operations :

Internal re-arrangement

It does not imply the manufacturing process, but it seems to be necessary to envisage a re-arrangement within the workshops, a thorough control of pollution points, and a sound management of water, raw materials and manufactured articles. These judicious measures have to be implemented in priority in all the establishments and therefore, before the purification process.

ISWA PROCEEDINGS
VOLUME 1 ISBN 0-12-058451-4

The modification of manufacturing process

It means clean technologies involving a significant pollution reduction. The clean technologies do not modify the very nature of the process : it brings an improvement in a less pollutant way.

Clean process

A new manufacturing method allowing to reduce or to avoid pollution, is implemented. The thus created **irreversibility** is the absolute guarantee of the final limitation of pollu-ion. In these rather rare cases, it is possible to speak of "clean technologies" in the **literal** meaning of this term.

SOME REFLECTIONS

To cure does not mean to eliminate pollution. It is in fact to remove it from an aqueous environment where it settles in a certain form in order "to dispose" it in another envi-ronment, in another form. If the implementation of these curative technologies makes the aqueous environment free from a part of this pollution, the risks of new degradations are to be considered because of the new state of pollution : Moreover, those risks are more serious because the **"pollutable" environment is much more sensible to each kind** of pollution. As the total reliability of a tool is practi-cally impossible, an aggregate of techniques is to be imple-mented in order "to make reliable" the whole equipments.
In order to implement a real policy of promotion of clean production technologies, it is thus necessary :

- to make all the people implied in the activity of the firm, aware of problems related to the fight against industrial nuisances. No durable result will be achieved as far as this wide ditch between "the productives" and "the functionals" responsible for reducing the nuisance generated by the firm will exist, as it is the case presently. The whole management services have to be actuated by the motivation not to create nuisance; why isn't it possible to attribute a reward for innovations as it already exists in certain establisments in the field of safety, working conditions... productivity improvement ? Therefore, in this field, a lot remains to be done.

- To make equipments manufacturers aware of the necessity to take into account water use notions on the production equipments design ; the energy savings have become a first-rate commercial argument, the water savings and the pollution reduction have to become an other one.

- To institute a real distribution of the research topics among the competent (qualified) organisms. As a matter of fact, isn't it paradoxical to know that Professional Technical Centers devote quite a significant part of their activity to extended studies on the various purification technologies for wastewater, whereas it exists several laboratories and specialized research Centers in this field ? As experts in the branch technology, the part they have to play is essentially to think over the new methods, as revolving as it may be and even if they have to face a great scepticism at the beginning; Universities and specialized schools can and have to contribute in this effort.

- To motivate industrials in the choice of these new technologies by a financial assistance that makes it possible to cover a great part of the risks involved by this new technology ;
- To inform and to train industrials that the pollution is to be rejected directly into the effluents or indirectly into

the landfill, but it does not mean to get rid of a "waste". At the utmost, better would be to replace this terme "waste" by the term of " secondary product", which in most cases, corresponds to the notion of possible integration within the consumption cycle. It is thus necessary to be imaginative in this field and to undertake a set of studies on this topic.

- To be at last very concrete and very pragmatic in all these fields ; it would be vain to find out the moon and stars or to modify with a fairy's wand both a technological aggregate sometimes very old, very "affected" and very efficient but also human habits as well sometimes old, affected and efficient.

CRITERIA OF EVALUATION OF A DEVELOPMENT STRATEGY IN THE FIELD OF CLEAN TECHNOLOGIES.

The implementation of clean technologies has set a limit concerning the matters discharge while improving the very conditions of production ; in order to obtain such results, to settle purification systems is not sufficient. It is necessary to modify entirely the production tool. Pollution corresponds generally to a discharge of lost material : raw materials, energy, intermediary products, finished products. It expresses the poor performance of a production tool. The **poorer this performance is, the more important the pollution** is. The additional purifier systems do not bring any improvement in these technologies although it makes it possible to stop the damages for the natural environment. The poor performance remains a source of wasting, therefore an economical charge to which must be added the purification. The solution is not satisfactory on the plan of profitability. The pollutant methods of production are not necessarily the most efficient on an economical point of view. It corresponds very frequently to a rather few advanced technologies, and to an unsatisfactory productivity. Acting on the effects and not on the causes leads to expensive methods of elimination that do not repair the defects of the system.

- when the production methods are renewed in order to improve the performance, to reduce the formation of pollutants, to recycle the fluids, it can be envisaged to achieve a better economical profitability and to reduce the ecologic charge. It is certain that the major interest to implement these new technologies is of economical nature because of the savings it could involve in the field of energy and working costs of a waste water treatment plant and for the valorization of the by-products that were formerly unused. In order to give an opinion on the economic interest of the promotion of these technologies, it is better to consider 2 levels: the firm level (or industrial activity) and the collectivity or nation level.

The economic interest of clean technologies to the level of the firm.

As for the industrial, the implementation of clean technologies, either by a better exploitation of the existing production tools, or by the implementation of control and separation equipments, allowing the recovery of products either by the substitution for a less pollutant process of the existing process, will imply in most cases an additional expenditure. Practically 2 situations can be observed

- the operation is not profitable for the industrial, i.e. the costs are not compensated by earnings but the introduction of clean technologies shows an improvement in the return cost/efficiency compared with the conventional external purification technologies.

- the operation seems to be profitable for the industrial ; the earnings are equal or higher than the costs.

Those situations will be illustrated below by concrete cases

- the implementation of clean technologies leads to a better balance cost/purification efficiency. The wool-scouring goes together with oily and earthy water disposal (4,5 m3 water for 1 ton wool-scouring, containing 135 kg suspended solids and 300 kg cod). The process makes it possible to separate **water from bituminous and oily pollutant, while using the** energy contained in these products in order to accomplish the essential operation of this separation ; purified water is recycled during the rinsing. On the environmental plan, this operation has made it possible to totally suppress the polluted water disposal ; from the economic point of view, this operation necessitated a 15 000 000 FF (1980) investment. The renewal of wool-washing equipments alone would have cost 2 500 000 FF (1980) but the waste water treatment plant building would have represented a 30 000 000 FF (1980) investment. As for the working costs, annual cost comes to 2 400 000 FF (1980) as against 600 000 FF (1980) only for the pollutant process formerly used. But the waste water treatment plant would have increased the cost of 2 000 000 FF (1980).

- The implementation of clean techniques makes it possible to earn money. Synthesis of ammonium nitrate is got by reaction of nitric acid on ammonia. This reaction shows the formation of vapours bearing away ammonia. The factory is equipped with a cleaner that collects the biggest part of this ammonia contained in the vapours, by washing them with nitric acid. It necessitated a 350 000 FF (1982) investment and it allowed a 300 000 FF (1982) savings on the ammonia purchase (cost of industrial ammonia in France in 1979 : 1 000 F/Ton), as against an additional expenditure of 40 000 FF for energy and 4 000 FF for labour. The investment is very rapidly made profitable.

The economic interest of clean technologies for the collectivity.

The economic interest of clean technologies is not to be considered from the micro-economical point of view alone, but compared with objectives that the national or international collectivity as a whole strive to reach. It leads to question about the cogency of these technologies promotion compared to criteria not exclusively based upon the immediate or short-term profitability but entering within the framework of a long-term global strategy aiming at :

- protecting the environment and reducing the social costs ;
- saving energy and raw materials ;
- valorizing a research potential and the installation of technologies likely to be integrated in a scheme of development assistance for other countries .

The development of clean technologies guarantees a better protection of the environment.

Compared to conventional purification technologies, clean technologies present a set of advantages revealed by the pollution reduction that can be measured by precise indicators (suspended solids, organic matters, toxicity ...) as shown by the above-mentioned examples. But it presents also the great advantage to contribute to the reduction of pollution, social costs and of nuisances for the collectivity, while reducing pollution and nuisances once for all upstream and reducing the accidental pollution risks and pollution transfer. When the pollution removal, or the non-production of pollution is directly integrated within the production tool, the efforts made for the manufacturing are devoted as well to have it run correclty ; moreover a bigger safety-security-reliability is assured.

The development of clean techniques may facilitate the implementation of a policy aiming at saving rare resources and energy.

Clean techniques are often going together with raw material savings : let's note the recovery of solvents on activated carbon, of chromium on ion exchange resins, of copper by electrolysis, of soda by concentration, hydrochloric acid by cracking. These savings can be significant. Concerning **the above-mentioned process, it has permitted some savings** of 67 % for solvents, 33 % for chromium, 60 % for soda, 90 % for hydrochloric acid on the utilized quantities. Those savings are profitable not only for the firm concerned but for the whole collectivity as it represents a cost reduction and therefore has an effect on the collectivity supply.

The development of clean techniques may encourage industrial innovation

It would be useful to consider the impact of a development strategy of clean technologies on industrial innovation. Each kind of innovation demonstrates a productivity earnings. It makes it possible for the plant to face more efficiently the international competition and to find out new markets. Thus **it is important to evaluate whether this action undertaken** to promote clean technologies goes together with patents by industrial or on the contrary with purchase licences.

REASONS EXPLAINING THE DIFFICULTIES OF INTRODUCTION OF CLEAN TECHNOLOGIES

Among the industrial investments made during the last years, only a small number has been devoted to the implementation of clean technologies. It seems that the economical difficulties that the various industrial activities had to face during this period, have led them to modernize or rationalize their production tool in a saving concern. This fact is particularly true for pulp and paper industries and leather industries. In an economically difficult context, prior actions are to be undertaken in the field of investment likely to increase the productivity. Thus, a choice has been made in favour of the less expensive solutions at first, i.e. either in favour of a simple modification of the conditions of the manufacturing process implementation or by the material or fluids recycling or recovery process. These facts have led to question about the reasons why it is so difficult to introduce clean technologies in industry. In which extent the diffusion of clean technologies implies a deep change in the mentalities of industrials when associating the investment concept to be achieved and money profit with the fight against pollution and nuisances ?
As a matter of fact, the implementation of clean technologies may generate a profit. So it is surprising that industrials have not thought over this idea sooner. This inaction can certainly be explained by the weight of habits and routine :
It can also be explained perhaps because the preoccupations related to the fight against pollution have not yet integrated the services of the firms in charge of research, products commercialization, and production activities.

Is there a possibility to avoid this situation by a sustained information and training action while demonstrating the very interest of such techniques ?
Industrial field entertains undoubtedly suspicions as regard the development of new technologies compared to the old one, used for a long time. Moreover, it can restrict the development of technologies not to communicate its "know-how" since very often the clean technologies are not a real material revolution but rather a specific way to take advantage of a given tool. It is as well absolutely certain that competition and cost price intervene a lot in this wait-and-see policy.
In which extent the organization of the fight against pollution until now mainly oriented towards purification does not constitute a restraint to the introduction of clean technologies. The first political generations of fight against pollution have been characterized by taking up actions aiming at reducing the pollution in the old facilities. Standards and regulations have been prescribed and a purification market has been developed. Is there a real incitation from the Governments to encourage the clean technologies development ? Is the promotion of clean technologies really preferred to the conventional de-pollution solutions. In which extent the working and financial assistance in the field of purification established by the Government does not turn the industrial away from the solutions aiming at using the clean technologies. What improvements could be brought to the financial assistance system established at the present time ? What efforts are to be undertaken in the field of the organization related to the products market and the recovery channels ? In which extent the establishment of national and international regulations or the institutional context encourages agreements between firms or plants and communities

and allowing a better valorization of by-products or waste thanks to clean technologies? Isn't it necessary to undertake an action on the consumption products quality, for which some finishing requirements are not always necessary because of their sometimes precarious use. This situation requires from the industrial to use some noble raw materials instead of certain by-products that might be recycled or re-used for the same utilization. It necessitated to extract a maximum quantity of impurities that sometimes do not constitute an inconvenient for the finished product quality but that involves an additional pollution.

CONCLUSIONS

- The main factors that determine the development and the implementation of clean technologies are for the most part economical factors. It is sure that the recent modifications of the world-wide economical context concerning both the general increase in raw materials and energy costs, associated with a more and more acute knowledge of natural available resources, will lead the industrial field and the whole governments to save as well the whole natural resources. It would be advisable to integrate within those economical and political concerns the notion of clean production techniques at each level of training, information and production.

- The re-use and re-cycle of by- and co-products of industrial manufacturing and community have to be strongly encouraged by carrying out the necessary research and studies proving that it is quite possible to substitute noble raw materials by those products. When the air and aqueous environment is degraded, the waste water treatment plant does not seem to be a forward solution and therefore it is necessary to do the utmost to avoid the introduction of what will have to be removed from this environment at great cost.

- All the political and economical powers have to encourage consumers and producers not to require systematically some sophisticated products, of which time of living is often short. Paper recycling, glass recycling, waste re-use coming from agro-industry for cattle feeding or for agriculture have to be highly encouraged. The industrial field has to perform evolutions in that way but it can be difficult to perform it alone in a national context necessarily affected by political and economical considerations that do not always facilitate research and innovation. International organizations might constitute some incentive factors in the development of such a policy.

GENERATOR'S REFLECTION FOR HAZARDOUS WASTES

Sachiho Naito, Prof., Kanto Gakuin Univ.,
Yokohama, Japan

PREFACE

The Ministry of Health & Welfare of the Japanese Government, in an attempt to reduce waste generation, decided to instruct manufacturers and other waste generators to "reflect on proper prior measures if the product concerned, when becoming waste, is expected to be difficult of treatment and/or disposal." (Such difficulty is to be referred to hereinafter as "untreatability".) Based on the decision, the Ministry established the following guideline and notified it on December 4th, 1987 to Chairmen of the Federation of Economic Organization and the Japan Waste Management Association. The guideline is shown below in whole, along with some explanatory notes.

CHAPTER 1 GENERAL PROVISIONS

1. Objectives of the Guideline

(1) The Waste Disposal & Public Cleansing Law provides that generators, in manufacturing, processing, selling or otherwise handling products, containers, etc. (hereinafter referred to collectively as "products"), shall ensure that the product concerned, when becoming waste, will not be difficult of proper treatment.

(2) To meet the requirement of (1), each generator shall reflect for itself beforehand (hereinafter referred to as "reflection") on the untreatability of the product concerned to ensure that the product, when becoming waste, will not be difficult of proper treatment.

(3) Reflection is a responsibility of each generator to be performed in its own manner, and untreatability to be evaluated depends on products.

ISWA PROCEEDINGS
VOLUME 1 ISBN 0-12-058451-4

Thus, the method of reflection should be determined by each generator according to the characteristics of the product concerned.
(4) This guideline aims at spreading and promoting reflection. Among other objectives, it is intended to be a guide for generators to establish their specific reflection methods for their own products, by specifying fundamental matters common to various products.

2. Definition of Terms

Basic terms used in this guideline are defined as follows:

(1) Waste: Waste as specified in the Waste Disposal & Public Cleansing Law(waste such as refuse and bulky refuse or disused articles); i.e., those materials or articles which have become unnecessary because of the inability of their occupants to use them or sell them to others;

(2) General waste: General waste as specified in the Waste Disposal & Public Cleansing Law, mainly refuse, bulky refuse, excrement, etc. generated from daily life of citizens. Also including waste generated from business activities except industrial waste;

(3) Generator: Generator as specified in the Waste Disposal & Public Cleansing Law, engaged, as business, in manufacturing, processing, selling or other handling of products to be disposed of as general waste;

(4) Proper treatment of waste: Prompt elimination of generated waste from livelihood space without posing problems to maintenance of living environments nor to public health, and treatment or disposal of such waste not to cause it to damage nature;

(5) Waste difficult of proper treatment: Waste of which treatment is impossible or difficult at municipal levels for technical, equipment or economical reasons. The concept of "difficult for technical reasons" includes cases where "equipment might be damaged" or "there are problems in terms of safety and health of workers."

3. Parties and Products Needing Reflection

(1) The provision of the Waste Disposal & Public Cleansing Law applies to all generators and pro-

ducts. Thus, reflection along this guideline is required, in principle, of all generators engaged in manufacturing, processing, selling and other handling of products to be disposed of as general waste, with respect to all of such products.

(2) However, with a view to avoiding redundant reflection and clarifying who is responsible, reflection on untreatability of a product is required of the generator engaged in the final manufacturing (or processing) activity for the product. Consequently, reflection is not necessary for those generators who only manufacture parts not offered for independent use.

(3) In the case of imported products, the importers concerned are required to make reflection. Socalled "sample imports" in small quantities do not need reflection.

(4) Products subject to reflection ("subject products") shall be those products which are on a market where free access is usually allowed to general consumers and which are newly manufactured from other products by changing the properties and other characteristics of the materials, parts, processes, etc. of the latter products in such a manner as is substantial in physical, chemical, biological or safety terms from the viewpoints of waste treatment.

4. Responsibilities of Generators and Cooperation of related Parties

(1) Generators assume the responsibility of ensuring in manufacturing, processing, selling or other handling of products that the products, when becoming waste, will not be difficult of proper treatment. To perform the responsibility, generators are required to make reflection.

(2) If reflection has resulted in expectation that a product will be difficult of treatment, the generator concerned shall consider corrective measures such as those for changing or improving the production method or other factors, developing or promoting appropriate treatment techniques, and establishing proper treatment systems. Reflection shall be newly made on changes in untreatability accompanying such corrective measures, to finally ensure that proper treatment will not be difficult.

(3) At the request of generators, municipalities and other related parties shall extend necessary cooperation such as information provision, and as the need arises, provide advice or guidance so that reflection would be made in a smooth and effective manner.

CHAPTER 2 REFLECTION

1. Principles

(1) All products will be disposed of as waste through our daily life or business activities. Thus, their proper treatment is a fundamental requirement for preserving our living environments and improving public health, as well as being an essential requirement for maintaining and expanding our socio-economic activities.

(2) The basic condition for ensuring proper treatment of waste is development of treatment techniques and establishment of treatment sytems for generated waste. It is also important to prevent generation of waste difficult of proper treatment, by promoting adequate measures on products themselves, in an effort to make waste treatment efficient and stable.

(3) From the above viewpoints, the Waste Disposal & Public Cleansing Law is intended to prevent generation of waste difficult of proper treatment, through reflection by the generators concerned, thus promoting proper treatment of waste.

2. Reflection Procedure and Others

(1) Time and scope of reflection

1) Generators shall make reflection as early as possible prior to manufacturing of the product concerned.

2) When experiences or other data demonstrate that the product concerned will not pose problems to proper waste treatment, the 2nd and subsequent steps specified in (2) are not necessary.

(2) Procedure and extent of reflection

1) Reflection shall be made in the following steps (see the slides in the Session). Details and specific methods are to be determined by each generator according to the characteristics of the product concerned, as specified in Chapter 3 of this guideline.

(3) Confirmation of related parties'views and examination of reflection results

1) In making reflection, generators shall collect information necessary for examining the existing treatment system at municipal and other levels.

2) In examining reflection results, generators shall seek advice or guidance, as necessary, from related parties.

3) The related parties shall meet the request for advice or guidance, and exercise due care to observe secrecy in respect to the matter.

4) Generators shall duly respect the advice or guidance of the parties concerned, and report reflection results to the parties.

5) Advice or guidance from related parties as well as reports by generators shall not be deemed to lessen the responsibility of the generator concerned with respect to its subsequent reflection on untreatability of the product.

(4) Keeping of reflection results

1) Generators shall keep reflection results.

2) Reflection results shall be kept until after a reasonable period from the time of termination of the manufacturing of the product concerned.

(5) Subsequent monitoring

1) Following reflection, generators shall be responsible for constant monitoring to confirm, among others, that the product concerned, when disposed of as waste, is not causing any treatment problem and that there is no change in the conditions of the product and treatment system having been considered at the last reflection.

2) When necessary as a result of monitoring, generators shall review the results of the last reflection, and make reflection again based on new conditions.

3) Each generator shall, as necessary, designate some of its personnel as liaisons with relevant municipalities and other related parties with a view to implementing the abovementioned activities in a smooth and effective manner. Each generator shall make efforts to train specialists in waste treatment.

CHAPTER 3 TECHNICAL MATTERS

Specific technical matters relevant to reflection shall be separately determined by each generator according to the characteristics of the product concerned. Such technical details shall be determined for each of the following matters.

1. Collection and examination of relevant information: Matters regarding treatment systems, treatment contractors, and other related conditions in the municipalities concerned.
2. Examination of product characteristics: Matters regarding product characteristics necessary for evaluating product attributes and quantifying untreatability.
3. Determination of evaluation items: Matters regarding evaluation items to be determined after examination of product characteristics and municipal treatment systems.
4. Determination and examination of evaluation method: Matters regarding evaluation method to be determined for evaluating the items of 3 abovementioned.
5. Evaluation of untreatability: Matters regarding evaluation of untreatability and specific effects expected on municipal treatment systems.
6. Consideration of corrective measures and overall evaluation: Matters regarding corrective measures to be taken as necessary for lessening untreatability of the product, and matters regarding overall evaluation.
7. Preparation of reflection reports: Matters regarding preparation of reflection reports.
8. Keeping of reflection reports: Matters regarding how to keep reflection reports.

RECOVERY OF CHEMICAL WASTES IN EUROPE

Kim Christiansen, Technological Institute, Copenhagen
Mogens Palmark, Chemcontrol, Copenhagen
Torben Hansen, Gendan

Introduction

For the Commission of the European Community and the OECD a data gathering on recovery of chemical wastes i.e. reuse or recycling of constituents of hazardous wastes has been made. The data are intended for use in development of policies to promote maximum recovery of constituents of potentially hazardous wastes with minimum risk to man and the environment.

The data were collected from central government and other central sources of knowledge in the following countries:
- Belgium, Denmark, Federal Republic of Germany, France, Italy, Luxemburg, the Netherlands and United Kingdom.

This was achieved by correspondence, questionnaires and personal visits.

Background

The background for the Europa-study was a project carried through by the Technological Institute and financed by the Technology Advisory Board of Denmark (Christiansen and Hansen, 1986), where the actual and potential recovery of chemical wastes by external recyclers in Denmark was investigated.

Not all data needed to fully complete the data sheets on the above mentioned recyclers and recoverable chemical waste types were available.

In summary the major conclusions of the Danish case study were as follows:

In gathering data the most efficient approach was personal contacts with key persons in industry and government.

There is a vast amount of literature (e.g. ERL(1985), Null et al. (1985), OTA (1985) UMPLIS (1982) and Sutter (1987)) dealing with recycling possibilities including laboratory research, pilot plant testing, on-site industry implementations, and commercialized by external recycling companies. However, such information is seldom critically evaluated and updated.

ISWA PROCEEDINGS
VOLUME 1 ISBN 0-12-058451-4

Danish industry produces approximately 120.000 tonnes of chemical wastes every year, excluding waste oils. About 75% is destructed at the Kommunekemi by incineration or inorganic detoxification followed by landfilling of any residues.

Around 5.000 tonnes of chemical wastes of which organic solvents comprise about 90% is recycled or reused by external contractors for secondary raw materials or by chemical companies as additional feedstocks.
Supplementary to external recycling activities in Denmark an additional 20.000 tonnes of chemical wastes are estimated to be exported. Reclamation of some fraction of these wastes occurs in other countries.
The most promising possibilities for an increase in external recovery of chemical wastes are associated with the following waste categories:

- recovery of iron, zinc, nickel and copper from filter cakes, waste waters and fly ashes and other metal containing wastes from surface plating industries, energy production etc.

- reuse of chromium pickling waste in formulating wood preservatives

- recovery of metallic copper from electric circuit print industry wastewater

- recovery of mercury from industrial wastes, batteries and dental wastes

- recyling of solvents from waste products such as paints

These approaches if combined with a suitable waste exchange system could lead to an increase in recovery of 4 to 5 times.
Concerning more general and political incentives to increase recovery of chemical wastes the feasibility of clean technologies/internal recovery, a selected tax on the most hazardous feedstocks, inclusion of minimum contents of secondary raw materials in new products and a tax on finial disposal were recommended for further investigations.

Results of the European study

The conclusions made in the Danish case study on the appropriate approach for gathering of data were confirmed through contacts to centers of knowledge e.g. central authorities and consultants in visited countries. The data collected and used in the data sheets evolves both in Denmark and in the Netherlands from a direct contact to waste producers (i.e. the industry), waste disposal and recycling facilities and local authorities. In Flanders (Belgium) and Luxembourg the data came from central authorities. but this seems to be an exception. In the Federal Republic of Germany, France, Italy and the United Kingdom, the information from central authorities is dealing only with selected industries or waste groups.

In table 1 the information concerning waste generation and recovery presented in one-page sheets in an annex to the study are summarized for Denmark, Belgium (Flanders) and the Netherlands. The OECD and UNEP preliminary code (Lieben and Huismans, 1987 and Yakowitz, 1984) has been used as order of listing. Any use of the data for further processing are not recommended as the data reliability has not been officially approved by the involved authorities.The figures only give a first impression of the importance of recovery for relatively few waste categories. Therefore no attempts have been made to add up data for several countries regarding specific constituents. The percentage of hazardous waste externally recovered in Belgium (Flanders), Denmark and the Netherlands, respectively, is nearly 20%, nearly 40% and nearly 10%.

Conclusions

The study (Hansen et al. 1987) showed that only few countries in the EEC possess central statistics on generation and fate of hazardous wastes. These countries are Denmark, Luxembourg, the Netherlands and the region Flanders in Belgium. Other countries i.e. the Federal Republic of Germany and France are planning to implement some type of control registration of generation, import/export, and disposal of hazardous waste. In other countries included in this study i.e. Italy and the United Kingdom, similar thoughts have been raised but practical fulfilment seems difficult due to a highly decentralized structure.
Only to a limited extent do these statistical registers include figures on resource recovery and recycling both externally and internally, and the actual figures are often uncertain and out of date.

Table 1 Waste Categories Included in the National Data Sheets
(Please note that all figures are cited without the reliability judgement used in the study)

CONSTITUENT EXPLICITLY COVERED BY SCOPE OF WORK	OECD CODE	DENMARK AMOUNT OF WASTE	BELGIUM AMOUNT OF WASTE	THE NETHERLANDS AMOUNT OF WASTE
OILY WASTES	-	5000 + 50000		N.D.
BERYLLIUM	1	N.D.	N.D.	N.D.
CHROMIUM (VI)	3	N.D. + 1000	N.D. + 80	30 + 0 + 5
COBALT AND OTHERS	4	N.D.	N.D.	0 + 0 + 1000
NICKEL	5	N.D.	N.D.	100 + N.D. 1200
COPPER	6	200 + 700[3] + 50	N.D. + 14	200 + 0 + 2200
ZINC	7	15000	N.D.	1500 + 250 + 800
ARSENIC	8	N.D.	N.D. + 2	N.D.
SELENIUM	9	N.D.	N.D.	N.D.
CADMIUM	11	N.D.	3 + 16	40 + 0 + 50
ANTIMONY	13	N.D.	650 + 765	N.D.
MERCURY	16	5 + 2000	N.D. + 1396	250 + 3 + 50
THALLIUM	17	N.D.	16 + 127	N.D.
LEAD	18	9450 + 1050	370 + 205	N.D. + 20000 + 400
LEAD/TIN		N.D.		N.D. + 5000 + 450
ACIDS	23	3000 + 3000	N.D. + 758[2]	40000 + 5500 + 3500
HALOGENATED SOLVENT	40	300 + 1600	3190 + 6508	10 + 16000 + 4000
NON-HALOGENATED SOLVENTS	41	4500 + 5000	INCLUDED ABOVE	INCLUDED ABOVE
PRODUCTS WITH	-	N.D. + 2000	N.D.	N.D.
HEAVY METAL MIXTURE	-	0 + 5000	N.D.	N.D.
PRECIPITATION METAL (Al, Fe)	-	N.D. + 2000	N.D.	N.D.
AMMONIA COOLING LIG.	-	20 + N.D.	N.D.	N.D.
TOXIC CHEMICALS	-	N.D.	N.D. + 215	N.D.
TOXIC FUME PRODUCING CHEMICAL WASTE WHEN IN CONTACT WITH WATER	-	N.D.	N.D. + 1254	N.D.
GOLD	-	N.D.		2
LABORATORY CHEMICALS	-	N.D.	26[1] + 7034	N.D.
PACKINGS FOR TOXIC WASTES	-	N.D.	N.D. + 26	N.D.
SILVER	10	500 + 1000 + 1000	N.D.	10 + 7500 + 65

TELLURIUM	14	N.D.	N.D.	N.D.
SULPHUR-CONTAINING WASTE	19	5000[3]		
FLUORIDE CONT. WASTE	20	N.D.	N.D. + 92	N.D.
INORGANIC CYANIDE	21	N.D.	N.D. + 51	N.D.
BASIC SOLUTIONS	24	N.D.		1500 + 2600 + 300
ASBESTOS	25	N.D.	N.D. + 235	N.D.
METAL CARBONYLS	27	N.D.	N.D.	N.D.
PEROXIDES, ETC.	28,29, 30	N.D.	N.D. + 19	N.D.
PCB	32	N.D.	N.D. + 97	N.D.
PESTICIDES ETC.	33 + 34	N.D.	N.D. + 302	N.D.
ORGANIC CYANIDE	37	N.D.	N.D. + 30	N.D.
PHENOLS	39	N.D. + 1000	N.D. + 124	N.D.
PAH	43	N.D.	N.D. + 4	N.D.
NITROGEN CONTAINING	44	400 + N.D.		
ETHERS	46	N.D.	N.D. + 41	N.D.
HALOGENATED ORGANICS	51	N.D.	536 + 3483	N.D.
TOTAL		45075 + 74701	4791 + 23010	43641+ 56855+ 14021

Legend: Denmark : First figures : external recovery
Second figures : other disposal methods (incl. heat recovery)

Belgium : as Denmark

Netherlands : First figures : internal recovery
Second figures : external recovery
Third figures : recovery abroad (export)

N.D. = No Data

Notes:
1. Solvents
2. Acid/basic waste from metal surface industry etc.
3. Exported for recovery

References: Denmark: Christiansen *et al*. (1986).
Belgium: Flanders Ministry of Environment, Mechelen; personal communication.
Netherlands: TAUW Infra Consults (1986).

Table 2 summarizes our first and very rough evaluation of the experiences with the gathering of information sought for in this study, which can be compared with tables in Van Veen and Mensink (1985) and Forester and Skinner (1987). In the study recommendations for further data gathering are included for each country.

Table 2. Evaluation of the availability of the information sought for in the visited countries.

Type of information	Existing at central authorities, consultants etc.	Existing at local authorities, industries etc.	Not available
Total generated quantity, import and export	B, L, (DK), (NL), (F)	NL, DK, FRG	UK, I
National capacity external recycling	(DK), (F)	B,NL	
Amounts actually recycled off-site	DK, NL, B		
Technologies used	NL, DK		
Use of recycled material	NL, DK		
Yield of recycling processes	DK	?	All
Types and quantities of residues from recycling		All	All
Fate of residues from recycling	DK, NL, B	FRG	
Amounts exported/imported for waste exchanges	FRG		
Amounts exported/imported for disposal	B, NL, FRG, DK		
Costs of recycling			All
Costs of disposal	DK	NL, FRG, B, F I, L	
Price available for end-products			All
Price of virgin materials	All		
Legislation/incentives for recycling	All		

(English abbreviations have been used for the countries)

References

Bentley, J. (1987): Developments in DOE Guidelines for the Proper Management of Hazardous Wastes. Safewaste Conference, Cambridge 31.3.- 2.4.1987

Christiansen, K. and Hansen, T. (1986): Resource Recovery, Reuse and Recycling of Constituents of Hazardous Wastes. Danish Case Study. Interim Report for the Commission of the European Communities. Technological Institute of Copenhagen, Denmark.

ERL (1985): Recycling of chemical wastes in the European Community. Prepared for the Commission of the European Communities. Directorate General for the Environment, Consumer Protection and Nuclear Safety by Environmental Resources Limited, London, United Kingdom

Forester, W.S. and Skinner, J.H. (ed.) (1987): International Perspectives on Hazardous Waste Management. Academic Press, London, United Kingdom.

Hansen, T., Palmark, M.I. and Christiansen, K. (1987): Resource Recovery, Reuse and Recycling of Constituents of Hazardous Wastes. Report for the CEC. Technological Institute of Copenhagen, Denmark.

Lieben, P. and Huismans, J.W. (1987): Identifying, Classifying and Describing Hazardous Wastes. Paper contributed by OECD Environment Directorate and UNEP International Register of Potentially Toxic Chemicals to the World Conference on Hazardous Waste, Budapest, October 1987.

Noll, K.E., Haas, C.N., Schmidt, C. and Kodukula, P. (1985):Recovery, Reuse and Recycle of Industrial Waste. Industrial Waste Management Series, Lewis Publishers, Inc., Mich., USA.

OTA (1983): Technologies and Management Strategies for Hazardous Waste Control. Office of Technology Assessment. Congress of the United States, Washington, USA.

Sutter, H. (1987): Vermeidung und Verwertung von Sonderabfallen - Grundlagen, Verfahren, Enrwicklungstendensen. Erich Schmidt Verlag, Berlin, Federal Republic of Germany.

TAUW Infra Consult (1986): Recycling of Hazardous Wastes, Generated in the Netherlands. Deventer, the Netherlands.

UMPLIS (1982): Recycling-Handbuch. Umweltbundesamt. Erich Schmidt Verlag, Berlin, Federal Republic of Germany.

Van Veen, F., Mensink, J.A. (1985): Brief Survey of Legislation and Arrangements for the Disposal of Chemical Waste in a Number of Industrialized Countries. Hazardous Waste & Hazardous Materials 2(1), 333-353.

Yakowitz, H. (1984): Recycling/Reclamation of Materials from Potentially Hazardous Wastes: Analysis, Data Needs and Policy Options. Organisation for Economic Cooperation and Development. Environment Committee, ENV/WMP/84.13. OELD, Paris, France.

FROM WASTE MANAGEMENT TO WASTE PREVENTION

CONCEPTS, VALUES AND APPROACHES ESSENTIAL FOR THE TRANSITION

D. Huisingh, T E M / University of Lund

J.C. van Weenen, University of Amsterdam

Introduction

In the years since the participants at "THE ONLY ONE EARTH" Environmental Conference held in 1972 in Stockholm, Sweden released their declaration and plan of action, the public's awareness about environmental pollution from human activities has increased dramatically. In response, governmental officials in almost every nation enacted a wide-array of POLLUTION CONTROL regulations. Most of these regulations, while instrumental in helping to reduce the pollution levels of some pollutants in some locations or situations, frequently served to merely transfer the pollutants from one medium or location to another. In many instances these materials had to be dealt with at a later time at greatly increased costs and risks. Hazardous waste polluted sites are being 'Cleaned-Up' in many nations, but the cost is prohibitive and there is doubt that many of the sites can ever really be 'Cleaned-Up' at any cost.

ISWA PROCEEDINGS
VOLUME 1 ISBN 0-12-058451-4

There is also growing concern about environmental problems such as the thinning of the Ozone Layer; the pollution of the oceans, destruction of tropical rainforests, acid deposition and with 'Accidents' like, Bhopal, Chernobyl and Sandoz. A review of these problems suggests that the goals of the participants of the Stockholm conference have not yet been achieved by the POLLUTION CONTROL LAWS AND THE POLLUTION CONTROL TECHNOLOGIES enacted and installed, to date.

Our lack of success in solving these problems appears to be symptomatic of a series of metaproblems that have their bases in the fundamental philosophy and life-style of humans that is best characterized by the phrase, "HUMANS VERSUS NATURE". The authors believe that if ecosystem dynamics and human sustainability are to be ensured, the "HUMANS WITH NATURE", metaphor must become the new paradigm through which ethical values based upon ecological values are established and upon which human activities are planned and developed. Societies and communities built upon such values will be cybernetically responsive; capable of quickly and sensitively exercising preventative action in balancing COMPETITIVITY WITH COOPERATIVITY.

In the sections which follow, approaches are suggested that should help societies change to the HUMANS WITH NATURE relationship that focuses upon WASTE PREVENTION as a central theme of this new harmonious relationship. Concepts for the design of ENVIRONMENTALLY FRIENDLY PRODUCTS and part of a model CODE OF ETHICS are presented as factors that can serve as guides for industrial, governmental and societal changes to the HUMANS WITH NATURE metaphor.

WASTE PREVENTION THROUGH GOVERNMENTAL AND INDUSTRIAL POLICIES AND PRACTICES.

During the last fifteen years, many of the nations of the world established a preferred ranking for the management of wastes that includes the following:

a. Waste prevention
b. Waste reuse within the production process
c. Waste reuse off-site
d. Utilization of the energy value of the wastes
e. Sanitary and/or secure landfilling of the wastes

However, during those fifteen years, the first option, PREVENTION, did not get much attention. Prevention was considered to be the best possible option but the one that would be possible only in the long-run. In the meantime 99% of the efforts were focused upon the end-of-pipe waste management option b through e. At the same time the generation of wastes, of all types, steadily increased and the composition of wastes became more complex; thereby posing increasingly severe threats upon the quality of present and future environments.

Fortunately, many companies, a few governments and some national and international environmental organizations led the way to more positive and environmentally sound products and production processes that capitalize upon the economic and environmental benefits of focusing upon prevention of the production of waste. These encouraging developments often reduced energy use and the amount and toxicity of wastes generated within industries and households.

Based upon the experiences of these leading organizations, there now appears to be sufficient knowledge and experience with waste prevention that we can present a technically feasible and ecologically sound list of preferred emphases upon WASTE PREVENTION for the 90's:

a. Change industrial processes for the production of CURRENTLY USED PRODUCTS so as to reduce production process and product burdens upon the environment. This implies the search for and incorporation of materials that are less toxic than the substances currently being used. It also implies making process modifications to reduce the probability of Bhopal-like accidents and to ensure that housekeeping and other aspects of the production processes thoroughly incorporate the best waste prevention practices available.
b. Change the planning of NEW products so as to incorporate ECO-SYSTEM AND HUMAN HEALTH considerations of the product's entire LIFE-CYCLE at the design phase of the new products. This implies that consideration must be given to the anticipated short- and long-term human health and environmental impacts of the production of the raw materials to be used in production of the product, to the impacts of the manufacturing phase of the product, to the impacts of consumer use of the product and to the manner in which the product are to be managed at the end of their useful lifetime. Such an integrated emphasis upon PREVENTION OF ENVIRONMENTAL PROBLEMS, already at the 'product design phase' should drastically reduce the environmental and human health threats posed by them.
c. Strive for an overall reduction of the use of substances and energy by making products more durable, more

readily repairable, more energy efficient and more readily recyclable.

d. Strive to select all raw materials for production so that the composition and properties of production wastes can be readily reused within the production processes (CLOSED-LOOP).

The authors believe that as society moves to implement these four emphases DESIGNED TO FOCUS UPON PREVENTION OF THE PRODUCTION OF WASTES, the amount and toxicity of waste that must be managed in environmentally sound ways will be drastically reduced. It is increasingly evident that such changes in approaches and practices are imperative since we are now, belatedly, recognizing the limitations to the currently prevalent, 'end-of-pipe', waste management practices.

The evolving new technologies like microelectronics, biotechnology, new materials technology and microsystems technology may present many new environmental and human health threats. But, if the types of preventative emphases suggested above are incorporated into their design, they may also provide many new ways for simultaneously meeting human needs and ensuring an environmentally safe and sustainable future for humans upon this planet.

CODE OF ETHICS TO HELP US MOVE TO THE HUMANS WITH NATURE SOCIETY

The World Federation of Engineering Organizations charged their committee on Engineering and Environment with the responsibility of developing a CODE OF ENVIRONMENTAL ETHICS FOR ENGINEERS. That group approved the code at their conference in New Delhi on November 5, 1985[1]. We quote from that code in order to emphasize that this group of environmental engineers recognizes that ethical values and ecological values are not abstractions but are concrete essentialities for survival that must be used as guidelines in professional and personal decisions.

We quote part of the Code as follows:

"1. Try with the best of your ability, courage, enthusiasm and dedication to obtain a superior technical achievement, which will contribute to and promote a healthy and agreeable surrounding for all men, in open spaces as well as indoors."

"2. Strive to accomplish the beneficial objectives of your work with the lowest possible consumption of raw materials and energy and the lowest production of wastes and any kind of pollution."

"7. Be aware that the principles of ecosystemic interdependence, diversity maintenance, resource recovery and interrelational harmony form the bases of our continued existence and that each of those bases poses a threshold of sustainability that should not be exceeded."

Such codes of ethics, when developed and utilized by professionals in industry, government and academia, could contribute much to the necessary transition to the HUMANS WITH NATURE relationship essential for human survival upon this planet.

---

[1] Code of Environmental Ethics for Engineers, printed in 1986 in Buenos Aires, headquarters of the Engineering and Environment Committee of the World Federation of Engineering Organizations.

The US EPA Waste Minimization Research Program

H. M. Freeman,
U.S. Environmental Protection Agency
Hazardous Waste Engineering Research Laboratory
Cincinnati, OH, USA

Introduction

In 1986 the US EPA submitted a Report to Congress on the Minimization of Hazardous Waste(1). In this report, which was submitted in response to the 1984 Hazardous and Solid Waste Amendments to the Resource Conservation and Recovery Act (RCRA), the Agency found that "aggressive action in favor of waste minimization is clearly needed...:" Subsequently, the Agency has undertaken to develop a program that includes among several elements, a waste minimization research program to encourage the development and adoption of production and recycling technologies that result in less waste being generated. A description of that program and a discussion of some of the projects in the program are the subject of this paper.

WMR Program Overview

The overall purpose of the waste minimization research (WMR) program is to facilitate the identification, development, and adoption of production and recycling technologies that result in reducing the generation of hazardous wastes. The program is intended to contribute and to support other waste minimization activities within the EPA that are designed to encourage waste minimization.

Although 1988 is the first year that the Agency has supported a program specifically identified as a waste minimization research program, it has in the past supported several projects that can be appropriately categorized as waste minimization research. In 1987 the HWERL published the results of a rather extensive project to document the advantages of incorporating selected recycling and treatment technologies in the printed circuit board industries(2). Two of the case studies evaluating the use of sodium borohydride reduction as a substitute for lime/ferrous sulfate precipitation, found that the technology was a

ISWA PROCEEDINGS
VOLUME 1 ISBN 0-12-058451-4

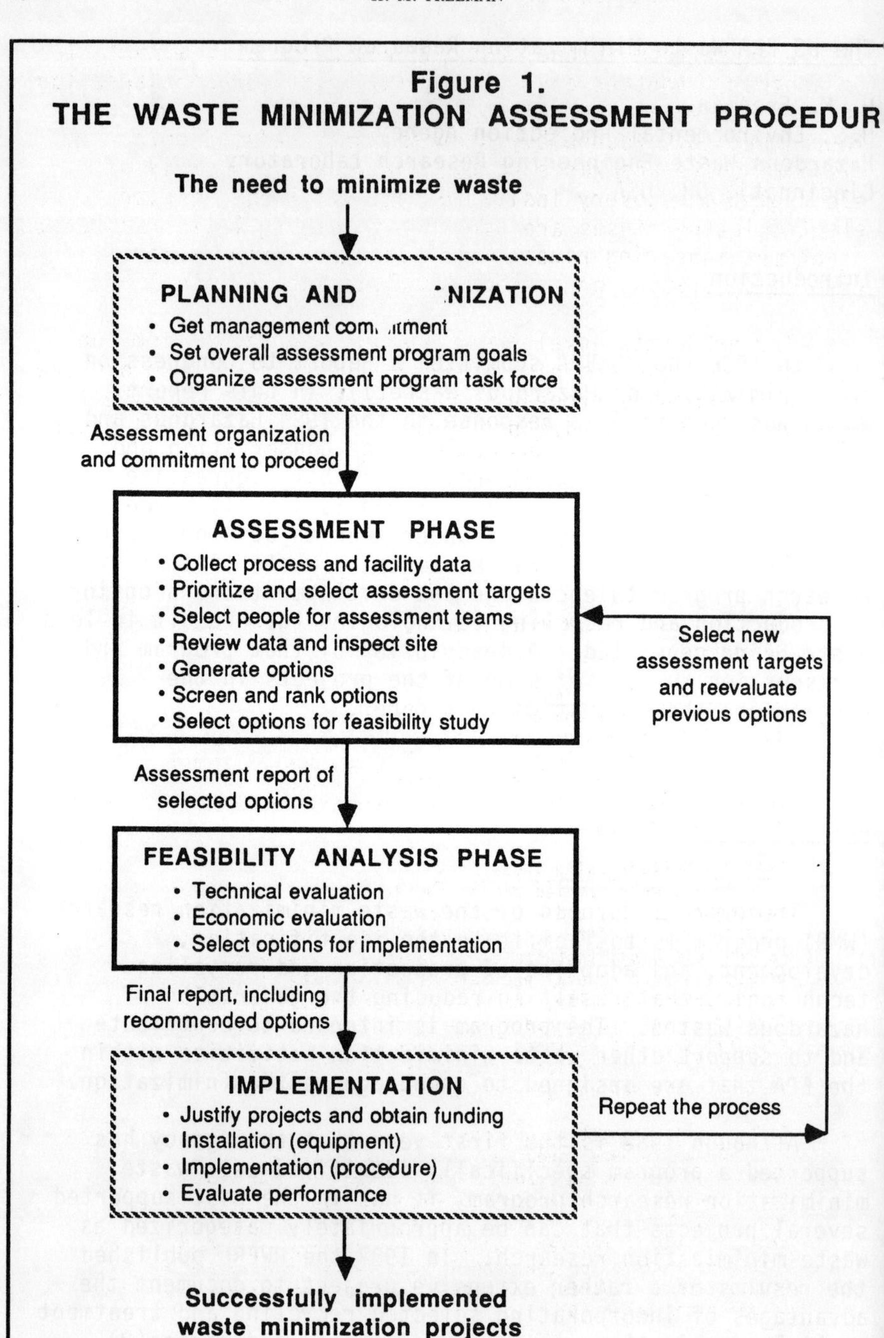
Figure 1.
THE WASTE MINIMIZATION ASSESSMENT PROCEDUR
The need to minimize waste
PLANNING AND ORGANIZATION
• Get management commitment
• Set overall assessment program goals
• Organize assessment program task force
Assessment organization and commitment to proceed
ASSESSMENT PHASE
• Collect process and facility data
• Prioritize and select assessment targets
• Select people for assessment teams
• Review data and inspect site
• Generate options
• Screen and rank options
• Select options for feasibility study
Select new assessment targets and reevaluate previous options
Assessment report of selected options
FEASIBILITY ANALYSIS PHASE
• Technical evaluation
• Economic evaluation
• Select options for implementation
Final report, including recommended options
IMPLEMENTATION
• Justify projects and obtain funding
• Installation (equipment)
• Implementation (procedure)
• Evaluate performance
Repeat the process
Successfully implemented waste minimization projects

Fig. 1

viable substitute in one case and was marginally acceptable in another. Another case study, involving carbon adsorption removal of organic contaminants from plating bath wastes, found that this technology significantly reduced both disposal costs and waste volume. A final case study of electrolytic recovery indicated that while acid copper electroplating rinses are amenable to electrolytic recovery, other metal-bearing rinses, such as those from solder (tin/lead) plating or etching are less appropriate.

The HWERL has developed a generic EPA manual for conducting waste minimization assessments. (3) An outline of the recommended procedure is shown in Figure 1. The agency supported the demonstration of the recommended procedure in selected waste generating facilities. Summaries of those studies are shown in Table 1.

Table 1

Selected Summaries of Waste Minimization Assessment Case Studies

Case 1 - Specialty steel making complex

A technically and economically feasible recycling option for the recovery of fluorspar (calcium fluoride) from a nitric/hydrofluric acid pickling waste water stream was identified. Use of the WM option would result in savings of $168,000 and a 30 percent reduction in final waste disposal volume.(4)

Case 2 - Solvent using industries

Audits were carried out at two facilities, an aircraft equipment facility and a ceramic capacitor manufacturer. At the first facility such source control options as counter current cleaning, installation of a level alarm, and periodic sludge removal had payouts of less than 3 years. At the second facility, the most cost effective options identified were to standardize solvents, to use metal mesh type filters, and to install a common batch solvent recovery still. This last option had a payout of - .9 years and reduced waste solvent generation by 50% annually.(5)

Case 3 - Electroplating operations

WM audits were conducted at two electroplating facilities generating cyanide - bearing wastes. At the first facility, WM options determined to have less than a 3yr payback were use of drain boards to reduce drag-out, still rinsing to recover drag-out and spray rinsing. Cost effective options for the second facility were determined to be use of drain boards, and bath life extension through use of deionized water.(6)

In cooperation with the California Department of Health Services the EPA is presently supporting two significant efforts in the area of waste minimization. The first is the development of industry-specific waste minimization assessment manuals (See Table 2), and the second is a project to estimate the potential for reducing halogenated organic waste streams in the Los Angeles area(7). The assessment manuals incorporate the format of the recommended EPA procedure for conducting waste minimization assessments. The manuals will be available by the fall of 1988, and the halogenated organics project will produce recommendations in the winter of 1989.

Table 2

US EPA/California DOHs Waste Minimization Assessment Manuals

. Pesticide Formulators	. Photographic Processing Labs
. Paints & Coatings	. Commercial Printing
. Circuit Board Manufacturers	. Educational, Scientific, & Research Institutions
. Automotive Repairs	. Hospitals
. Automotive Paint Shops	. Fiberglass & Rigid Plastic Products

Current Expanded Program

Reflecting the increased emphasis being given to waste minimization in the U.S., the HWERL is expanding the WM program. This expanded program is being built on five newly established subprograms as outlined below.

WRITE

The Waste Reduction Innovative Technology Evaluation (WRITE) Program - a program to identify, evaluate and/or demonstrate new ideas and technologies that lead to waste reduction.

The WRITE Program is a program to involve the EPA with private industry to encourage the development and/or demonstration of effective techniques and technology for hazardous waste minimization. The program is based upon the model currently being used in the Agency's Superfund Innovative Technology Evaluation (SITE) Program, i.e., the EPA generally provides funds to support only the evaluation of the demonstration and the generator assumes the costs of carrying out the demonstration.

Typical projects come through contacts with industry associations, state governments, professional associations, or public solicitations. The EPA's contributing support for evaluation expenses depends upon the projects having a wide applicability and a high chance of success, or upon its innovativeness. Efforts are made to maintain active projects in all of the major waste generating industry sectors.

The WRITE program is based upon the Agency's perception that a significant disincentive to the acceptance of new processes is the lack of credible technical information on the processes. The 1986 EPA Report to Congress states, "Finally the most significant technical barrier to waste minimization may often be a lack of suitable information on source reduction and recycling techniques."

WRAP

The Waste Reduction Assessments Program (WRAP) - a program to demonstrate the EPA Waste Minimization Opportunity Assessment Procedure and encourage the use of waste minimization assessments;

The WRAP program is designed to encourage the use of waste minimization assessment as a tool for identifying operations for reducing waste. It supports activities to demonstrate the procedure in various industrial settings. During 1988 the program includes two major projects, one a cooperative agreement with a research institute to utilize the faculties at several engineering schools in carrying out assessments for small generators, and the other a cooperative agreement with the New Jersey Department of Environmental Protection to support demonstration assessments in many different types of facilities.

WREAFS

The Waste Reduction Evaluations at Federal Sites (WREAFS)- a program to develop cooperative waste minimization technology demonstration projects with other Federal agencies.

The various services within the Department of Defense are among the more active organizations currently pursuing waste minimization as a broad management strategy. Other Federal agencies such as the Department of Energy and the Tennessee Valley Authority are also involved in waste minimization. The manufacturing facilities operated by these agencies are not unlike manufacturing facilities operated by the private sector. Thus, there is significant potential for transfer of technology from Federal sites to industrial organizations employing the same manufacturing operations.

WREAFS provides a low-cost structure for involving the EPA in a wide variety of projects already being funded by other Federal agencies. The EPA plays the role of information broker and/or advisor for other agencies on project selection, evaluation methodologies and assessment of potential environmental benefits. It is planned to convene a conference in the fall of 1988 to provide a forum for discussion of waste minimization efforts and programs thoughout the federal sector.

Sb WRITE

The Small Business WRITE (Sb WRITE) program- a program to encourage the introduction of waste minimization technology at the facilities of small quantity generators. This program is to be done in cooperation with state waste reduction agencies.

The Sb WRITE Program is designed to involve the EPA and cooperating state agencies in efforts to promote the demonstration of waste reduction techniques that are of particular interest to companies that generally produce less tha 1000Kg per month of waste. A typical Sb-Write project consists of the demonstration of a new source reduction technique at a small business, such as an electroplating shop. The state agency would be responsible for the project award and monitoring. It is envisioned that funds provided by the Agency would be leveraged considerably by the state and the involved company. Current plans are to initiate at least one Sb WRITE program in 1988. However, funding of the Sb WRITE Program is still under discussion within the Agency.

WISE - Waste Reduction Institute for Senior Executives - An institute comprised of senior individuals knowledgeable in the principles and practices of waste minimization to act as advisors to the EPA and as liaisons to those generators with whom the Agency is working cooperatively to encourage new waste reduction technologies.

Conclusion

There has been established a program for the EPA to provide technical data to promote a national waste minimization policy, and to provide a technical foundation for furthering the acceptance of technologies that reduce hazardous waste generation. As the U.S. continues to evolve toward production systems that involve more low and nonwaste technologies, it is planned that the Agency's waste minimization research activities will continue to contribute to progress.

References

1. US EPA. Report to Congress: Waste Minimization. 530-SW-86-033. Washington, D.C., 1986

2. Nunno, T. S. Case Studies of Minimizing Plating Bath Wastes. (1987) APCA Journal. Vo 137, No. 6. Pittsburgh, PA USA

3. U.S. EPA. The EPA Manual for Waste Minimization Opportunity Assessments Available Summer 1988. HWERL. Cincinnati, OH.

4. U.S. EPA. Wm Audit Report: Case Studies of Corrosion and Heavy Metal Waste Minimization at a Specialty Steel Manufacturing Complex. EPA 600/52-87/055. 1987. Cincinnati, OH

5. U.S. EPA. Wm Audit Report: Case Studies of Minimization of Solvent Wastes from Parts Cleaning and from Electronic Capacitor Manufacturing Operations. EPA 600/52-87/057. 1987. Cincinnati, OH

6. U.S. EPA. Wm Audit Report Case Studies of Minimization of Cyanide Waste from Electroplating Operations. EPA 600/2-87/056. 1987. Cincinnati, OH

7. Wolf. Source Reduction Research Partnership. Draft Work Plan. 1987.

TRADITIONAL THINKING LIMITS WASTE REDUCTION

by

Kirsten U. Oldenburg and Joel S. Hirschhorn[1]

Any comprehensive environmental protection strategy should consist of three phases: prevention, control, and cleanup.[2] On a national basis in the United States, the second phase has been in place for over fifteen years and the third was initiated in 1980. Meanwhile the first, most fundamental step--not generating pollution in the first place--is struggling for widespread recognition.

There are two fundamental reasons why we in the United States are reluctant to deal directly with sources of pollution. The first step is a deep-seated concern about how far government should penetrate into and affect operations within the private sector; the second is a tendency for government to react rather than be proactive. The result for environmental protection is that a pollution control culture is firmly in place. Moving to prevention must overcome this relatively young tradition, as well as find its place in the much older philosophical tradition of proper role of government.

Why Three Approaches?

As with many multifaceted problems, a systematic approach --a set of solutions--is often better than any one solution. Pollution is a result of industrial operations, transportation, farming and everyday use of products. For any one of these, the environment and public health are protected with greatest

ISWA PROCEEDINGS
VOLUME 1 ISBN 0-12-058451-4

certainty by preventing the generation of pollutants in the first place. Pesticides which are not applied to land cannot run off and contaminate water supplies. Manufacturing processes that do not produce toxic air pollutants cannot emit them into the air, where they can ultimately cause effects difficult to predict.

But, attaining zero level of pollutants across all sources everywhere is not technically or economically possible. The second approach to environmental protection is to control pollutants that are produced--to manage them in ways that reduce the risks of their being placed into the air and water and on the land. Control techniques are not foolproof--inevitable errors occur--and, as society's concept of risk changes and knowledge of risk increases, today's benign waste can become tomorrow's toxic waste. Then environmental clean up--the third option--becomes necessary.

Historical Development and Consequences

In the United States, environmental problems have been dealt with over time, one by one, since the late 1960s. Each piece of the puzzle was crafted when a crisis caused us to recognize the consequences of a single class of pollutants: dirty air, undrinkable water, land overburdened with trash. And, the consequences--rather than the sources--became the problem to solve. Focusing on consequences meant that government did not have to act inside the production sphere. The result is a series of individual laws that attempt to manage pollutants by media after they are created by regulating how and in what quantities they can end up in the water, air, or be deposited on the land.

People were, of course, aware that attacking the problem of pollutants at their source would provide the most certain environmental protection. In 1970 a report from the President's Council on Environmental Quality (CEQ) discussed the interrelated, interchangeable nature of environmental pollutants and the limits of the pollution control approach.[3] And, Rachel Carson's Silent Spring in 1964, which galvanized public support for environmental protection, took a preventive stance. In 1976, before the new hazardous waste law was completed, the Environmental Protection Agency published a "preferred strategy" of waste management options with prevention at the top.[4]

The environmental laws enacted in the 1970s and revised in the 1980s have not foreclosed prevention, but they have not often advanced it either. As the regulatory structure grew, controlling pollution through end-of-pipe treatment plants and collection devices became the standard approach. The result is a complex web of regulations that differ by environmental medium in terms of the substances covered, standards used, operations effected, enforcement authority, and information collected. Pollution control has become a culture, an established, traditional way of thinking. Huge investments have been made in political, bureaucratic, environmental, and industrial careers and in technology to control pollution. In 1985 along $73.8 billion was spent by the government and industry on pollution control.[5]

A Revival for Pollution Prevention

Pollution prevention has seen a revival of sorts in the last few years. Why? First, there is disappointment with what pollution control has been able to accomplish, substance by substance, medium by medium. According to environmentalist Barry Commoner, "...the few instances of environmental success ...occur only when the relevant technologies of production are changed to eliminate the pollutant."[6] And, the CEQ in 1987, "The system is marked by substantial noncompliance, delay, consent decrees, and the other apparatus of legalistic combat, rather than by steady reduction of toxics."[7]

Second, the costs of pollution control--growing even for gross pollutants like oxygen demand in water bodies--may escalate exponentially as control techniques are applied to minute quantities of hundreds of toxic substances from countless sources. Moreover, pollution control may be reaching its technical limits. Lee Thomas, the administrator of EPA has called pollution control the "strategy of the cork" and has said:

> ...At first the corks may be somewhat loose and some pollution escapes. But with advances in technology they can be pushed in tighter. Of course, as we have seen, the pollution will tend to squirt out in new and unexpected places. The solution is a new set of corks, and the process of jamming them in begins all over again...[8]

The U.S. Congress has taken two significant actions in response to the concerns about the effectiveness of pollution control. In 1984, while revising the hazardous waste law that applies to those substances traditionally disposed on the land, Congress declared:

> ...it to be the national policy of the United States that, wherever feasible, the generation of hazardous waste is to be reduced or eliminated as expeditiously as possible.[9]

Later, while revising Superfund legislation in 1986, Congress mandated a new way of collecting comprehensive information on industrial plant emissions, one purpose of which is to assess waste reduction progress.[10]

At the request of Congress, the Office of Technology Assessment (OTA) studied pollution prevention and issued two reports, Serious Reduction of Hazardous Waste and From Pollution to Prevention: A Progress Report on Waste Reduction.[11] The overall conclusion of these reports was that waste reduction is not yet a serious pursuit in the United States, despite its potential for increasing the competitiveness of manufacturing firms, decreasing the cost of complying with and enforcing environmental regulations, and lowering risks to public health and the environment.

The OTA reports suggested that major and rapid movement toward the comprehensive adoption of waste reduction by American industry required a focal point in the national government, such as an Office of Waste Reduction in EPA, to counterbalance the national pollution control institutions. OTA also found that one of the most important things to do, at this early stage of waste reduction, is to fund technical assistance programs within each State. The pollution control culture has created jobs for environmental professionals who make sure that their firms stay in compliance with environmental regulations. They cope with the pollutants that production units discharge and often do not have the authority or expertise to reach back and modify production processes and operations. The main purpose of a government technical assistance program, as discussed by OTA, is not to dictate to plant production people how to change their production systems to reduce waste but to educate and motivate them to recognize the full costs of waste streams and the feasibility and benefits of reduction.

EPA, in a report mandated by Congress, reached conclusions similar in many respects to OTA's.[12] However, the EPA report stated that increasing costs of the regulation-driven marketplace provided motivation for American industry to adopt waste reduction. Thus, OTA found a need for a fundamental shift in thinking before waste reduction would be widely adopted; EPA's report saw an ongoing, incremental process.

Current Status

Most current waste reduction activity lies at the State government level, pushed by a few dedicated people. But most efforts are small, especially compared to expenditures on pollution control, and they usually cover only hazardous wastes as defined by one set of regulations--thus leaving out air or water pollutants. Annual individual State government spending on waste reduction ranges from a million dollars to nothing.

Notable actions have recently occurred at the national level. Six waste reduction bills are under consideration in Congress, and for 1988 Congress appropriated $4.5 million to initiate a multimedia waste reduction and recycling program within EPA. Most of the money is to be spent on technical assistance and an information clearinghouse. For 1987 the agency converted unused hazardous waste training funds into waste reduction and recycling grants for State governments. EPA's 1989 budget request, however, does not include funds to continue or expand this program or the technical assistance program. OTA estimated that an aggressive, highly visible national waste reduction program would cost $255 million over five years. In comparison, EPA's operating budget is now $1.6 billion per year.

Why Isn't It Happening?

Shifting from pollution control to prevention requires a paradigm change. That means that the existing culture has to accept its limitations and build new, parallel institutions dedicated to prevention. In the absence of strong national government support for a prevention strategy, individual members of the pollution control culture are on their own.

Some major corporations have moved dramatically when key people in top management learn that prevention benefits the corporation. But internal barriers are strong and a strong corporate policy and a goal-oriented prevention program are necessary for success.

The actions of some corporations, however, are not representative of American industry. Thousands of firms behind the leading edge need help to change the way they and the rest of the community think about environmental matters. And despite the RCRA national policy statement, only one of the numerous studies on specific pollution problems released by EPA in the last two years considers prevention as a possible solution.[13] That is, regulators do not yet understand that supporting prevention will not eliminate their jobs but can make them easier.

Environmentalists have fought intensively for over fifteen years for the current environmental statutes and regulations. A number of small, grassroots groups are enthusiastic supporters of prevention and have helped get State programs started. But, no major national group has prevention high on its agenda. They are not yet ready to divert resources from familiar battles even though it would enhance longstanding goals.

Policymakers have spent fifteen years juggling the benefits of protecting the environment and the costs to industry. Most do not yet see that a prevention strategy will benefit both the public <u>and</u> industry. And, finally, the general public is not making the shift because they are not receiving the "good news" message of prevention. The bad news of regulatory failure is more sensational.

Conclusions

A national, longterm commitment to waste reduction, as a serious complement to traditional pollution control, is not at hand. No one is against waste reduction in the United States. At the same time, while there are a number of dedicated individual supporters, there is no organized force to advance of waste reduction. Until there is, waste reduction will continue to move sporadically and unevenly through American industry.

References:

1. The authors are analyst and senior associate, respectively, with the U.S. Congress, Office of Technology Assessment (OTA). The views expressed, however, are theirs and not necessarily those of OTA.
2. In this paper, the terms pollution prevention and waste reduction or prevention and reduction are synonymous. Pollutants is used as a general term for substances that can cause harm to public health or the environment.
3. Council on Environmental Quality, Environmental Quality 1970, p. 295.
4. The EPA Position Statement was published in the Federal Register, vol. 41. No. 61, August 16, 1976; the law, the Resource Conservation and Recovery Act (Public Law 94-580) was enacted on October 21, 1976.
5. U.S. Department of Commerce, Bureau of Economic Analysis, 1987.
6. Barry Commoner, "A Reporter at Large: The Environment," The New Yorker, June 15, 1987, p. 57.
7. Council on Environmental Quality, Environmental Quality 1985, 1987.
8. Lee M. Thomas, "A Systems Approach: Challenge for EPA," EPA Journal, September 1985, p. 22.
9. Resource Conservation and Recovery Act as amended by the Hazardous and Solid Waste Amendment Act of 1984 (Public Law 98-616).
10. Superfund Amendments and Reauthorization Act of 1986 (Public Law 99-499), Title III, Section 313.
11. U.S. Congress, Office of Technology Assessment, Serious Reduction of Hazardous Waste, September 1986; and From Pollution to Prevention: A Progress Report on Waste Reduction, June 1987.

12. U.S. Environmental Protection Agency, Report to Congress: Minimization of Hazardous Waste, EPA/530-SW-033, October 1986.
13. U.S. Environmental Protection Agency, Agricultural Chemicals in Ground Water: Proposed Pesticide Strategy, December 1987.

Session 13

International Activities in Solid Waste Management

Session 13

International Activities in Solid Waste Management

THE INTERNATIONAL ORGANIZATIONS AND THE MANAGEMENT OF SOLID AND HAZARDOUS WASTES

Jacqueline Aloisi de Larderel, Director
UNEP Industry and Environment Office

Gérard P. Loiseau, Senior Industry Consultant
UNEP Industry and Environment Office

INTRODUCTION

There is an increasing public awareness and concern about the growing amount of wastes, and in particular, hazardous wastes. The immediate and long-term environmental and health problems caused by improper management of these wastes are now recognized, and an increasing number of countries are in the process of formulating policies and strategies to manage urban and industrial wastes.

There is also recognition that the institutional, administrative and technical experience in resolving the problems faced by the industrialized countries could be adapted and used by the developing countries. This should minimize unnecessary repetition of trials and errors, which would be costly and consume scarce resources.

UNEP, during the past decade, has been undertaking a number of activities to assist countries to develop hazardous waste management strategies.

UNEP is also collaborating with other UN organizations such as the World Health Organization (WHO), the United Nations Industrial Development Organization (UNIDO), the International Labour Office (ILO) and the World Bank, international and regional organizations (like the Organization for Economic Co-operation and Development - OECD), industry and trade associations, university and research institutions and non-governmental organizations.

This paper presents the activties of UNEP that deal mainly with hazardous wastes and summarizes the programme of other international organizations.

I. UNEP ACTIVITIES

I.1 Exchange and Transfer of Information

One of UNEP's objectives is to stimulate the exchange of information and the sharing of information through the world.

ISWA PROCEEDINGS
VOLUME 1 ISBN 0-12-058451-4

For example, a publication entitled "Management of Hazardous Waste"(1) was published under the joint sponsorship of UNEP and WHO in 1983. This document provides guidance on the main elements to be considered in formulating a policy for the management of hazardous waste, and on the more technical aspects to be considered in implementing this policy. The information provided intends to assist policy and decision-makers in governments, control authorities and industry, to develop and organize hazardous waste management schemes appropriate to their specific needs.

A special issue of the UNEP/IEO review "Industry and Environment" devoted to "Industrial Hazardous Waste Management"(2) was published in 1983. This issue gave information on the industrial hazardous waste management policies implemented in developing and developed countries. It contained articles, contributed by different authors, on the status and trends on management of industrial hazardous waste in different regions and deals with topics that include the legal, economic, health and safety aspects of handling hazardous wastes. The need for adequate national legislations and international legal agreements to control the disposal of hazardous wastes was stressed by many of the authors.

A second issue of the UNEP review "Industry and Environment" in 1988 will be concentrating again on Hazardous Waste Management. The articles will endeavour to identify the new trends and developments in hazardous waste management policies, strategies and actions.

The UNEP/International Register of Potentially Toxic Chemicals has set up in 1985, an information system on chemicals entitled: Treatment and Disposal Methods for Waste Chemicals(3). This data profile provides brief information on recommendable methods for the treatment and disposal of chemicals when they are presented in concentrated form.

A Technical Manual for the Safe Disposal of Hazardous Wastes with Special Emphasis on the Problems and Needs of Developing Countries(4) has been prepared in co-operation with the World Health Organization and the World Bank. This manual, which is principally intended to provide guidance to lead administrative and technical staff in developing countries, deals with the various aspects of hazardous wastes including the status of management, health and environmental effects, definitions and classification, planning and infrastructure for management and waste reduction and recycling. The available treatment techniques as well as the economic, financial and institutional considerations are also reviewed in detail in the manual which is further illustrated by case studies drawn mainly from developing countries. This manual, to be published in 1988, will be a major input for regional training workshops.

UNEP/INFOTERA published "Wastes and their Treatment: Information sources and Bibliography"[5] in 1986. This Directory is designed to provide planners, researchers and decision makers with a working reference tool on wastes and their treatment. This book brings together almost 6000 cross-index sources.

Query-response services able to respond to questions related to environmentally sound management of industrial wastes are carried out in close co-operation by the UNEP/International Register of Potentially Toxic Chemicals and the UNEP Industry and Environment Office. The data bases of both the UNEP Offices are computerized and have been in operation for the past decade.

In addition, the UNEP Industry and Environment Office is promoting the development of Low and Non-Waste Technologies, in order to prevent waste production[6].

I.2 Technical and Legislative Guidelines and International Conventions

The Cairo Guidelines and Principles for the Environmentally Sound Management of Hazardous Wastes.

These Guidelines which were adopted by the 14th UNEP Governing Council in 1987 were prepared with a view to assist governments in the process of developing policies for the environmentally sound management of hazardous wastes. They cover the management of hazardous wastes from their generation to their final disposal and, in particular, the problem of transfrontier movements of such wastes which calls for international co-operation between exporting and importing countries in light of their joint responsibility for the protection of the global environment. Also, they provide a framework for effective and environmentally sound hazardous waste management policies in developing countries. Comprehensive aspects of hazardous waste management are covered by this text which contains: general provisions including definition, generation and management, monitoring, remedial action and record-keeping, safety and contingency planning, transport, liability and compensation of hazardous wastes.

Global Convention on the Control of the Transboundary Movements of Hazardous Wastes

The UNEP Governing Council has entrusted an ad hoc Working Group of Legal and Technical Experts with a mandate to prepare a Global Convention on the Control of the Transboundary Movement of Hazardous Wastes. The second session of this ad hoc Working Group is to be held in Caracas in early June 1988. The aim of this convention, which should be adopted in 1989, is to allow importation-exportation of wastes with the necessary provisions safeguarding the environment and human health in the importing country.

I.3 Training Programmes

In 1986, a first training workshop was organized in Singapore by the UNEP Regional Office for Asia and the Pacific (ROAP), with a view to training government officials responsible for developing hazardous waste management policies. Further to this first experience and in order to foster the implementation of the Cairo Guidelines, a training programme is being designed by UNEP with a specific support of the Federal Republic of Germany. The aim of this Programme is to create increased awareness and understanding of the problems and issues related to hazardous wastes and of the ways and means for their control and abatement amongst policy makers and industrialists in Asia and Latin America and to promote the use of environmentally sound hazardous wastes treatment practices. Under this Programme, two training workshops will be organized in 1988 in co-operation with WHO and other organizations. One will be on development of policies and strategies for managing hazardous wastes in Latin America. The other will be a training workshop for Asia and the Pacific countries and will focus specifically on the Siting, Design, Construction and Operation of Hazarduous Waste Landfills.

National training workshops are also organized with UNEP's co-operation by some countries: for example, the Côte d'Ivoire organized a seminar entitled "Valorisation et élimination des déchets d'origine industrielle et domestique en Côte d'Ivoire" in Abidjan in 1987.

It is UNEP's intention to continue to support such activities by providing advice, documentation and resource persons within its available budget.

II. OTHER INTERNATIONAL ORGANIZATIONS' ACTIVITIES

Many international organizations have activities in the field of waste management. The following quickly summarizes some of them.

II.1 World Health Organization (WHO)

Within WHO, two environmental health programmes are dealing with the problems of solid wastes:

- Environmental Health in Rural and Urban Development and Housing (solid wastes in general);
- Control of Environmental Health Hazards (hazardous wastes).

The main concern regarding solid wastes is due to the implications which improper disposal practices have on the spread of communicable diseases, accidents and visual degradation of housing areas. WHO technical co-operations have emphasized development of appropriate technologies including recycling of wastes, training, strengthening of national and local institutions, community participation, and formulation of projects for national and international support. Solid waste management is inter-related with other environmental health considerations such as community water supply and sanitation, drainage, and rural and urban development and housing.

With regard to hazardous wastes, WHO's programme emphasizes the evaluation of health risks of exposure to chemicals in the human environment. This is done primarily through the International Programme on Chemical Safety which is co-sponsored jointly with UNEP and ILO (see § II.3) The programme of WHO will be more extensively presented by Mr. Suess.

II.2 U.N. Industrial Development Organization (UNIDO)

UNIDO, of which the primary objective is to promote and accelerate industrial development in developing countries, aims at minimizing the environmental risk from waste generation and disposal while ensuring maximal resource conservation as, for example, in a number of projects promoting the recycling of waste materials.

In dealing with solid waste, UNIDO assists developing countries in composting municipal solid waste and sewage sludge. These efforts concentrate on assisting municipal authorities in mainly arid regions where there is a need for organic fertilizers and where municipal solid waste is a cause of pollution and its disposal is costly. UNIDO's current activities in this area include the provision of assistance to Bahrain in selecting a contracter for the turn-key installation of a compost plant, and to China in designing compost facilities.

II.3 International Programme on Chemical Safety (IPCS)

Environmental Health Criteria documents and different types of health and safety guides and cards on chemicals are regularly issued by the International Programme on Chemical Safety which is co-sponsored by WHO, UNEP and ILO. Such information, evaluated by international groups of experts, is already available for nearly 70 priority chemicals and more are in various stages of completion.

II.4 Organization for Economic Co-Operation and Development (OECD)

OECD's direct involvement with waste management questions dates back at least to March 1974, when the Waste Management Policy was established by the Environment Committee. A review of relevant Member countries' policies at that time provided the basis for a Recommendation on a Comprehensive Waste Management Policy, which the OECD Council adopted in September 1976. This Recommendation defines the three main principles of waste management as:

(i) reducing at source the total quantity of waste generated;

(ii) extracting the maximum benefit from waste through material reclamation and energy recovery; and

(iii) ensuring the safe transportation and disposal of waste, expecially for toxic and hazardous wastes.

During the period 1976-1980, the activities focussed mainly on household wastes, with a view to implementing in practical cases the provisions of the Recommendation. A number of studies were carried out on the re-use and recycling of beverage containers, waste paper recovery, the management of used tyres, the use of economic instruments in solid wastes management, product durability, and separate collection of household waste.

At the end of the 1970's, the problems raised by abandoned sites where industrial waste had been inadequately disposed of in the past, became a new focus of public and government concern. This concern was not only to have these sites cleaned up but to prevent the creation of new indadequately controlled dumps. International co-operation to control exportations of industrial waste was clearly seen to be necessary by OECD Member countries.

Thus, on 1 February 1984, the OECD Council adopted a Decision and Recommendation on Transfrontier Movements of Hazardous Waste. The Council decided that Member countries shall control the transfrontier movements of hazardous waste and, for this purpose, shall ensure that the competent authorities of the countries concerned are provided with adequate and timely information concerning such movements. Experts of hazardous waste to a destination outside the OECD area have now been covered by a Decision adopted by the OECD Council in June, 1986, which requires that the appropriate authorities of the importing country have give their written consent before the exportation can take place.

Work is presently underway on the development of an international system for effective control of transfrontier movements of hazardous wastes, which will include appropriate OECD instruments such as further Acts of the Council covering notification, identification and control of such transfrontier movements, as well as an international agreement of a legally binding character. Such work, which should be completed by the end of 1988, includes:

- the development of an agreed list of hazardous wastes whose transfrontier movements shall be controlled, and the finalization of a system for cross-referencing wastes listed as hazardous in different countries;
- the establishment of an OECD-wide system of notification, identification and monitoring of tranfrontier movements;
- the development of adequate procedures for transfrontier movements of waste materials intended for recycling.

Other aspects of hazardous waste management are also being addressed by OECD, albeit at a lower level of effort at the present time. These aspects include:

- management of small quantities of hazarous wastes from multiple sources;
- legal and technical aspects of PCB waste disposal;
- release of toxic substances (e.g. dioxin) from waste incineration;
- assessment of policies relating to uncontrolled hazardous waste sites;
- reduction of hazardous waste generation and methods of rendering wastes less hazardous.

CONCLUSION

The transport, handling and disposal of hazardous wastes was selected by UNEP as one of its three priority areas of work when the Montevideo Programme was adopted in 1981. Since then, by promoting exchange and transfer of information between developed and developing countries, by proposing Guidelines and Conventions to assist states in the process of developing appropriate bilateral, regional and multilateral agreements and national legislation, by supporting training programmes for policy makers and technical personnel, UNEP, in close co-operation with other UN Agencies, will continue to collaborate with governments, industry, academic and research institutions in improving the safe management of hazardous wastes.

REFERENCES

1. Suess, M. J. and Huismans, J.W., Management of Hazardous Waste, Policy Guidelines and Code of Practice, WHO Regional Publications, European Series N° 14, Copenhagen 1983.

2. Industrial Hazardous Waste Management, Industry and Environment review, N° 4, 1983 and N° 11, 1988, Industry and Environment Office, Tour Mirabeau, 39-43 quai André Citröen, 75739 PARIS CEDEX 15.

3. Technical Manual for the Safe Disposal of Hazardous Wastes with Special Emphasis on the Problems and Needs of Developing Countries, WHO/UNEP/World Bank, (to be published).

4. Treatment and disposal methods for waste chemicals, International Register of Potentially Toxic Chemicals/UNEP, United Nations Publications, Geneva, 1985, Palais des Nations, CH-1211 Geneva (Switzerland).

5. Wastes and their Treatment: Information Sources and Bibliography. UNEP, P.O. Box 30552, Nairobi (Kenya) 1986.

6. Low- and Non-Waste Technologies, UNEP/IEO, 1986 - 1987.

Integrated Resource Recovery: Optimizing Waste Management (The UNDP/World Bank Global Program in Developing Countries)

F. Wright, C. Bartone, S. Arlosoroff[1]
Water and Sanitation Division
Infrastructure and Urban Development Department
The World Bank, 1818 H St., N.W. Washington, D.C. 20433 USA

Introduction

The management of municipal solid waste is a growing concern in the developing world. Despite large sums spent, service levels are very low in terms of both coverage and the degree of environmental and health protection afforded. Increasingly, resource recovery and recycling is being seen as an important strategy for improving waste management practices. The rationale for waste recycling rests on a number of factors: (a) the rising costs of waste management and conventional pollution control; (b) the need for municipal self-sufficiency; (c) the depletion and degradation of natural resources, including water, soil nutrients, raw materials, energy and space; (d) the growing scarcity of foreign exchange in many of the developing countries.

In response to these concerns a global research and demonstration project on integrating waste management with resource recovery and recycling, financed by the UNDP and executed by the World Bank, was initiated by the UNDP's Governing council in August 1981 as a component of the International Drinking Water Supply and Sanitation Decade. The aim of the Project is to evaluate, document, research, demonstrate and promote low-cost technologies for waste management that incorporate resource recovery and recycling technologies in order to generate financial resources that can offset the costs of waste management.

---

[1] The views and interpretations in this paper are those of the authors and should not be attributed to the World Bank, The UNDP, or their affiliated organizations

ISWA PROCEEDINGS
VOLUME 1 ISBN 0-12-058451-4

This paper reports on some of the past, present, and planned activities of the project related to municipal solid wastes in particular.

Project Background

During the initial phase, the Project surveyed the state-of-the-art in resource recovery and recycling, carried out technological, financial and economic research studies on municipal waste management, recovery and recycling practices, and evaluated the economic potential for resource recovery of both municipal solid and liquid wastes. This led to the identification of the following areas of concentration: (a) solid waste recovery and recycling; (b) co-composing of domestic solid and human wastes; (c) large scale community biogas plants; (d) landfill gas recovery; (e) effluent irrigation; and (f) development of sewage or septage fed aquaculture. Studies have shown that each of these technologies is highly site specific, and their feasibility must be determined on a site by site basis.

To help discover ways to effectively deal with the specific problems of individual cities, the Project has also conducted surveys and prefeasibility studies for assessing resource recovery activities and opportunities in a number of developing country municipalities. These include Abidjan, Colombo, Dakar, Douala, Khartoum, Mexico City, and others. In-depth case studies of several specific resource recovery and recycling systems have also been conducted. These include the (a) Shanghai Resource Recovery and Utilization Company; (b) The Cairo Zabaleen community, (c) and landfill gas recovery in Sao Paulo, Brazil, and others. In addition several rapid technology assessment case studies have been carried out in Bangkok, Manila, Lima, India, Nepal, Israel, Europe and the United States.

Some of the research results have been reported in several forms, as World Bank Technical Papers, articles, conference papers, or simply as unpublished reports and case studies which are available upon request.

Findings to date

Studies of individual cities indicate that the costs of solid waste management in cities of developing countries typically consumes between 30 and 50% of municipal operating budgets, yet coverage rarely reaches more than 70% of the population. This figure masks wide swings in service levels throughout the year, especially for those residing in lower income or spontaneous neighborhoods which often receive no service at all. Poor quality service is the result of a range of deficiencies, from the absence of a waste management strategy and organization planning, and the inadequate number of public containers, to the insufficient numbers of vehicles for pick up and collection, equipment downtime due to poor maintenance and the lack of spare parts, poor pay and low job status. Typically 90-95% of solid waste management costs in developing countries are those associated with collection and transport. The uncontrolled expansion of the cities' boundaries is lengthening the distance to the disposal areas, increasing the need to recover and recycle materials in order to reduce overall transport costs through volume reduction and to optimize the location of disposal sites and introduce transfer stations.

Field studies by the Project also indicate that much of what is recoverable from the solid waste stream is currently being recovered in all the countries but often not very efficiently. Much of this recycling is done by the informal sector or by municipal collectors to supplement their wages. Detailed information on current recycling practices has been assembled from the Dakar study, where a high level of well organized recovery and recycling occurs via "packs" (any place where one or more individuals are involved directly or indirectly in resource recovery or recycling), and at the landfill, where some 500 people reside and eke out a subsistence living as scavengers. It is estimated that the packs provide direct employment for at least 4600 people. Unfortunately, there are no quantitative measurements of the contribution recovery and recycling make to reducing the volume or cost of waste to be transported and disposed. At the same time, sampling at the dump site show that considerable recoverable material remains at the end of the waste chain, suggesting that more could be recovered at the source and further reduce waste management costs. For example scavengers at the landfill recover and sell about 250 tons of ferrous scrap metal per month, while an estimated 730 tons are actually picked up and transported to the landfill.

The situation in Douala is similar to Dakar. In Douala it is estimated that about 10% of the total amount of waste generated is recovered, or about 40 tons of residential and 80 tons of industrial, commercial and institutional wastes. In Colombo, the chain of collectors, middlemen, and end users is also well organized. Most of the end users are located outside the city limits and the middlemen have come into business mainly to provide a warehousing and transporting facility. In Abidjan, informal domestic solid waste resource recovery operates well and appears to be concentrated upstream in the waste chain, with less scavenging occurring at the dump. Industries also tend to recovery and reuse their wastes. One item that is not extensively recycled is waste oil, only 15% of which is recovered and reused, the bulk of which is being poured into the sewer system or elsewhere in the environment. It is estimated that in Abidjan alone, the quantity of discarded oils is more that 5,000 tons per year.

Informal sector scavenging is an important economic activity in many large cities of Mexico. In Mexico City alone it is estimated that more than 10,000 scavengers are involved in separation, sorting and selling of refuse materials which arrive at three landfill sites. Approximately 10 percent of the waste stream is recovered, with most of the recovered materials being sold to industry. On average each scavenger recovers 60 kg of materials daily and earns 1,260 peso (about US$1.00), less than half the prevailing minimum daily wage. Within each disposal site there exists a well established internal organization controlled by a leader, through whom all transactions are centralized. Individuals have little control over working conditions or the unit prices received from the leader who in turn resells to industry at three to five times the price paid to scavengers. In an effort to determine ways of improving the situation of these scavengers, a detailed study was made of the Juarez City Cooperative of Materials Recoverers. The Cooperative was established in 1975, and currently has 211 members, and has a concession from the municipality of Juarez to operate the landfill site. Activities include separation, sorting, weighing, packing and sale of materials to industry. Approximately five percent of the waste stream is recovered, and sales of recovered materials in 1984 were estimated at 70 million pesos versus operating costs of 14.1 million pesos, which include social costs such as education and health care. There is considerable room for improving the efficiency of recovery operations, especially through the introduction of light machinery for intermediate processing. Profits are distributed among the Cooperative members. The Juarez City Cooperative model is one which should be considered for implementation in Mexico City.

In many of the cities studied by the Project the only ultimate disposal option is landfilling, or more typically open dumping. Pressure to close open dumps, convert them into sanitary landfills or open new landfills further away from population centers are all combining to raise the costs of disposal. Landfill capacity was found to be a major problem in the city of Colombo, where the project assisted in the identification of additional disposal cites and will be assisting them in demonstration of reclaiming low lying lands with municipal waste in both a low cost and environmentally sound way.

Also during the field studies the problem of industrial solid wastes was raised by the studied cities as a priority problem. One of the main problems encountered in addressing this problem was the general lack of data on industrial waste generation and the almost total lack of special handling and disposal procedures. It was clear that in many cities these wastes present substantial environmental and human health hazards. Again using Dakar as an example, there is extensive recovery and recycling by industry. A number of commercial enterprises deal exclusively in recovered items. Others save on foreign exchange by recovering and reusing imported raw materials used in production.

A notable example of a successful formal recycling system is the Shanghai Resource Recovery and Utilization Company, which has retrieved more than 20 million tons of waste metal, paper, cloth, plastics, rubber and other materials since it began operations in 1957. The company handles both household and industrial wastes, separating out materials for recycling before they become mixed with general municipal refuse.

In Mexico city the Project identified industrial wastes as a significant component of the waste stream where recycling can be increased. The Project is assisting authorities in the design and implementation of an industrial waste census, using a computerized data management system, and the development of an industrial waste exchange program.

Future Activities

The Project is now shifting its concentration from technological and generic research to work on the implementation of knowledge already developed. Future research activities will be focus on key issues surrounding municipal solid waste management including the need to strengthen the planning and management capacity of solid waste institutions; define affordable service levels;

technology options and delivery systems for low income groups; better utilization of the informal sector; and provision of environmentally safe disposal options.

Highest priority will be placed on technical support to waste management/resource recovery demonstration and investment projects that are being assisted by bilateral and multilateral agencies including the World Bank. Further research will be focused on obtaining a better understanding of the software issues, institutional factors and other constraints that affect the implementation of resource recovery programs. As a follow up on the findings of the early research and survey work of the Project, implementation of a demonstration project is underway in Lima, and other demonstration projects are now in preparation in Colombo, Dakar, Douala, Guatemala, and Mexico City with bilateral cost sharing support.

The project will also focus on policy issues and the role governments must play in incorporating resource recovery into national environmental and natural resource management policies. This work will complement the expanded role the World Bank is taking in the protection of the environment. Recently the World Bank has set up two positions one dealing specifically with urban waste management issues in the Infrastructure and Urban Department and one dealing with toxic and hazardous wastes in the Environment Department.

Additional Information

Cointreau, S., Gunnerson, C., Huls, J., Seldman, N. (1984). Recycling From Municipal Refuse; A State-of-the-Art Review and Annotated Bibliography, World Bank Tech. Pub. No. 30.

Lund, R. (1984). Remanufacturing: The Experience of the United States and Implications for Developing Countries, World Bank Technical Paper No. 31.

Abert, J. (1985). Municipal Waste Processing in Europe: A Status Report on Selected Materials and Energy Recovery Projects, World Bank Technical Paper No. 37.

Gunnerson, C., Stuckey, D. (1986). Anaerobic Digestion: Principles and Practices for Biogas Systems, WBTP No. 49.

Obeng, L., Wright, F. (1987). The Co-Composting of Domestic Solid and Human Wastes, World Bank Technical Paper No. 57.

ISWA INTERNATIONAL ACTIVITIES
ORGANIZATION, PROJECTS AND PROGRAM, 1988-92

By
Dipl.-Ing. Werner Schenkel

Director and Professor
Federal Environmental Agency

and

Chairman
Scientific and Technical Committee
International Solid Waste and Public Cleansing
Association

INTRODUCTION

While the waste management problem has many dimensions, it originates as a problem basic to our economic system. Professor Georgescou Roegen, an American economist, has pointed out, "It is the nature of our industrial production to transform precious raw materials into worthless good and wastes." Never before in the world's history has a generation of human beings transformed such large amounts of raw materials into wastes. At the same time, we have never before produced such large amounts of products and goods so cheaply. We have been characterized as living in a "throw-away society" where it is cheaper to throw away than it is to recycle, or repair, or reuse.

The problem with this attitude, of course, is that there are not enough places to throw away all the wastes we are capable of producing. Siting of waste management facilities is a problem endemic to all industrial countries. Thus, we are faced with the need to develop new, environmentally sound solutions for solving our waste management problems.

We have reached a point in world history where we are reminded constantly that each product will change in character, after its use, into a waste product. At the same time, we realize that this pattern from raw material, to product, to waste can be changed by adopting innovative recycling/

ISWA PROCEEDINGS
VOLUME 1 ISBN 0-12-058451-4

reuse processes. To solve such problems, we must focus our efforts on designing modern industrial systems that utilize fully what we know about product production, recycling, waste reduction, disposal and environmental protection.

The need for innovative problem solving, however, does not end here. The waste management problem is global in nature, affecting nonindustrial countries as well as industrial ones. It is becoming apparent that the industrial countries, which also produce goods for export to nonindustrial countries, must assist their consumers in nonindustrial countries in disposing of the used product, if the full problem is to be addressed.

Also, we must consider the special problem of how better to handle our hazardous wastes. Hazardous wastes can and do cross national boundaries, significantly impacting the oceans and the stratosphere as well as the land. The scope of the problem is so far-reaching that international cooperation is imperative if we are to devise adequate solutions. It is to foster this imperative for international cooperation that ISWA accepts as its charge for the next four years and beyond.

ISWA ACTIVITIES, 1984-87

At the 1984 ISWA International Solid Wastes Congress and Exhibition in Philadelphia, the association took an important step in broadening its sphere of activities. It formally included hazardous wastes management as an area of concern along with such traditional concerns as collection practice, public cleansing, disposal practice, and incineration and resource recovery. The first of ISWA's working groups, the ISWA Working Group on Hazardous Wastes, along with a three-year work plan, was approved by the Governing Board at this meeting.

During the four years since the 1984 Congress, the working group in conjunction with the Scientific and Technical Committee considered the following topics in detail:

- The classification of hazardous wastes;
- Manifest and trip-ticket systems;
- Environmental standards;
- Siting of facilities;
- Long-term care at land disposal facilities; and
- Treatment and disposal techniques.

The working group ended its three-year work plan in June, 1987, with its successful "Hazardous Waste Symposium" in Tokyo, Japan, and the publication of its report, "International Perspectives on Hazardous Waste Management", by Academic Press in London.

As we turn to the future, ISWA had identified the following problem areas where international information exchanges are needed:

- Transboundary transport of hazardous wastes;
- The import and export of wastes and recyclable goods;
- The establishment of minimum technical standards;
- The codisposal of wastes in landfills, and emissions at old tips;
- Ending the practice of disposal at sea;
- Standardizing national treatment practices for highly toxic wastes;
- Development treatment systems for special hazardous wastes;
- Developing standards for long-term storage of wastes; and
- Codes of waste practice for firms working in developing countries.

ISWA believes it is in finding answers to the above problems that we best can serve our national member countries.

ISWA OUTPUTS AND ACTIVITIES

ISWA utilizes the following three types of outputs and activities to provide its membership with the most current waste management information:

- The organization and management of official technical meetings, such as seminars, symposia, and congresses;

- The creation of working groups, commissions, and committees, such as the working groups for hazardous wastes and sanitary landfills, and the ISWA Commission for European Affairs;
- The publication of technical and policy documents such as the ISWA "Policy and Professional Activities, 1987-90", the "ISWA Guide", the ISWA journal, and the ISWA newsletter.

I will discuss these outputs and activities in greater detail.

Seminars, Conferences, and Congresses

Figure 1 provides an overview of seminar, conference and congress activities with which ISWA was involved during the past four years. We were able through our Scientific and Technical Committee to provide program support to a large number of organizations. We did this by developing conference programs for them and recommending experienced international speakers for their meetings.

Two problems face us as we continue to provide this information for the international waste management community. One is to define exactly what form ISWA collaboration should take. We need to decide whether we wish only to co-sponsor meetings arranged by other organizations, to develop full technical programs for other organizations, or to manage our own meetings through our Scientific and Technical Committee, Executive Committee, and Governing Board. We must find a balance between local and regional activity needs and ISWA's responsibility to produce high quality international programs. I believe we need to develop specific procedures for providing this program support, and to define precisely our conditions for working together with other groups.

The second problem concerns how we distribute information such as the announcement of meetings and the proceedings after the meetings are completed. While we have all had experience organizing our respective national member organization meetings, it is quite another thing to organize international meetings.

We are made quite aware of the great amount of work there is to be done as we prepare for the Copenhagen Congress. Another example of this type of international work was the specialized ISWA seminar, "Emission of Trace Organics from Municipal Solid Waste Incinerators", held in Copenhagen January 20-22, 1987. The proceedings were published by Academic Press, and thus made available to individuals and libraries worldwide. Another good example of publishing success was the working group report, "International Perspectives on Hazardous Waste Management". We must continue to make sure such materials are published, and thus made available to those in need of our information.

The ISWA Working Groups

Currently, there are two ISWA working groups, the Working Group on Hazardous Wastes and the Working Group on Sanitary Landfills. A third working group, to deal with incineration and energy recovery problems is under consideration.

Since its formation at the 1984 ISWA Congress, the hazardous waste group has been responsible for a wide-range of activities in the hazardous waste field. In addition to the "Tokyo Hazardous Waste Symposium" last summer, the working group sponsored a major conference in Washington, D.C., in March 1986, in conjunction with the U. S. Environmental Protection Agency and the Organization for Economic Cooperation and Development (OECD). Also, contacts have been established with other international organizations, such as the World Health Organizations (WHO), and the United Nations Environmental Programme (UNEP), that have proven beneficial to all concerned.

The major output from this group was, of course, its final report, which was published in book form as, "International Perspectives on Hazardous Waste Management".

What, however, are the main issues for the next four years? To start with, our work in the hazardous waste field will be more detailed and specific. Our group of international experts on the Working Group on Hazardous Wastes will continue its ambitious program by developing reports in four areas:

- Waste Minimization/Clean Technologies;
- Safe Management Practices/Practical Treatment Technology;
- Household Hazardous Wastes/Small Quantity Generators; and
- Developing Countries/Special Wastes.

The Working Group on Sanitary Landfills, on the other hand, will continue to consider all aspects of land disposal of solid wastes, with particular focus on ways to stabilize wastes to ensure that landfills remain secure even after closure. The working group held a successful international symposium, "Process, Technology, and Environmental Impact of Sanitary Landfills", in Cagliari, Sardinia, in October, 1987, and will conduct a second symposium in Amsterdam following the Copenhagen congress.

The new working group now being established will deal with the problems of thermal destruction of wastes and energy recovery. With this group, the important topics of trace metals and organics in stack emissions, dealt with by our Danish National Member in specialized seminars, likewise will continue.

Many waste management problems are not task force-type problems. A task force by definition is set up to deal with a specific problem during a set period of time. Other problems are on-going in nature and require a more or less permanent organizational structure. These "Commissions (or Councils) are authorized by ISWA Statutes, and one such group, the ISWA Commission (Council) for European Affairs, was approved by the Governing Board at its Budapest meeting in October, 1987.

Another such commission or council is needed in the area of developing countries. Many goods produced in developed countries are exported, and eventually become wastes, in developing countries that lack the waste management infrastructure to handle the waste goods. Another aspect of this problem is that developing countries, which may be in financial need, may agree to accept shipped hazardous wastes that it cannot handle, simply for the income it may provide. These problems need to be addressed, and an ISWA commission or council would be an effective vehicle to do this. Representatives from the working groups could assist the commissions when and as they are needed, thus allowing us to draw from a broad spectrum of expertise to solve problems of mutual interest.

Policy and Professional Activities, 1987-90

This document, developed through the Scientific and Technical Committee, provides an overview of ISWA's future intentions. The authors point out that new working fields have come into existence and there is a need to get new, professional-quality information to our members.

Three problems are identified. First, international travel is costly, and these costs for members will increase as ISWA becomes more active. Second, all ISWA work is voluntary, and thus outputs are limited to what its members and organizations have the time and resources to donate. Third, the ISWA Secretariat needs to be reinforced. With the exception of the ISWA Journal Office, all Secretariat assistance is donated. There has been discussion concerning the establishment of a full professional office with a paid staff hired to look after ISWA on-going needs. All of this is very costly, but if we are interested in increased ISWA activities we must find ways to finance them.

The ISWA Guide

The ISWA Guide is a new publication in the form of a directory that provides detailed waste management information on ISWA countries. The country reports are presented in similar format thus allowing for quick cross-references. We hope these figures, addresses, and descriptions will fill a need in our organization, and will be used often to obtain information and to assist members in contacting persons and organizations in other countries.

The ISWA Journal

A successful international organization must have good publications. ISWA has two, the ISWA journal and the ISWA newsletter. The journal, "Waste Management and Research", is published four times a year. It is an excellent journal already containing first-rate scientific and technical articles on management by encouraging managers and directors of waste management articles worldwide to write down their experiences and submit them to the journal editors.

Currently, there are two problems concerning the journal. First, there is insufficient income due to insufficient advertising and subscription sales. While the journal provides an excellent forum to inform waste management professionals, its continued publication depends upon our increasing its revenue. Currently, it requires almost two-thirds of ISWA's income to fund the operation of the Journal Office.

The second problem is language. Sales are very low in non-English speaking countries. In some other, nonprogressive countries there also is an absence of interest in waste management problems.

The ISWA Newsletter

The ISWA Newsletter could be the basic information instrument for the association, if members would contribute on a regular basis. The newsletter does a good job of reporting on ISWA meetings, and the activities of the ISWA committees, working groups and the Governing Board. It could benefit, however, if more national member organizations would submit news and articles. Some thought is being given to change the newsletter to magazine form, but this will require far more financial resources than currently are avilable to ISWA.

COLLABORATION WITH OTHER ORGANIZATIONS AND COUNTRIES

Contacts betwѳen ISWA and other countries such as Turkey, Greece, and the Peoples Republic of China, demonstrate something vѳry important to ISWA's future. These countries are vitally interested in what they can learn from ISWA. They are characterized by having developing economies, and as they assume western industrialized lifestyles they produce more and more wastes. They, thus, are becoming more and more interested in solving the collection, transport, disposal, and hazardous waste problems developed countries are facing.

The problem is that these countries lack the funds to participate in ISWA. The first problem concerns membership. ISWA currently does not have a membership category these countries can afford. In most cases these countries do not have the waste management organization or committee necessary to represent their country in ISWA. We therefore need to change the ISWA Statutes to allow a special membership category for such countries.

At the same time, we need to recognize that the membership fee is only a small part of the cost of belonging to an international organization. Travel monies are needed to send members to working group and other meetings. Also, it is expensive to translate working group reports and other documents into home-country languages. Therefore, we need a new arrangement for funding our working groups. The Working Group on Hazardous Wastes secretariat is funded jointly by the American Public Works Association and the U. S. Environmental Protection Agency. The Working Group on Sanitary Landfills, however, does not yet have such funding, although there was an attempt to obtain some funding from the European Community Commission. As ISWA finds new sources of revenue during the next four years, we must devise a more equitable way to fund our working groups.

While our contacts with other international organizations, such as WHO and UNEP have served to strengthen our own organization, these too have been costly, and deserve our close attention. We are asked to attend meetings or to serve on special taks forces. Travel funds seldom are provided to attend these sessions. Thus, the costs must be assumed either by the ISWA member or by the ISWA Secretariat. Also, we are asked to provide papers for conferences or to critique reports. This must be done by individual members, as we cannot afford to have it professionally done by a consultant. All these new challenges, as our organization continues to mature, point to the desirability of expanding our secretariat income so that we can achieve the full potential inherent in our organization.

In closing, I will end on a positive note. Our work with other international organizations has benefited ISWA in many ways. It puts us in contact with our professional colleagues throughout the world, and enables us to learn from their successful practices. ISWA actually belongs to two international organizations, the UATI and the International Public Works Federation. We not only receive their publications, but share information and contacts. We are convinced of the value of international relations, and look eagerly to an ISWA future of growth and expanded waste management benefits for its membership.

Session 14

Local, Regional and National Solid Waste Management

Session 15

ISWA 88 Closure